Lecture Notes in Artificial Intelligence 7969

Subseries of Lecture Notes in Computer Science

LNAI Series Editors

Randy Goebel
University of Alberta, Edmonton, Canada
Yuzuru Tanaka
Hokkaido University, Sapporo, Japan
Wolfgang Wahlster
DFKI and Saarland University, Saarbrücken, Germany

LNAI Founding Series Editor

Joerg Siekmann
DFKI and Saarland University, Saarbrücken, Germany

Sarah Jane Delany   Santiago Ontañón (Eds.)

Case-Based Reasoning Research and Development

21st International Conference, ICCBR 2013
Saratoga Springs, NY, USA, July 8-11, 2013
Proceedings

 Springer

Volume Editors

Sarah Jane Delany
Dublin Institute of Technology, School of Computing
Kevin Street, Dublin 8, Ireland
E-mail: sarahjane.delany@dit.ie

Santiago Ontañón
Drexel University, Department of Computer Science
3141 Chestnut Street, Philadelphia, PA 19104, USA
E-mail: santi@cs.drexel.edu

ISSN 0302-9743   e-ISSN 1611-3349
ISBN 978-3-642-39055-5   e-ISBN 978-3-642-39056-2
DOI 10.1007/978-3-642-39056-2
Springer Heidelberg Dordrecht London New York

Library of Congress Control Number: 2013940603

CR Subject Classification (1998): I.2.1, I.2.3-4, I.2.6, I.2.8, I.2.11, H.3.3-5, H.2.8, H.4, H.5.3, J.1, J.3

LNCS Sublibrary: SL 7 – Artificial Intelligence

Typesetting: Camera-ready by author, data conversion by Scientific Publishing Services, Chennai, India

Printed on acid-free paper

Springer is part of Springer Science+Business Media (www.springer.com)

Preface

This volume contains the papers presented at ICCBR 2013: the 21st International Conference on Case-Based Reasoning (http://www.iccbr.org/iccbr13/) held during July 8–11, 2013, in Saratoga Springs, USA. There were 39 submissions and each one was reviewed by at least three Program Committee members using the criteria of relevance, significance, originality, technical quality, and presentation. The committee accepted 17 papers for oral presentation and nine papers for poster presentation at the conference.

The International Conference on Case-Based Reasoning (ICCBR) is the preeminent international meeting on case-based reasoning (CBR). Previous ICCBR conferences have been held in Sesimbra, Portugal (1995), Providence, USA (1997), Seeon Monastery, Germany (1999), Vancouver, Canada (2001), Trondheim, Norway (2003), Chicago, USA (2005), Belfast, UK (2007), Seattle, USA (2009), Alessandria, Italy (2010), London, UK (2011), and most recently in Lyon, France (2012).

Day 1 of ICCBR 2013 included the Industry Day and the Doctoral Consortium, which were very kindly hosted by General Electric on their premises at GE Global Research in Niskayuna, New York. The industry day consisted of eight presentations on applications of CBR. Concurrently with the industry presentations was the annual Doctoral Consortium that involved presentations by seven research students in collaboration with their respective senior CBR research mentors.

Day 2 was dedicated to workshops on current and interesting aspects of CBR including CBR in Social Web Applications, CBR in Health Sciences and Experience Reuse: Provenance, Process Orientation and Traces.

Days 3 and 4 consisted of scientific paper presentations on theoretical and applied CBR research as well as invited talks from two distinguished scholars: Ashok K. Goel, Professor of Computer and Cognitive Science in the School of Interactive Computing at Georgia Institute of Technology in Atlanta, USA, and Igor Jurisica, Senior Scientist at the Ontario Cancer Institute and Professor in the Departments of Computer Science and Medical Biophysics at the University of Toronto, Canada. Ashok Goel gave a keynote address on exploring and exploiting case-based reasoning in biologically inspired design – the invention of new sustainable technologies by analogy to biological systems. He described how we can view nature as a large library of sustainable designs and use human case-based reasoning for designing novel technological products, processes, and systems. Igor Jurisica discussed scaling up case-based reasoning for "big data" applications, describing two main systems. The first involved estimating a job at runtime during scheduling functional regression tests for the IBM DB2 Universal Database on an internal, heterogeneous computer grid, and the second was a computational biology application for managing protein crystallization experiments.

The presentations and posters covered a wide range of CBR topics of interest both to researchers and practitioners including case retrieval and adaptation, similarity assessment, case base maintenance, knowledge management, recommender systems, multiagent systems, textual CBR, and applications to healthcare and computer games. Day 3 also hosted the Computer Cooking Contest, the aim of which is to encourage the use of AI technologies such as case-based reasoning, semantic technologies, search, and information extraction.

As part of the main conference, David Aha and Odd Erik Gundersen conducted and presented a Reproducibility Process for Case-Based Reasoning that involved confirming the results for voluntarily provided empirical studies reported at ICCBR 2013. It is hoped that this will become a fixture of the ICCBR conferences of the future.

Many people participated in making ICCBR 2013 a success. William Cheetham, GE Research, USA, served as the Conference Chair with Sarah Jane Delany, Dublin Institute of Technology, Ireland, and Santiago Ontañón, Drexel University, USA, as Program Co-chairs. We would like to thank Michael Floyd, Carleton University, Canada, and Jonathan Rubin, PARC, USA, who acted as Workshop Chairs as well as William Cheetham, GE Research, USA, and Frode Sørmo, Verdande Technology, Norway, who coordinated the Industry Day. Our thanks also go to Thomas Roth-Berghofer, University of West London, UK, and Rosina Weber, Drexel University, USA, for organizing the Doctoral Consortium. We thank Michel Manago, Kiolis, France, who was responsible for the Computer Cooking Competition and to David Aha, who acted as Sponsorship Chair. We are very grateful also to all our sponsors, which at the time of printing included General Electric, AAAI, Knexus Research Corporation, Empolis, *Artificial Intelligence Journal* and Drexel University.

We thank the Program Committee and the additional reviewers for their timely and thorough participation in the reviewing process. We appreciate the time and effort put in by the local organizers at GE Global Research, particularly Aisha Yousuf. Finally, we acknowledge the support of EasyChair in the submission, review, and proceedings creation processes and thank Springer for its continued support in publishing the proceedings of ICCBR.

May 2013   Sarah Jane Delany
                                                                 Santiago Ontañón

Organization

Program Chairs

Sarah Jane Delany — Dublin Institute of Technology, Ireland
Santiago Ontañón — Drexel University, USA

Conference Chair

William Cheetham — GE Global Research, USA

Workshop Chairs

Michael Floyd — Carleton University, Canada
Jonathan Rubin — PARC, A Xerox Company, USA

Industry Day Coordinators

William Cheetham — GE Global Research, USA
Frode Sørmo — Verdande Technology, Norway

Doctoral Consortium Chairs

Thomas Roth-Berghofer — University of West London, UK
Rosina Weber — Drexel University, USA

Sponsorship Chair

David W. Aha — Naval Research Laboratory, USA

Program Committee

David W. Aha — Naval Research Laboratory, USA
Klaus-Dieter Althoff — DFKI / University of Hildesheim, Germany
Kevin Ashley — University of Pittsburgh, USA
Ralph Bergmann — University of Trier, Germany
Isabelle Bichindaritz — State University of New York at Oswego, USA
Derek Bridge — University College Cork, Ireland
William Cheetham — GE Global Research, USA
Amélie Cordier — LIRIS, France

Additional Reviewers

Sponsoring Institutions

ICCBR 2013 was supported by General Electric, AAAI, Knexus Research Corporation, Empolis, *Artificial Intelligence Journal* and Drexel University.

Table of Contents

goal—an explicit target that must be achieved—and this paradigm is still dominant in AI problem solving. But as application domains become more complex and realistic, it is apparent that the dichotomic notion of a goal, while adequate for certain puzzles, is too crude in general. The problem is that in many contemporary application domains [...] the user has little knowledge about the set of possible solutions or feasible items, and what she typically seeks is the best that's out there. But since the user does not know what is the best achievable plan or the best available document or product, she typically cannot characterize it or its properties specifically. As a result, she will end up either asking for an unachievable goal, getting no solution in response, or asking for too little, obtaining a solution that can be substantially improved."

In [1], we made a first step toward preference-based CBR by addressing the important part of case-based inference, which is responsible for predicting a "contextualized" preference relation on the solution space. More specifically, the latter consists of inferring preferences for candidate solutions in the context of a new problem, given knowledge about such preferences in similar situations. In this paper, we go one step further by embedding this inference procedure in a more general, search-based problem solving framework. In this framework, case-based problem solving is formalized as a search process, in which a solution space is traversed through the application of adaptation operators, and the choice of these operators is guided by case-based preferences.

...inder of the paper is organized as follows. By way of back... ...f P...f CBR and Sect...

Preference-Based CBR:
A Search-Based Problem Solving Framework

Amira Abdel-Aziz, Weiwei Cheng, Marc Strickert, and Eyke Hüllermeier

Department of Mathematics and Computer Science
Marburg University, Germany
{amira,cheng,strickert,eyke}@mathematik.uni-marburg.de

Abstract. Preference-based CBR is conceived as a case-based reasoning methodology in which problem solving experience is mainly represented in the form of contextualized preferences, namely preferences for candidate solutions in the context of a target problem to be solved. This paper is a continuation of recent work on a formalization of preference-based CBR that was focused on an essential part of the methodology: a method to predict a most plausible candidate solution given a set of preferences on other solutions, deemed relevant for the problem at hand. Here, we go one step further by embedding this method in a more general search-based problem solving framework. In this framework, case-based problem solving is formalized as a search process, in which a solution space is traversed through the application of adaptation

Preference-Based CBR:
A Search-Based Problem Solving Framework

Amira Abdel-Aziz, Weiwei Cheng, Marc Strickert, and Eyke Hüllermeier

Department of Mathematics and Computer Science
Marburg University, Germany
{amira,cheng,strickert,eyke}@mathematik.uni-marburg.de

Abstract. Preference-based CBR is conceived as a case-based reasoning methodology in which problem solving experience is mainly represented in the form of contextualized preferences, namely preferences for candidate solutions in the context of a target problem to be solved. This paper is a continuation of recent work on a formalization of preference-based CBR that was focused on an essential part of the methodology: a method to predict a most plausible candidate solution given a set of preferences on other solutions, deemed relevant for the problem at hand. Here, we go one step further by embedding this method in a more general search-based problem solving framework. In this framework, case-based problem solving is formalized as a search process, in which a solution space is traversed through the application of adaptation operators, and the choice of these operators is guided by case-based preferences. The effectiveness of this approach is illustrated in two case studies, one from the field of bioinformatics and the other one related to the computer cooking domain.

1 Introduction

A preference-based approach to case-based reasoning (CBR) has recently been advocated in [1]. Building on general ideas and concepts for preference handling in artificial intelligence (AI), which have already been applied successfully in other fields [2–4], the goal of preference-based CBR, or Pref-CBR for short, is to develop a coherent and universally applicable methodological framework for CBR on the basis of formal concepts and methods for knowledge representation and reasoning with preferences.

In fact, as argued in [1], a preference-based approach to CBR appears to be appealing for several reasons, notably because case-based experiences lend themselves to representations in terms of preference relations quite naturally. Moreover, the flexibility and expressiveness of a preference-based formalism well accommodate the uncertain and approximate nature of case-based problem solving. In this sense, the advantages of a preference-based problem solving paradigm in comparison to the classical (constraint-based) one, which have already been observed for AI in general, seem to apply to CBR in particular. These advantages are nicely explained in [5]: "Early work in AI focused on the notion of a

S.J. Delany and S. Ontañón (Eds.): ICCBR 2013, LNAI 7969, pp. 1–14, 2013.

goal—an explicit target that must be achieved—and this paradigm is still dominant in AI problem solving. But as application domains become more complex and realistic, it is apparent that the dichotomic notion of a goal, while adequate for certain puzzles, is too crude in general. The problem is that in many contemporary application domains [...] the user has little knowledge about the set of possible solutions or feasible items, and what she typically seeks is the best that's out there. But since the user does not know what is the best achievable plan or the best available document or product, she typically cannot characterize it or its properties specifically. As a result, she will end up either asking for an unachievable goal, getting no solution in response, or asking for too little, obtaining a solution that can be substantially improved."

In [1], we made a first step toward preference-based CBR by addressing the important part of case-based inference, which is responsible for predicting a "contextualized" preference relation on the solution space. More specifically, the latter consists of inferring preferences for candidate solutions in the context of a new problem, given knowledge about such preferences in similar situations. In this paper, we go one step further by embedding this inference procedure in a more general, search-based problem solving framework. In this framework, case-based problem solving is formalized as a search process, in which a solution space is traversed through the application of adaptation operators, and the choice of these operators is guided by case-based preferences.

The remainder of the paper is organized as follows. By way of background, Section 2 recapitulates the main ideas of Pref-CBR, and Section 3 briefly recalls the case-based inference procedure of [1]. Although these two sections are to some extent redundant, they are included here to increase readability of the paper and to make it more self-contained. In Section 4, we introduce and detail our search-based problem solving framework. In Section 5, two case studies are presented to illustrate the effectiveness of this approach, one from the field of bioinformatics (molecular docking, drug discovery) and the other one related to the computer cooking domain. The paper ends with some concluding remarks and an outlook on future work in Section 5.

2   Preference-Based CBR

2.1   Conventional CBR

Experience in CBR is most commonly (though not exclusively) represented in the form of problem/solution tuples $(x, y) \in X \times Y$, where x is an element from a problem space X, and y an element from a solution space Y. Despite its generality and expressiveness, this representation exhibits some limitations, both from a knowledge acquisition and reuse point of view.

- *Existence of correct solutions*: It assumes the existence of a "correct" solution for each problem, and implicitly even its uniqueness. This assumption is often not tenable. In the cooking domain, for example, there is definitely not a single "correct" recipe for a vegetarian pasta meal. Instead, there will be many possible alternatives, maybe more or less preferred by the user.

– *Verification of optimality*: Even if the existence of a single correct solution for each problem could be assured, it will generally be impossible to verify the optimality of the solution that has been produced by a CBR system. However, storing and later on reusing a suboptimal solution y as if it were optimal for a problem x can be misleading. This problem is less critical, though does not dissolve, if only "acceptable" instead of optimal solutions are required.

– *Loss of information*: Storing only a single solution y for a problem x, even if it can be guaranteed to be optimal, may come along with a potential loss of information. In fact, during a problem solving episode, one typically tries or at least compares several candidate solutions, and even if these solutions are suboptimal, preferences between them may provide useful information.

– *Limited guidance*: From a reuse point of view, a retrieved case (x, y) only suggests a single solution, namely y, for a query problem x_0. Thus, it does not imply a possible course of action in the case where the suggestion fails: If y is not a good point of departure, for example since it cannot be adapted to solve x_0, there is no concrete recommendation on how to continue.

Table 1. Notations

notation	meaning
$\mathbb{X}, x$	problem space, problem
$\mathbb{Y}, y$	solution space, solution
CB	case base (storing problems with preferences on solutions)
S_X, Δ_X	similarity/distance measure on $\mathbb{X}$
S_Y, Δ_Y	similarity/distance measure on $\mathbb{Y}$
$\mathcal{N}(y)$	neighborhood of a solution y
$\mathfrak{P}(\mathbb{Y})$	class of preference structures on $\mathbb{Y}$
$\mathcal{P}(x)$	set of (pairwise) preferences associated with a problem
CBI	case-based inference using ML estimation (see equation (4))

2.2 Preference-Based Knowledge Representation

To avoid these problems, preference-based CBR replaces experiences of the form "solution y (optimally) solves problem x" by information of the form "y is better (more preferred) than z as a solution for x". More specifically, the basic "chunk of information" we consider is symbolized in the form $y \succeq_x z$ and suggests that, for the problem x, the solution y is supposedly at least as good as z.

This type of knowledge representation obviously overcomes the problems discussed above. As soon as two candidate solutions y and z have been tried as solutions for a problem x, these two alternatives can be compared and, correspondingly, a strict preference in favor of one of them or an indifference can be expressed. To this end, it is by no means required that one of these solutions is optimal. It is worth mentioning, however, that knowledge about the optimality of a solution y^*, if available, can be handled, too, as it simply means that $y^* \succ y$

for all $y \neq y^*$. In this sense, the conventional CBR setting can be considered as a special case of Pref-CBR.

The above idea of a preference-based approach to knowledge representation in CBR also suggests a natural extension of the case retrieval and inference steps, that is, the recommendation of solutions for a new query problem: Instead of just proposing a single solution, it would be desirable to predict a *ranking* of several (or even all) candidate solutions, ordered by their (estimated) degree of preference:

$$y_1 \succeq_x y_2 \succeq_x y_3 \succeq_x \cdots \succeq_x y_n \tag{1}$$

Obviously, the last problem mentioned above, namely the lack of guidance in the case of a failure, can thus be overcome.

In order to realize an approach of that kind, a number of important questions need to be addressed, including the following: How to represent, organize and maintain case-based experiences, given in the form of preferences referring to a specific context, in an efficient way? How to select and access the experiences which are most relevant in a new problem solving situation? How to combine these experiences and exploit them to infer a solution or, more generally, a preference order on a set of candidate solutions, for the problem at hand?

2.3   Formal Setting and Notation

In the following, we assume the problem space $\mathbb{X}$ to be equipped with a similarity measure $S_X : \mathbb{X} \times \mathbb{X} \to \mathbb{R}_+$ or, equivalently, with a (reciprocal) distance measure $\Delta_X : \mathbb{X} \times \mathbb{X} \to \mathbb{R}_+$. Thus, for any pair of problems $x, x' \in \mathbb{X}$, their similarity is denoted by $S_X(x, x')$ and their distance by $\Delta_X(x, x')$. Likewise, we assume that the solution space $\mathbb{Y}$ to be equipped with a similarity measure S_Y or, equivalently, with a (reciprocal) distance measure Δ_Y. While the assumption of a similarity measure on problems is common in CBR, the existence of such a measure on the solution space is often not required. However, the latter is neither less natural than the former nor more difficult to define. In general, $\Delta_Y(y, y')$ can be thought of as a kind of adaptation cost, i.e., the (minimum) cost that needs to be invested to transform the solution y into y'.

In Pref-CBR, problems $x \in \mathbb{X}$ are not associated with single solutions but rather with preferences over solutions, that is, with elements from a class of preference structures $\mathfrak{P}(\mathbb{Y})$ over the solution space $\mathbb{Y}$. Here, we make the assumption that $\mathfrak{P}(\mathbb{Y})$ is given by the class of all weak order relations $\succeq$ on $\mathbb{Y}$, and we denote the relation associated with a problem x by $\succeq_x$; recall that, from a weak order $\succeq$, a strict preference $\succ$ and an indifference $\sim$ are derived as follows: $y \succ y'$ iff $y \succeq y'$ and $y' \not\succeq y$, and $y \sim y'$ iff $y \succeq y'$ and $y' \succeq y$.

More precisely, we assume that $\succeq_x$ has a specific form, which is defined by an "ideal" solution $y^* \in \mathbb{Y}$ and the distance measure Δ_Y: The closer a solution y to $y^* = y^*(x)$, the more it is preferred; thus, $y \succeq_x y'$ iff $\Delta_Y(y, y^*) \leq \Delta_Y(y', y^*)$. Please note that, when starting from an order relation $\succeq_x$, then the existence of an "ideal" solution is in principle no additional assumption (since a weak order has a maximal element, at least if the underlying space is topologically closed).

Instead, the additional assumption we make is that the order relations $\succeq_{\boldsymbol{x}}$ and $\succeq_{\boldsymbol{x}'}$ associated with different problems $\boldsymbol{x}$ and $\boldsymbol{x}'$ have a common structure, which is determined by the distance measure Δ_Y. In conjunction with the regularity assumption that is commonly made in CBR, namely that similar problems tend to have similar (ideal) solutions, this property legitimates a preference-based version of this assumption: *Similar problems are likely to induce similar preferences over solutions.*

3 Case-Based Inference

The key idea of Pref-CBR is to exploit experience in the form of previously observed preferences, deemed relevant for the problem at hand, in order to support the current problem solving episode; like in standard CBR, the *relevance* of a preference will typically be decided on the basis of problem similarity, i.e., those preferences will be deemed relevant that pertain to similar problems. An important question that needs to be answered in this connection is the following: Given a set of observed preferences on solutions, considered representative for a problem $\boldsymbol{x}_0$, what is the underlying preference structure $\succeq_{\boldsymbol{x}}$ or, equivalently, what is the most likely "ideal" solution $\boldsymbol{y}^*$ for $\boldsymbol{x}_0$?

3.1 Case-Based Inference as Probability Estimation

We approach this problem from a statistical perspective, considering the true preference model $\succeq_{\boldsymbol{x}_0} \in \mathfrak{P}(\mathbb{Y})$ associated with the query $\boldsymbol{x}_0$ as a random variable Z with distribution $\mathbf{P}(\cdot \,|\, \boldsymbol{x}_0)$, where $\mathbf{P}(\cdot \,|\, \boldsymbol{x}_0)$ is a distribution $\mathbf{P}_\theta(\cdot)$ parametrized by $\theta = \theta(\boldsymbol{x}_0) \in \Theta$. The problem is then to estimate this distribution or, equivalently, the parameter θ on the basis of the information available. This information consists of a set $\mathcal{D}$ of preferences of the form $\boldsymbol{y} \succ \boldsymbol{z}$ between solutions.

The basic assumption underlying nearest neighbor estimation is that the conditional probability distribution of the output given the input is (approximately) locally constant, that is, $\mathbf{P}(\cdot \,|\, \boldsymbol{x}_0) \approx \mathbf{P}(\cdot \,|\, \boldsymbol{x})$ for $\boldsymbol{x}$ close to $\boldsymbol{x}_0$. Thus, if the above preferences are coming from problems $\boldsymbol{x}$ similar to $\boldsymbol{x}_0$ (namely from the nearest neighbors of $\boldsymbol{x}_0$ in the case base), then this assumption justifies considering $\mathcal{D}$ as a representative sample of $\mathbf{P}_\theta(\cdot)$ and, hence, estimating θ via maximum likelihood (ML) by

$$\theta^{ML} = \arg\max_{\theta \in \Theta} \mathbf{P}_\theta(\mathcal{D}) \ . \tag{2}$$

An important prerequisite for putting this approach into practice is a suitable data generating process, i.e., a process generating preferences in a stochastic way.

3.2 A Discrete Choice Model

Our data generating process is based on the idea of a discrete choice model as used in choice and decision theory [6]. Recall that the (absolute) preference for a solution $\boldsymbol{y} \in \mathbb{Y}$ supposedly depends on its distance $\Delta_Y(\boldsymbol{y}, \boldsymbol{y}^*) \geq 0$ to an "ideal" solution $\boldsymbol{y}^*$, where $\Delta(\boldsymbol{y}, \boldsymbol{y}^*)$ can be seen as a "degree of suboptimality" of $\boldsymbol{y}$. As explained in [1], more specific assumptions on an underlying (latent) utility function on solutions justify the *logit* model of discrete choice:

$$\mathbf{P}(\boldsymbol{y} \succ \boldsymbol{z}) = \left(1 + \exp\left(-\beta(\Delta_Y(\boldsymbol{z}, \boldsymbol{y}^*) - \Delta_Y(\boldsymbol{y}, \boldsymbol{y}^*))\right)\right)^{-1} \qquad (3)$$

Thus, the probability of observing the (revealed) preference $\boldsymbol{y} \succ \boldsymbol{z}$ depends on the degree of suboptimality of $\boldsymbol{y}$ and $\boldsymbol{z}$, namely their respective distances to the ideal solution, $\Delta_Y(\boldsymbol{y}, \boldsymbol{y}^*)$ and $\Delta_Y(\boldsymbol{z}, \boldsymbol{y}^*)$: The larger the difference $\Delta_Y(\boldsymbol{z}, \boldsymbol{y}^*) - \Delta_Y(\boldsymbol{y}, \boldsymbol{y}^*)$, i.e., the less optimal $\boldsymbol{z}$ in comparison to $\boldsymbol{y}$, the larger the probability to observe $\boldsymbol{y} \succ \boldsymbol{z}$; if $\Delta_Y(\boldsymbol{z}, \boldsymbol{y}^*) = \Delta_Y(\boldsymbol{y}, \boldsymbol{y}^*)$, then $\mathbf{P}(\boldsymbol{y} \succ \boldsymbol{z}) = 1/2$. The coefficient β can be seen as a measure of precision of the preference feedback. For large β, the probability (3) converges toward 0 if $\Delta_Y(\boldsymbol{z}, \boldsymbol{y}^*) < \Delta_Y(\boldsymbol{y}, \boldsymbol{y}^*)$ and toward 1 if $\Delta_Y(\boldsymbol{y}', \boldsymbol{y}^*) > \Delta_Y(\boldsymbol{y}, \boldsymbol{y}^*)$; this corresponds to a deterministic (error-free) information source. The other extreme case, namely $\beta = 0$, models a completely unreliable source reporting preferences at random.

3.3   Maximum Likelihood Estimation

The probabilistic model outlined above is specified by two parameters: the ideal solution $\boldsymbol{y}^*$ and the (true) precision parameter $\beta^* \in \mathbb{R}_+$. Depending on the context in which these parameters are sought, the ideal solution might be unrestricted (i.e., any element of $\mathbb{Y}$ is an eligible candidate), or it might be restricted to a certain subset $\mathbb{Y}_0 \subseteq \mathbb{Y}$ of candidates.

Now, to estimate the parameter vector $\theta^* = (\boldsymbol{y}^*, \beta^*) \in \mathbb{Y}_0 \times \mathbb{R}^*$ from a given set $\mathcal{D} = \{\boldsymbol{y}^{(i)} \succ \boldsymbol{z}^{(i)}\}_{i=1}^N$ of observed preferences, we refer to the maximum likelihood (ML) estimation principle. Assuming independence of the preferences, the log-likelihood of $\theta = (\boldsymbol{y}, \beta)$ is given by

$$\ell(\theta) = \ell(\boldsymbol{y}, \beta) = -\sum_{i=1}^N \log\left(1 + \exp\left(-\beta(\Delta(\boldsymbol{z}^{(i)}, \boldsymbol{y}) - \Delta(\boldsymbol{y}^{(i)}, \boldsymbol{y}))\right)\right). \qquad (4)$$

The maximum likelihood estimation (MLE) $\theta_{ML} = (\boldsymbol{y}^{ML}, \beta^{ML})$ of θ^* is given by the maximizer of (4):

$$\theta_{ML} = (\boldsymbol{y}^{ML}, \beta^{ML}) = \arg\max_{\boldsymbol{y} \in \mathbb{Y}_0, \beta \in \mathbb{R}_+} \ell(\boldsymbol{y}, \beta)$$

The problem of finding this estimation in an efficient way is discussed in [1].

4   CBR as Preference-Guided Search

Case-based reasoning and (heuristic) search can be connected in various ways. One idea is to exploit CBR in order to enhance heuristic search, which essentially comes down to using case-based experience to guide the search behavior [7–9].

The other way around, the CBR process itself can be formalized as a search process, namely a traversal of the space of potential solutions [10]. This idea is quite appealing: On the one side, it is close to practical, human-like problem solving, which is indeed often realized as a kind of trial-and-error process, in which a candidate solution is successively modified and improved until a satisfactory solution is found. On the other side, this idea is also amenable to a proper formalization and automation, since *searching* is what computers are really good at; besides, heuristic search is one of the best developed subfields of AI.

Needless to say, both directions (enhancing search through CBR and formalizing CBR as search) are not mutually exclusive and can be combined with each other. In our approach, this is accomplished by implementing case-based problem solving as a search process that is guided by preference information collected in previous problem solving episodes. The type of application we have in mind is characterized by two important properties:

- *The evaluation of candidate solutions is expensive.* Therefore, only relatively few candidates can be considered in a problem solving episode before a selection is made. Typical examples include cases where an evaluation requires time-consuming simulation studies or human intervention. In the cooking domain, for example, the evaluation of a recipe may require its preparation and tasting. Needless to say, this can only be done for a limited number of variations.
- *The quality of candidate solutions is difficult to quantify.* Therefore, instead of asking for numerical utility degrees, we make a much weaker assumption: Feedback is only provided in the form of pairwise comparisons, informing about which of two candidate solutions is preferred (for example, which of two meals tastes better). Formally, we assume the existence of an "oracle" (for example, a user or a computer program) which, given a problem x_0 and two solutions y and z as input, returns a preference $y \succ z$ or $z \succ y$ (or perhaps also an indifference $y \sim z$) as output.

We assume the solution space $\mathbb{Y}$ to be equipped with a topology that is defined through a *neighborhood structure*: For each $y \in \mathbb{Y}$, we denote by $\mathcal{N}(y) \subseteq \mathbb{Y}$ the neighborhood of this candidate solution. The neighborhood is thought of as those solutions that can be produced through a single modification of y, i.e., by applying one of the available adaptation operators to y (for example, adding or removing a single ingredient in a recipe). Since these operators are application-dependent, we are not going to specify them further here.

Our case base **CB** stores problems x_i together with a set of preferences $\mathcal{P}(x_i)$ that have been observed for these problems. Thus, each $\mathcal{P}(x_i)$ is a set of preferences of the form $y \succ_{x_i} z$. As will be explained further below, these preferences are collected while searching for a good solution to x_i.

We conceive preference-based CBR as an iterative process in which problems are solved one by one; our current implementation of this process is described in pseudo-code in Algorithm 1. In each problem solving episode, a good solution for a new query problem is sought, and new experiences in the form of preferences are collected. In what follows, we give a high-level description of a single problem solving episode (lines 5–23 of the algorithm):

- Given a new query problem x_0, the K nearest neighbors[1] $x_1, \ldots, x_K$ of this problem (i.e., those with smallest distance in the sense of Δ_X) are retrieved from the case base **CB**, together with their preference information $\mathcal{P}(x_1), \ldots, \mathcal{P}(x_K)$.
- This information is collected in a single set of preferences $\mathcal{P}$, which is considered representative for the problem x_0 and used to guide the search process (line 8).
- The search for a solution starts with a initial candidate $y^* \in \mathbb{Y}$ chosen at random (line 9) and iterates L times. Restricting the number of iterations by an upper bound L reflects our assumption that an evaluation of a candidate solution is costly.
- In each iteration, a new candidate y^{query} is determined and given as a query to the oracle (line 15), i.e., the oracle is asked to compare y^{query} with the current best solution y^* (line 16). The preference reported by the oracle is memorized by adding it to the preference set $\mathcal{P}_0 = \mathcal{P}(x_0)$ associated with x_0 (line 17), as well as to the set $\mathcal{P}$ of preferences used for guiding the search process. Moreover, the better solution is retained as the current best candidate (line 18).
- When the search stops, the current best solution y^* is returned, and the case $(x_0, \mathcal{P}_0)$ is added to the case base.

The preference-based guidance of the search process is realized in lines 9 and 14–15. Here, the case-based inference method (referred to as CBI in the pseudo-code) described in Section 3 is used to find the most promising candidate among the neighborhood of the current solution y^* (excluding those solutions that have already been tried). By providing information about which of these candidates will most likely constitute a good solution for x_0, it (hopefully) points the search into the most promising direction. Please note that in line 15, case-based inference is not applied to the whole set of preferences $\mathcal{P}$ collected so far, but only to a subset of the J preferences $\mathcal{P}^{nn}$ that are closest (and hence most relevant) to the current search state y^*; here, the distance between a preference $y \succ z$ and a solution y^* is defined as

$$\Delta (y^*, y \succ z) = \min \{\Delta_Y (y^*, y), \Delta_Y (y^*, z)\} , \qquad (5)$$

i.e., the preference is considered relevant if either y is close to y^* or z is close to y^*. This is done in order to allow for controlling the locality of the search: The smaller J, the less preferences are used, i.e., the more local the determination of the direction of the search process[2] becomes (by definition, CBI returns a random element from $\mathbb{Y}^{nn}$ if $\mathcal{P}^{nn} = \emptyset$, i.e., if $J = 0$). Note that, if $J = 1$, then only the preference that has been added in the last step is looked at (since this

---

[1] As long as the case base contains less than K cases, all these cases are taken.
[2] The term "direction" is used figuratively here; if $\mathbb{Y}$ is not a metric space, there is not necessarily a direction in a strictly mathematical sense.

Algorithm 1. Pref-CBR Search(K, L, J)

Require: K = number of nearest neighbors collected in the case base
L = total number of queries to the oracle
J = number of preferences used to guide the search process

1: $\mathbb{X}_0 \leftarrow$ list of problems to be solved $\triangleright$ a subset of $\boldsymbol{X}$
2: $Q \leftarrow [\cdot]$ $\triangleright$ empty list of performance degrees
3: $\mathbf{CB} \leftarrow \emptyset$ $\triangleright$ initialize empty case base
4: **while** $\mathbb{X}_0$ not empty **do**
5:     $\boldsymbol{x}_0 \leftarrow$ pop first element from $\mathbb{X}_0$ $\triangleright$ new problem to be solved
6:     $\{\boldsymbol{x}_1, \dots, \boldsymbol{x}_K\} \leftarrow$ nearest neighbors of $\boldsymbol{x}_0$ in $\mathbf{CB}$ (according to Δ_X)
7:     $\{\mathcal{P}(\boldsymbol{x}_1), \dots, \mathcal{P}(\boldsymbol{x}_K)\} \leftarrow$ preferences associated with nearest neighbors
8:     $\mathcal{P} \leftarrow \mathcal{P}(\boldsymbol{x}_1) \cup \mathcal{P}(\boldsymbol{x}_2) \cup \dots \cup \mathcal{P}(\boldsymbol{x}_k)$ $\triangleright$ combine neighbor preferences
9:     $\boldsymbol{y}^* \leftarrow \mathrm{CBI}(\mathcal{P}, \mathbb{Y})$ $\triangleright$ select an initial candidate solution
10:    $\mathbb{Y}^{vis} \leftarrow \{\boldsymbol{y}^*\}$ $\triangleright$ candidates already visited
11:    $\mathcal{P}_0 \leftarrow \emptyset$ $\triangleright$ initialize new preferences
12:    **for** $i = 1$ to L **do**
13:       $\mathcal{P}^{nn} = \{\boldsymbol{y}^{(j)} \succ \boldsymbol{z}^{(j)}\}_{j=1}^J \leftarrow J$ preferences in $\mathcal{P} \cup \mathcal{P}_0$ closest to $\boldsymbol{y}^*$
14:       $\mathbb{Y}^{nn} \leftarrow$ neighborhood $\mathcal{N}(\boldsymbol{y}^*)$ of $\boldsymbol{y}^*$ in $\mathbb{Y} \setminus \mathbb{Y}^{vis}$
15:       $\boldsymbol{y}^{query} \leftarrow \mathrm{CBI}(\mathcal{P}^{nn}, \mathbb{Y}^{nn})$ $\triangleright$ find next candidate
16:       $[\boldsymbol{y} \succ \boldsymbol{z}] \leftarrow \mathrm{Oracle}(\boldsymbol{x}_0, \boldsymbol{y}^{query}, \boldsymbol{y}^*)$ $\triangleright$ check if new candidate is better
17:       $\mathcal{P}_0 \leftarrow \mathcal{P}_0 \cup \{\boldsymbol{y} \succ \boldsymbol{z}\}$ $\triangleright$ memorize preference
18:       $\boldsymbol{y}^* \leftarrow \boldsymbol{y}$ $\triangleright$ adopt the current best solution
19:       $\mathbb{Y}^{vis} \leftarrow \mathbb{Y}^{vis} \cup \{\boldsymbol{y}^{query}\}$
20:    **end for**
21:    $q \leftarrow$ performance of solution $\boldsymbol{y}^*$ for problem $\boldsymbol{x}_0$
22:    $Q \leftarrow [Q, q]$ $\triangleright$ store the performance
23:    $\mathbf{CB} \leftarrow \mathbf{CB} \cup \{(\boldsymbol{x}_0, \mathcal{P}_0)\}$ $\triangleright$ memorize new experience
24: **end while**
25: return list Q of performance degrees

---

preference involves $\boldsymbol{y}^*$, and therefore its distance according to (5) is 0. Thus, search will move ahead in the same direction if the last modification has led to an improvement, and otherwise reverse its direction. In general, a larger J increases the bias of the search process and makes it more "inert". This is advantageous if the preferences coming from the neighbors of $\boldsymbol{x}_0$ are indeed representative and, therefore, are pointing in the right direction. Otherwise, of course, too much reliance on these preferences may prevent one from searching in other regions of the solution space that might be more appropriate for $\boldsymbol{x}_0$.

Although we did not implement this alternative so far, let us mention that a stochastic component can be added to our search procedure in a quite natural way. To this end, the case-based inference procedure CBI simply returns one of the candidate solutions $\boldsymbol{y} \in \mathbb{Y}^{cand}$ with a probability that is proportional to the corresponding likelihood degrees of these solutions (instead of deterministically choosing the solution with the highest likelihood).

5 Case Studies

5.1 Drug Discovery

The function of a protein in a living organism can be modulated by ligand molecules that specifically bind to the protein surface and thereby block or enhance its biochemical activity. This is how a drug becomes effective: By docking to a protein and changing its activity, it (hopefully) interrupts a cascade of reactions that might be responsible for a disease.

The identification and selection of ligands targeting a specific protein is of high interest for de-novo drug development, and is nowadays supported by computational tools and molecular modeling techniques. Molecular docking is an *in silico* technique to screen large molecule databases for potential ligands. Using the spatial (three-dimensional) structure and physicochemical properties of proteins, it tries to identify novel ligands by estimating the binding affinity between small molecules and proteins. However, since docking results are not very reliable, they need to be controlled by human experts. This is typically done through visual inspection, i.e., by looking at the docking poses predicted by the software tool and judging whether or not a molecule is indeed a promising candidate. Needless to say, this kind of human intervention is costly. Besides, a human will normally not be able to score a docking pose in terms of a numerical (affinity) degree, whereas a comparison of two such poses can be accomplished without much difficulties. Therefore, the search for a ligand that well interacts with a target protein is a nice example of the kind of problem we have in mind.

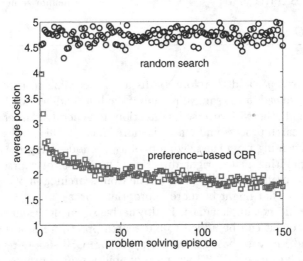

Fig. 1. Average performance of Pref-CBR and random search on the drug discovery problem in the first 150 problem solving episodes

We conducted experiments with a data set consisting of 588 proteins, which constitute the problem space $\mathbb{X}$, and 38 molecules, which correspond to the solution space $\mathbb{Y}$; this data set is an extension of the data used in [11]. For each protein/molecule pair, the data contains an affinity score (pairwise binding energy) computed by a docking tool. We make use of these scores in order to mimic a human expert, i.e., to realize our oracle: Given a protein and two candidate molecules, the oracle can provide a preference by looking at the corresponding affinity scores. As a similarity S_X on problems (proteins), we used the measure that is computed by the CavBase database; this measure compares proteins in terms of the spatial and physicochemical properties of their respective binding sites [12]. For the solutions (ligands), a similarity S_Y was determined based on molecular fingerprints derived from the SMILES code using a molecular operating environment. These fingerprints were used to create a graph representation of the molecules, for which the Tanimoto similarity was determined [13]. Both similarities S_X and S_Y were normalized to the unit interval, and corresponding distances Δ_X and Δ_Y were defined as $1 - S_X$ and $1 - S_Y$, respectively.

We applied Algorithm 1 with $\mathbb{X}_0$ as a random order of the complete problem space $\mathbb{X}$. Since the solution space is quite small, we used a global neighborhood structure, i.e., we defined the neighborhood of a solution y as $\mathcal{N}(y) = \mathbb{Y} \setminus \{y\}$. As a performance q of a proposed solution y^* for a problem x_0 (line 21), we computed the position of this solution in the complete list of $|\mathbb{Y}| = 38$ ligands ranked by affinity to x_0 (i.e., 1 would be the optimal performance). To stabilize the results and make trends more visible, the corresponding sequence of $|\mathbb{X}| = 588$ performance degrees produced by a single run of Algorithm 1 was averaged over 1000 such runs.

As a baseline to compare with, we used a search strategy in which the preference-guided selection of the next candidate solution in line 15 of Algorithm 1 is replaced by a random selection (i.e., an element from $\mathbb{Y}^{nn}$ is selected uniformly at random). Although this is a very simple strategy, it is suitable to isolate the effect of guiding the search behavior on the basis of preference information. Fig. 1 shows the results for parameters $K = 3$, $L = 5$, $J = 15$ in Algorithm 1 (other settings let to qualitatively similar results). As can be seen, our preference-based CBR approach shows a clear trend toward improvement from episode to episode, thanks to the accumulation and exploitation of problem solving experience. As expected, such an improvement is not visible for the random variant of the search algorithm.

5.2   The Set Completion Problem

In a second experiment, we considered a set completion problem that is similar to the problem solved by the Bayesian set algorithm proposed in [14]. Given a (small) subset of items as a seed, the task is to extend this seed by successively adding (or potentially also removing) items, so as to end up with a "good" set of items. As a concrete example, imagine that items are ingredients, and itemsets correspond to (simplified) representations of cooking recipes. Then, the problem is to extend a seed like {noodles, chicken}, suggesting that a user wants a meal including noodles and chicken, to a complete and tasty recipe.

More formally, both the problem space and the solution space are now given by $\mathbb{X} = \mathbb{Y} = 2^{\mathcal{I}}$, where $\mathcal{I} = \{\iota_1, \ldots, \iota_N\}$ is a finite set of items; thus, both problems and solutions are itemsets. We define the distance measures Δ_X and Δ_Y in terms of the size of the symmetric difference Δ, i.e.,

$$\Delta_X(x, x') = |x \, \Delta \, x'| = |x \setminus x'| + |x' \setminus x| \ .$$

Let $\mathbb{Y}^* \subset \mathbb{Y}$ be a set of reference solutions (e.g., recipes of tasty meals). For a $y \in \mathbb{Y}$, define the distance to $\mathbb{Y}^*$ as

$$d(y) = \min_{y^* \in \mathbb{Y}^*} |y \, \Delta \, y^*| \ .$$

Moreover, for a problem $x \in \mathbb{X}$, we define a preference relation on $\mathbb{Y}$ as follows: $y \succ z$ if either $c(y \,|\, x) < c(z \,|\, x)$ or $c(y \,|\, x) = c(z \,|\, x)$ and $|y| < |z|$, where

$$c(y \,|\, x) = \begin{cases} d(y) & \text{if } y \supseteq x \\ \infty & \text{otherwise} \end{cases}$$

Thus, the worst solutions are those that do not fully contain the original seed. Among the proper extensions of the seed, those being closer to the reference solutions $\mathbb{Y}^*$ are preferred; if two solutions are equally close, the one with less items (i.e., the less expensive one) is preferred to the larger one. For a candidate solution y, we define the neighborhood as the set of those itemsets that can be produced by adding or removing a single item:

$$\mathbb{Y}^{nn} = \{ y' \,|\, \Delta_Y(y, y') = 1 \} \ .$$

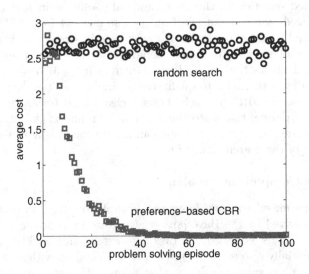

Fig. 2. Average performance of Pref-CBR and random search on the set completion problem in the first 100 problem solving episodes

Finally, for a given problem x_0, we define the performance of a found solution y^* in terms of $c(y^* \mid x_0)$.

We applied this setting to a database of pizzas extracted from the website `allrecipes.com`, each one characterized by a number of toppings (typically between 6 and 10). Seeds (problems) were produced at random by picking a pizza and removing all except three toppings. The task is then to complete this seed by adding toppings, so as to produce a tasty pizza (preferably one of those in the database, which plays the role of the reference set $\mathbb{Y}^*$). Again, we compared Algorithm 1 with the random search variant as a baseline. The results for parameters $K = 5$, $L = 10$, $J = 50$, shown in Fig. 2, which are qualitatively similar to those of the previous study.

6   Conclusion

In this paper, we have presented a general framework for CBR in which experience is represented in the form of contextualized preferences, and these preferences are used to direct an adaptive problem solving process that is formalized as a search procedure. This kind of preference-based CBR is an interesting alternative to conventional CBR whenever solution quality is a matter of degree and feedback is only provided in an indirect or qualitative way. The effectiveness of our generic framework has been illustrated in two concrete case studies.

For future work, we plan to extend and generalize our framework in various directions. First, the search procedure presented here can essentially be seen as a preference-based variant of a simple hill-climbing method. Needless to say, the idea of using preferences for guiding the search process can be applied to other, more sophisticated search methods (including population-based stochastic search algorithms) in a quite similar way. Second, since the number of preferences collected in the course of time may become rather large, effective methods for case base maintenance ought to be developed. Third, as already mentioned, the similarity (distance) measure in the solution space has an important influence on the preference relations $\succeq_x$ associated with problems $x \in \mathbb{X}$ and essentially determines the structure of these relations (cf. Section 2.3). Therefore, a proper specification of this measure is a prerequisite for the effectiveness of our preference-guided search procedure. It would hence be desirable to allow for a data-driven adaptation of this measure, that is, to enable the CBR system to adapt this measure whenever it does not seem to be optimal. The method for learning similarity measures from qualitative feedback proposed in [15] appears to be ideally suited for this purpose.

Acknowledgments. This work has been supported by the German Research Foundation (DFG). We are grateful to Peter Kolb and Denis Schmidt for providing us the data used for in the drug discovery experiment.

References

1. Hüllermeier, E., Schlegel, P.: Preference-based CBR: First steps toward a methodological framework. In: Ram, A., Wiratunga, N. (eds.) ICCBR 2011. LNCS, vol. 6880, pp. 77–91. Springer, Heidelberg (2011)
2. Doyle, J.: Prospects for preferences. Comput. Intell. 20(2), 111–136 (2004)
3. Goldsmith, J., Junker, U.: Special issue on preference handling for Artificial Intelligence. Computational Intelligence 29(4) (2008)
4. Domshlak, C., Hüllermeier, E., Kaci, S., Prade, H.: Preferences in AI: An overview. Artificial Intelligence (2011)
5. Brafman, R.I., Domshlak, C.: Preference handling–an introductory tutorial. AI Magazine 30(1) (2009)
6. Peterson, M.: An Introduction to Decision Theory. Cambridge Univ. Press (2009)
7. Kraay, D.R., Harker, P.T.: Case-based reasoning for repetitive combinatorial optimization problems, part I: Framework. Journal of Heuristics 2, 55–85 (1996)
8. Grolimund, S., Ganascia, J.G.: Driving tabu search with case-based reasoning. European Journal of Operational Research 103(2), 326–338 (1997)
9. Hüllermeier, E.: Focusing search by using problem solving experience. In: Horn, W. (ed.) Proceedings ECAI 2000, 14th European Conference on Artificial Intelligence, Berlin, Germany, pp. 55–59. IOS Press (2000)
10. Bergmann, R., Wilke, W.: Towards a new formal model of transformational adaptation in case-based reasoning. In: Prade, H. (ed.) ECAI 1998, 13th European Conference on Artificial Intelligence, pp. 53–57 (1998)
11. Karaman, M.W., et al.: A quantitative analysis of kinase inhibitor selectivity. Nature Biotechnology 26, 127–132 (2008)
12. Schmitt, S., Kuhn, D., Klebe, G.: A new method to detect related function among proteins independent of sequence and fold homology. Journal of Molecular Biology 323(2), 387–406 (2002)
13. Stock, M.: Learning pairwise relations in bioinformatics: Three case studies. Master's thesis, University of Ghent (2012)
14. Ghahramani, Z., Heller, K.A.: Bayesian sets. In: Proceedings NIPS 2005 (2005)
15. Cheng, W., Hüllermeier, E.: Learning similarity functions from qualitative feedback. In: Althoff, K.-D., Bergmann, R., Minor, M., Hanft, A. (eds.) ECCBR 2008. LNCS (LNAI), vol. 5239, pp. 120–134. Springer, Heidelberg (2008)

Applying MapReduce
to Learning User Preferences in Near Real-Time

Ian Beaver and Joe Dumoulin

NextIT Corporation,
421 W. Riverside Ave, Spokane WA 99201 USA
{ibeaver,jdumoulin}@nextit.com
http://www.nextit.com

Abstract. When computer programs participate in conversations, they can learn things about the people they are conversing with. A conversational system that helps a user select a flight may notice that a person prefers a particular seating arrangement or departure airport. In this paper we discuss a system which uses the information state accumulated during a person-machine conversation and a case-based analysis to derive preferences for the person participating in that conversation. We describe the implementation of this system based on a MapReduce framework that allows for near real-time generation of a user's preferences regardless of the total case memory size. We also show some preliminary performance results from scaling tests.

Keywords: case-based reasoning, dialogue systems, natural language.

1 Introduction

Next IT is a company in Spokane WA, USA that builds natural language applications for the worldwide web and for mobile device applications. Recently we have been working on features to improve the performance of machine directed conversations. This paper describes one of those enhancements, the development of a scalable system for learning user preferences from past experience in near real-time.

The system we developed is closely tied to our existing natural language system (NLS) so this paper will begin with a discussion of the behaviour of this system and how it interacts with people. The following sections discuss the architecture, operation, and performance testing of our CBR-based learning system.

1.1 User-Directed and Machine-Directed Conversations

The NLS in question enables designers to craft user-directed and machine-directed conversation templates. Users of the system can initiate these templates by starting a conversation with the system. The conversation can take the form of a chat or a task to complete. For example, we can assume that the application is running in an airline domain. A person can ask the system "how much does it

S.J. Delany and S. Ontañón (Eds.): ICCBR 2013, LNAI 7969, pp. 15–28, 2013.

cost to check an extra bag?" and the system may respond with a simple answer like "$10." This is a user-directed conversation.

In contrast, a machine-directed conversation turns the user's request into a task to be completed, and the system asks the user a series of subsequent questions to support that task. The system keeps track of what information has been gathered and what is still required to complete the task in a *form*, or conversation state object. For example, suppose the person wants to book a flight. They may start by saying "I need to fly to Seattle on Tuesday." The system can then ask a series of questions of the user to fill in the remaining information needed to complete the task.

There may be a validation step to perform for each form component, or *slot*, to ensure that the user has supplied valid information and to re-prompt the user in the case of invalid data. When the user has filled in all of the slots in the form through conversation, the system has enough information to complete the task by booking the flight the user requested.

To summarize, a user-directed conversation is conducted by the user asking questions of the system. A machine-directed conversation is conducted by a system asking questions of a user. Systems which perform both of these tasks are sometimes called mixed-initiative systems. Mixed-initiative dialogue systems have been proposed as far back as the 1970's[1] and have been shown to both perform well[2] and adapt well to different domains[3].

1.2   Machine-Directed Conversations and Personal Preferences

Designing a machine directed conversation includes a number of steps. The first step is creating the tasks the system will be capable of performing (defining the set of available forms). For each task in the system, the next step is determining the data needed to complete the task (defining the slots in the form to be filled). Once the task is well-defined, a final optimization step is looking at ways to minimize the number of questions the system must ask in order to complete the task. This is especially important if a person is expected to return to the system and complete this task many times, as may be the case when booking travel.

Something we noticed while investigating the need to minimize the number of questions being asked is that, for a given person, the answers to certain questions are consistently the same. For example, if a person is a frequent traveller and books flights through the system, there is a high likelihood that the flight's departure airport will be the home airport of the person. In another example, suppose the person prefers aisle seats. This will become evident after only a small number of bookings. We concluded that in order to shorten the task completion turns, we could learn a person's preferences based on previous conversations with the user.

The remainder of this paper describes the system that we constructed to learn user preferences in the scope of any machine-directed conversation and to make that learned behaviour accessible to subsequent sessions with the same person. This learned preferences system can then be applied to any of the domains in which our natural language system is deployed. In Section 2 we outline the implementation goals and the component architecture used to achieve them. In Section 3 we describe how we applied case-based reasoning methods to address some of the implementation goals. Section 4 describes the testing methodology we used to ensure that our system could perform adequately when installed on production hardware and under load. Finally, Section 5 outlines our Conclusions.

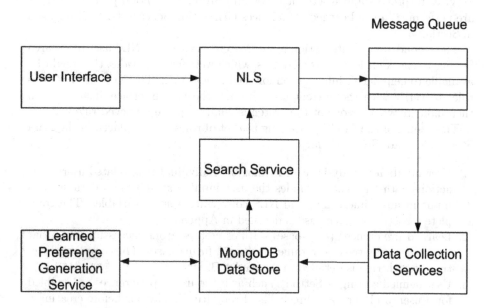

Fig. 1. A component view of the NLS and the CBR Learning system and the services that support it

2 Implementation Goals and Architecture

There were several general properties that we hoped to obtain from adding a learning component to our existing NLS. We will later show how the case-based reasoning (CBR) system we developed satisfied each of these properties:

- Reduce the number of turns required to complete a task for a returning user, thereby increasing the user's satisfaction with the system.
- Allow the user to adjust the level of personalization the system displays to them.
- Store the learned preferences in a way that they can be quickly and easily retrieved and displayed to the end user.
- Create preferences from user input in near real-time.
- Confirm new preferences with the user to prevent unexpected user experiences.

- Allow users to change their preferences at any time.
- Scale easily as the number of users and input history grows.

Figure 1 shows a simplified component view of the entire NLS with the CBR Learning System. There are two integration points for the CBR Learning System:

The first is where the NLS queues user responses for the Data Collection Services to process and insert into the Data Store. The actual processing steps are performed in the Learned Preferences Generation Services (LPGS) outlined in Section 3.

The second area of integration is the Search Service. The NLS uses the Search Service to check for learned preferences, which we refer to as *rules*, that might be available to help derive information for the in-focus task. The NLS also retrieves rules to verify with the current user through the user interface. The user can then indicate whether or not the system should keep the retrieved rule.

The results of the LPGS processing is a set of rules that is placed in the Data Store. The Data Store contains:

1. User inputs for analysis (case memory) - Individual task-related user interactions with the NLS. Includes the user input text and meta data such as input means, timestamp, and NLS conversation state variables. The complete structure of each case is detailed in Appendix A.
2. Learned preferences (case-base) - Rules created from successful cases that have been analysed and retained for use in future cases. The complete structure of each rule is detailed in Appendix B.
3. User defined settings - Settings such as if the use of preferences are enabled for a user, and per user thresholds of repetitive behaviour before creating a preference solution.

3 Application of Case-Based Reasoning

The primary design idea for the learning system was to look at entire user histories and group their NLS interactions by task. If a user consistently provides the same value for a given slot, we can assume it instead of prompting in future conversations. This was a perfect fit for a CBR methodology since we are looking at specific instances in a user's history and reusing that information to solve a future problem, namely minimizing the number of steps required to repeat the task in the future.

Similar systems have been proposed to learn user preferences for the purpose of creating user recommendations [4]([5] compares many such approaches) or generating a user profile[6,7]. The primary difference in our system is that instead of learning preferences for the purpose of filtering or ordering information presented to the user, we are attempting to anticipate responses to specific NLS prompts. Another key difference is that we require these rules to be constructed in near real-time as the user is aware of exactly when a rule should exist based on their personal settings. Our system must be able to maintain this near real-time property as the number of users and cases grow, as we have many existing natural language systems deployed to large companies with very large numbers of users.

Taking the scale into consideration, we decided on a MapReduce programming model as it allows functions to be applied to large datasets in an automatically parallelized fashion[8]. As the dataset grows it can be partitioned across more resources as needed to maintain the near real-time requirement without additional complexity or code changes.

3.1 Case Structure

The major issue that had to be addressed was how to structure the case feature set so that it was general enough that unique information like timestamps and sequences would not prevent a match with a similar conversation, but specific enough to maintain a distinction between conversations with different outcomes.

We overcame this problem by requiring the task designers to specify the set of features in the conversation state that were important for each slot in a task. Since the task designer is the knowledge expert that is defining the ontology of the system, they know ahead of time what feature variables exist in the conversation state and which ones are important to the slot that is being prompted for. When the set of available forms is generated for the NLS system to consume, a separate file is created that defines these relationships, which the learning system uses to identify features for each case.

This solution can still be error prone when feature variable names are too generic or are reused for several different values. For example, suppose we are trying to learn a preferred email address for the user and the designated variable is named *EmailAddr*. If Sally were to ask the system to "Send an email to Fred", and the same variable name is re-used for both source and destination addresses, it can lead to a rule never being generated for Sally's preferred email address because the addresses contained in *EmailAddr* will never agree. After a few iterations of conversation template deployment and testing most of these variable name issues became apparent and were resolved.

3.2 Case Storage

Every task-related user interaction with the NLS is saved as a case. Simple user-directed interactions are not retained as they are not a multi-step process that we can attempt to optimize for repeat users.

MongoDB was chosen as the Data Store for its schema flexibility[9] and its ability to easily scale as the number of cases increases[10]. It also includes a MapReduce mechanism which allows for complex parallel searching of the case memory. Any MapReduce framework and distributed file system could have been used for the learning system such as Hadoop or Disco, but we chose MongoDB as it supported our requirements in a single simple to deploy package.

3.3   Case Retrieval

When a user begins a task in the NLS, before the NLS prompts the user for each slot in the form (the new case to be solved), it first sends the new case features to the Search Service to see if there is an existing preferential value to place in the slot from the case-base. Fig. 2 shows an example of preferring learned preferences over asking the user.

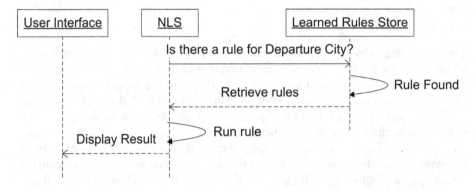

Fig. 2. Flow diagram for a portion of 'Book A Flight' dialogue. The system is resolving whether or not rules exist in order to avoid an explicit prompt for 'Departure City'.

The features of this new case are defined by:

1. A Universally Unique Identifier (UUID) generated for this new case.
2. The current UserID.
3. The array of variable-value pairs that make up the current conversation state, for example: [(IsLoggedIn,True), (LatLong,[47.656918,-117.418356])]
4. The active task the user has initiated in the NLS.

By ordering the array of context pairs by variable name, and then generating a SHA-1 hash of the string representation of the array, a search key is generated. Using this search key, the UserID, and TaskID, we query the case-base for an existing preference rule. If no preference is found in the case-base, the NLS will then prompt the user to fill in the slot.

3.4   Case Reuse

If a preference is found that is both active and valid, it will be used to fill in the slot value, and the NLS will then repeat this process to try and fill in the next slot in the form.

3.5   Case Revision

If a preference is matched and it has been inactivated, it will not be used. If it is active but has not been validated by the user yet, the NLS will ask the user if they would like to use the historical value from now on. If so, the learned preference will be validated and used without prompting the user in the future. Otherwise, it will be marked inactive and will not be returned by future queries. Fig. 3 shows how the verification event takes place during a conversation. This example was taken from a live system and shows how the user's repeat task is optimized by the learning system, satisfying our primary goal of reducing the number of steps for repetitive tasks. Note that the initial conversation will need to be repeated as many times as this specific user has defined in their preferences before a rule will be constructed.

Initial Conversation(s)	Verification of New Rule	Rule in Use

User: text Fred
NLS: I found more than one phone number for 'Fred'. Which do you mean?
 	
    Home - ###
    Mobile - ###
    Work - ###
 	
User: Home number
NLS: What do you want the text to say?
	User: text Fred
	NLS: I notice that in the past when you've asked me to send a text message to Fred you chose to use the phone number ###. Would you like me to assume you mean ### now and in the future?
	User: Yes.
	NLS: Great! I'll make a note of it. What do you want the text to say?
		User: text Fred
		NLS: What do you want the text to say?

Fig. 3. Conversations from the same user over time using the learning system

3.6   Case Retainment: Learning Preferences from Case Memory

The NLS forwards all attempts to fill slots to the Learning System by placing them in the Message Queue where they will be stored in case memory by the Data Collection Service. The NLS does this even when a learned preference rule was used to fill the slot.

The LPGS receives events that new cases are available and does reanalysis on the affected user's entire case memory. The LPGS is looking for repetitive behaviours by the specific user that lead to the same outcome in completing a slot for a given task. This is accomplished by two MapReduce jobs applied to the user's history in the case memory collection.

First MapReduce Job. The first MapReduce job compresses continuous user inputs that are trying to complete the same slot within the same task. It may take the user several interactions with the NLS to resolve a specific slot. The user may give incorrect or incomplete data, or respond to the system prompt with a clarifying question of their own.

The compression is done by keeping track of when the user was first prompted for the slot, and when the slot was either filled in or abandoned. The system evaluates a few scenarios to determine the correct cases to use for the learning step.

In some scenarios, we know that the slot was not successfully completed because the NLS re-prompted the user for the same slot variable. Multiple prompts are merged into a single case. The earlier case's slot value and starting context are combined with the later case's slot value and ending context.

This reduction process continues until either 1) the slot was satisfied or abandoned and the NLS prompted for a new slot; or 2) no more cases are found in that sequence of interactions, meaning that either the user is currently in the middle of the conversation or they abandoned the entire conversation without completing the task.

When this first job completes, its combined results are stored with the single cases where the slot was resolved in a single interaction.

Second MapReduce Job. The LPGS starts a second MapReduce job on the first job's results. This job attempts to count all of the slot outcomes for this user that are equivalent. The job first groups all of the cases where the end context was the same. When it finds two cases that meet this criteria, it checks to see if the final answers matched. If that is not the case, later analysis will be done to determine which answer (if any) is most appropriate to learn.

Finalize. After this step completes, the LPGS starts a finalize function over the resulting groups. It filters out any group whose slot was never satisfied. For example, if the starting value for the slot is "null" and, after all of the attempts to resolve it, the value remains "null", we know that the slot was never satisfied.

The finalize step also filters out any groups where the count of observed cases is less than an adjustable threshold. This threshold is in place to define how many times a user needed to repeat a behaviour before it was considered a preference.

For example, the very first time a user completes a slot, it should not be considered a "preference" since there is not enough historical data to support that they will do it again. However, each user may have a different expectation of how many times they need to repeat an answer before it is saved.

Since one of our system goals is to allow a user to adjust the level of personalization, we exposed this threshold to them in a "Settings" screen within the application. This setting is saved with their user preferences and looked up when that user's inputs are analysed by the LPGS. The finalize function filters out any answers that do not meet the user's threshold for repeat behaviours. There are pre-set defaults so that users are not required to configure the system before using it and we impose a lower bound of two, for the above mentioned reason. Currently this setting is used for all forms but we have considered implementing form "groups" of similar tasks and allowing users to control the settings per group. We do not expect that this would add significant load to the LPGS since it would only involve adding a group key to the user setting query that it is already performing. It would, however, add more complexity to the application interface, and possibly add confusion to users so more research needs to be done on the benefits of adding this feature.

Analysis of Results. At this point we have a second temporary collection that contains cases where the slot was satisfied, and the count of how many times each answer was observed in the user's history. The LPGS now analyses these cases to determine if an answer can be assumed.

For the purposes of this analysis, the LPGS runs a set of functions over the data. Each of these functions can output rules, but they use different criteria to determine the rules. Currently, we have two functions in the set but we anticipate using more in the future.

The first function looks for a user-specified number of contiguous cases with identical values for the slot in a group. If found, a rule is created for the slot using the value of the case.

The second function checks for a percentage of final slot values to be the same. The specific percentage used is user-defined.

The value of this second function is that a user does not have to repeat the same answer many more times if they had a single different answer for some reason. For example, consider a frequent traveller. Most of the time they book a flight from their home airport, but sometimes they are in a different city and they book a flight leaving from there. We can still learn which airport they prefer since the majority of the time they choose their home airport.

More specifically, suppose they have their percentage set to 75% and the sequential threshold set to 3 for example. If their answers for the DepartureAirport slot from case memory ordered by time looks like [SEA,SEA,LAX,SEA], we can still assume that they prefer to leave from Seattle, even though the sequential function could not assume that.

After a function analyses the composite cases, any new preferences found by it are saved into the case-base. New preferences will be verified or ignored the next time the user initiates the task. The two temporary collections are then deleted and the MapReduce jobs are applied to another slot in the user's task history. This re-analysis process is completed for each user that has added new cases since the last time the LPGS fetched updated users.

4   Testing and Performance

4.1   System Evaluation

The primary measure of success for the learning system is reducing the number of steps required for the user to complete a task in the future. To evaluate this measure we needed to ensure that when a user repeats a task the same way their configured amount of times, a rule is created and that rule is found on the next attempt to complete the task. The evaluation was done following these steps:

1. Create a new user account
2. Choose custom threshold settings or use system defaults
3. Walk through a task in the system conversationally
4. Repeat the conversation enough times to meet the set thresholds
5. Assert that on the next attempt to complete the task a prompt to validate a learned preference appears
6. Assert that on the next attempt to complete the task no prompt appears but the task is completed using the learned preference

Once the system was shown to be working correctly for a single user, we released access to the UI in the form of a mobile application to a limited group of 35 testers. The testers had the ability to enable and disable the use of the learned preferences during their conversation to compare the change in experience. In our limited release testing user feedback was very positive. One user commented that "Using the application without learning enabled is annoying", compared to the experience with it enabled. This was due to the decrease in prompting by the NLS on repeat uses with the learning system enabled. An example of a conversation collected from this evaluation was shown in Fig. 3 above.

4.2   System Performance

The system was functioning as intended, but we had to ensure that the solution would be able to scale to a production capacity. Since the majority of the analysis work is done within the MapReduce jobs, the ability for the LPGS to scale is closely tied to the ability for the MapReduce engine (MongoDB) to scale. One of the goals of the system is that the creation of new preferences for a specific user must happen in near real-time from when a user input is received.

The definition of near real-time in this context is driven purely by user experience. There is an expectation by the user that, for example, after the third time booking a flight it will not ask them for their departure airport that they have given the last three times in a row. This would happen if this user has their learned threshold setting at three. If the system were then to ask them for that information, their expectations would not be met. In this example the definition of near real-time must be less than a realistic window of time before the user would repeat this task.

In this domain of booking flights, several hours may be an acceptable time frame since it is rare that users would book multiple flights in a several hour period leaving from the same airport. There may be domains where the same tasks are completed many times a day, as in a personal assistant domain where the user wants the system to learn that a nickname is associated to a specific contact they write text messages to often. In this domain the acceptable time frame may be only a matter of minutes.

Therefore we recognize that since this acceptable time frame varies by domain and expectations of the user base, we can only show how the system performs with the testing hardware available to us and know there will be larger computing capacity needed to cover domains with fast preference availability expectations.

To test that the system was capable of scaling to large numbers of cases, we needed to create a test data set in incremental sizes and show how performance degrades. We measure how long it took to analyse a single user for the set of tasks in the NLS and the total time it took the system to analyse all users in the data set for each case memory size. Since the LPGS only works on users that added new cases since the last time it ran, running against all users would be a test of the worse case scenario in the system.

4.3   Test Dataset Creation

In order to create a data set to test with, we used actual conversations from users of an existing NLS in the personal assistant domain. These conversations were inserted into the case memory directly, truncated in a way that the number of inputs or cases per user would form a Normal Distribution where $\mu = 365, \sigma = 168$ with negatives remapped as

$$f(n) = \begin{cases} \chi & \chi \geq 1 \\ \chi \in 1 \leq \mathbb{Z} \leq 10 & \chi < 1 \end{cases}$$

Where $1 \leq \mathbb{Z} \leq 10$ is generated at random. This larger distribution between $1 \ldots 10$ is to simulate users that try the NLS out of curiosity with no intention of accomplishing any task and then abandon it. We chose μ and σ values based on projected usage expectation in the personal assistant domain after reviewing historical NLS usage in current production environments. A custom Data Collection Service was used that simply marked all of the users as having new cases available instead of reading off the Message Queue and marking users with queued messages. This way the LPGS would have to look at all users at the same time, creating the maximum load on the system.

4.4   Testing Environment

The MongoDB cluster was constructed with 8 homogeneous servers with 2xE5450 CPUs, 16GB RAM and 2x73GB 15k rpm drives with RAID0. The system OS is Ubuntu Server 12.04LTS and the database and MapReduce system is MongoDB v2.2.3-rc1.

MongoDB was configured as 4 shards of 2-node replica sets. In the case memory and case-base collections, the ID was used as the shard key. In the user settings collection the UserID was used as the shard key. The learning service itself was running on a workstation with an Intel i7-3930K CPU and 64GB RAM and was configured to use 32 worker threads, meaning 32 users would be worked on in parallel. This number is configurable based on the computing power of the machine the learning system is running on. Multiple instances of the learning system can be started on multiple machines in order to reduce the total analysis time.

4.5   Performance Results

Table. 1 shows the results of running the LPGS against all of the users in the database. After each run more users were imported in case memory and the database cluster was fully restarted in order to clear out any cached data that may skew the benchmarks. The *Total Cases* column shows the number of cases in the case memory collection, all of which would have to be looked at if all user histories are analysed. The *Analysis Time* column is the wall-clock time from when the LPGS started to the time it completed the last user. The *Avg. User Time* column is the wall-clock time it took to complete all of the historical analysis on a single user.

Table 1. Performance (MongoDB v2.2.3-rc1)

Total Cases	Users	Avg. User Cases	Analysis Time (H:M:S)	Avg. User Time (S)
369,536	1,000	369	0:07:22	2.839
743,719	2,000	371	0:14:39	6.279
3,622,196	10,000	362	0:47:12	8.416
7,200,767	20,000	360	1:34:00	8.407

MongoDB handles the load of 32 parallel MapReduce jobs on completely separate (meaning uncached) data very well. The total time it takes to process 20,000 users would be acceptable in most domains without needing to use multiple instances of the LPGS. The *Avg. User Time* meets our definition of near real-time given the size of the data we tested with. This is also testing the absolute worst case, in a real world case where 20,000 unique users would need to be reviewed every 1.5 hours would in all likelihood mean there was a great deal more total users in the system. Notice that the *Avg. User Time* does not change when the case memory size is doubled from 3.6M to 7.2M cases. Once MongoDB has reached a stable load, the job time appears to flat line, at least until the indexes would no longer fit in memory[11]. Further testing is needed with larger data sets to know if this assumption holds true and where the limitations of our testing hardware are.

In order to compare how the performance changes with a different M/R engine, the same cluster was rebuilt using a developer preview version of MongoDB 2.4 (2.3.2) using the V8 JavaScript engine[12], which supports multi-threaded MapReduce execution compared to the single-threaded execution in the SpiderMonkey engine used in versions 2.2 and below. This release proved to be still unstable, with several memory leaks observed and occasional long pauses in MapReduce job execution. In spite of this, as Table. 2 shows[1], the results are very promising as the average user analysis time was sped up between 50-62%[2].

---

[1] The apparent discrepancy on Table. 2 in the 2,000 user test between the *Avg. User Time* and the *Analysis Time* appears to be due to one of the jobs hanging for several minutes before returning. This seemed to be repeatable but since this was a developer preview version of MongoDB and the job did eventually return successfully it was not a cause for concern, other than making an inconsistent benchmark.

[2] Because of the stability issues, we were not able to reproduce the 20K data set size to compare.

Table 2. MongoDB v2.3.2 Performance

Total Cases	Users	Avg. User Cases	Analysis Time (H:M:S)	Avg. User Time (S)
754,748	2,000	377	0:12:44	2.364
4,130,057	10,000	413	0:25:50	4.175

5   Conclusion

We have shown how we applied CBR to the problem of automatically learning user preferences for repeat users. We have also shown how this system satisfied all of our initial goals, as well as shown that it has the ability to scale very well to millions of cases and tens of thousands of users. Given that our test cluster used 4 shards when MongoDB supports up to 1,000[13], we are confident that the solution we presented would continue to scale several orders of magnitude more than our test data size.

Appendix A   Case Memory Document Structure

ID - The case identifier.
UserID - System wide unique user identifier.
PreviousPrompt - The slot variable the user was previously prompted for.
JustPrompted - The slot variable the NLS just prompted the user for after their answer.
Context - A JSON object holding pairs of conversation state variable names and values at the time the user was prompted for the slot variable.
SearchContext - A case-normalized form of the **Context** stored as an array of [name,value] pairs sorted by variable names.
Answer - The value of the slot variable named by **PreviousPrompt** that was filled in from the users input.
Order - Sequence number used to order inputs for this user.
TimeStamp - Time from web server when user input occurred.

Appendix B   Case-Base Document Structure

ID - The learned preference identifier.
Active - Flag used to determine if this specific preference is available for use.
Prompt - The slot variable name the user was prompted for.
EndContext - A JSON object holding pairs of conversation state variable names and values representing the state of the system after the user had successfully filled in the slot variable contained in **Prompt**.
Type - The function type that discovered this learned preference.
UserID - System wide unique user identifier.
Verified - Flag used to determine if the user has verified that this preference is acceptable.

StartContext - A JSON object holding pairs of conversation state variable names and an array of values observed at the time the user was prompted for the slot variable contained in **Prompt**.

SearchKeys - Array of SHA-1 hex strings computed for each combination of name:value pairs in **StartContext**.

$$length(SearchKeys) = \prod_{var}^{vars} length(StartContext(var))$$

TaskID - The task that this preference relates to.

SlotFeatures - The set of variable relationships defined for this **Prompt** in this **TaskID** that was created by the domain knowledge expert when the task was defined. It is saved with the learned preference built from it for reporting and system auditing purposes.

Entries - A list of case IDs that contributed to this preference.

References

1. Bobrow, D.G., Kaplan, R.M., Kay, M., Norman, D.A., Thompson, H., Winograd, T.: GUS, a frame-driven dialog system. Artificial Intelligence 8(2), 155–173 (1977)
2. Levin, E., Narayanan, S., Pieraccini, R., Biatov, K., Bocchieri, E., Di Fabbrizio, G., Walker, M.: The AT&T-DARPA Communicator mixed-initiative spoken dialog system. In: Proc. of ICSLP, vol. 2, pp. 122–125 (October 2000)
3. Bohus, D., Rudnicky, A.I.: The RavenClaw dialog management framework: Architecture and systems. Computer Speech & Language 23(3), 332–361 (2009)
4. Saaya, Z., Smyth, B., Coyle, M., Briggs, P.: Recommending case bases: applications in social web search. Case-Based Reasoning Research and Development, 274–288 (2011)
5. Bridge, D., Göker, M.H., McGinty, L., Smyth, B.: Case-based recommender systems. The Knowledge Engineering Review 20(03), 315–320 (2005)
6. Sugiyama, K., Hatano, K., Yoshikawa, M.: Adaptive web search based on user profile constructed without any effort from users. In: Proceedings of the 13th International Conference on World Wide Web, pp. 675–684. ACM (May 2004)
7. Schiaffino, S.N., Amandi, A.: User profiling with case-based reasoning and bayesian networks. In: IBERAMIA-SBIA 2000 Open Discussion Track, pp. 12–21 (2000)
8. Dean, J., Ghemawat, S.: MapReduce: simplified data processing on large clusters. Communications of the ACM 51(1), 107–113 (2008)
9. Berube, D.: Encode video with MongoDB work queues, http://www.ibm.com/developerworks/library/os-mongodb-work-queues
10. Bonnet, L., Laurent, A., Sala, M., Laurent, B., Sicard, N.: Reduce, You Say: What NoSQL Can Do for Data Aggregation and BI in Large Repositories. In: 2011 22nd International Workshop on Database and Expert Systems Applications (DEXA), pp. 483–488. IEEE (August 2011)
11. Horowitz, E.: Schema Design at Scale. Presentation, MongoSV (2011), http://www.10gen.com/presentations/mongosv-2011/schema-design-at-scale
12. MongoDB 2.4 Release Notes, http://docs.mongodb.org/manual/release-notes/2.4/#default-javascript-engine-switched-to-v8-from-spidermonkey
13. Horowitz, E.: The Secret Sauce of Sharding. Presentation, MongoSF (2011), http://www.10gen.com/presentations/mongosf2011/sharding

Case-Based Goal Selection Inspired by IBM's Watson

Dustin Dannenhauer and Héctor Muñoz-Avila

Department of Computer Science and Engineering, Lehigh University, Bethlehem PA 18015, USA

Abstract. IBM's Watson uses a variety of scoring algorithms to rank candidate answers for natural language questions. These scoring algorithms played a crucial role in Watson's win against human champions in Jeopardy!. We show that this same technique can be implemented within a real-time strategy (RTS) game playing goal-driven autonomy (GDA) agent. Previous GDA agents in RTS games were forced to use very compact state representations. Watson's scoring algorithms technique removes this restriction for goal selection, allowing the use of all information available in the game state. Unfortunately, there is a high knowledge engineering effort required to create new scoring algorithms. We alleviate this burden using case-based reasoning to approximate past observations of a scoring algorithm system. Our experiments in a real-time strategy game show that goal selection by the CBR system attains comparable in-game performance to a baseline scoring algorithm system.

1   Introduction

This work presents a new solution to the problem of goal selection within a goal-driven autonomy agent. Goal-driven autonomy (GDA) is a reasoning model in which an agent selects the goals it will achieve next by examining possible discrepancies between the agent's expectations and the actual outcome of the agent's actions. GDA agents explain these discrepancies and generate new goals accordingly [10, 13–15]. The computer program Watson developed by IBM achieved fame when it defeated two previous (human) winners from the United States television show *Jeopardy!*. Watson's use [16, 17] of a variety of scoring algorithms to rank answers from evidence snippets can be applied to a game playing agent for ranking which goal to achieve next. We present two goal selection implementations: a baseline system inspired by Watson's answer scoring algorithms and a case-based reasoning system that approximates this baseline system.

In RTS games, players manage armies of units to defeat an opponent. Actions in the game are executed in real time (i.e., players do not wait for the opponent to make a move). RTS games follow a combat model where units of a certain kind are particularly effective against units of some other kind but particularly vulnerable against units of a third kind. For this reason RTS games are frequently used in case-based reasoning research [9]. A challenge of using CBR (or any

S.J. Delany and S. Ontañón (Eds.): ICCBR 2013, LNAI 7969, pp. 29–43, 2013.

other AI technique) in these domains is the large amount of state information. With such large state spaces, AI systems' state representations must exclude many details about the state (e.g., by using state abstraction techniques [20]). A primary motivation for borrowing the scoring algorithm technique from Watson is to make use of potentially all information in the game state.

A major drawback of this evidence-scoring technique is the significant knowledge engineering effort required to create new scoring algorithms. Each scoring algorithm contains heuristic-like knowledge that relates evidence to an answer by means of a numerical score. Such effort is apparent in Watson, which made use of thousands of scoring algorithms (sometimes referred to as evidence scoring strategies, in this paper we refer to them as evidence scorers). At least some (it is not clear how many) scoring algorithms used by Watson were NLP-based and were easily available due to research in the NLP community [19]. However, for other domains, including RTS games, it may not be the case that evidence scoring functions are readily available. Our work is partly motivated by the fact that creating evidence scorers may be too high of a knowledge engineering burden but past observations (e.g., made by experts) are available. Case-based reasoning can approximate evidence scorers by capturing and reusing the results from previous observations (e.g., scores given by human experts in previous episodes). We found that a CBR system with a simple state representation, a straightforward similarity function, and nearest neighbor retrieval approached the performance of the baseline evidence scoring system.

This paper is organized as follows: Section 2 discusses IBM's Watson and the core technique relevant to the work in this paper. Section 3 discusses the baseline Watson-inspired goal selection component using evidence scorers. Section 4 discusses an implementation of the CBR system using the ideas described in Section 3. We present our experiments and results in Section 5, followed by Related Work in Section 6, and finally end with Conclusions and Future Work in Section 7.

2   IBM's Watson's Evidence Scoring Algorithms

Jeopardy! is a game in which 3 competitors are given clues in natural language about some information that must be guessed and the first person to answer the information correctly wins. A wrong answer carries a penalty, so good players must be highly confident of their answers before choosing to respond. Successful play requires rapid understanding of English sentences and substantial background knowledge on a variety of topics [16, 17]. The heart of Watson is an extensible software architecture named DeepQA [16]. DeepQA is best thought of as a pipeline, where the question is given at the start and an ⟨answer,confidence-score⟩ pair is produced at the end (a full diagram of the pipeline can be found in [16]). This pipeline has many phases, we are only interested in the final stage of the pipeline, which ranks potential answers based on the evidence scores provided by the evidence scorers [19].

When the DeepQA pipeline reaches the final stage it has accumulated a list of candidate answers along with supporting pieces of evidence for each answer. An

. example of a supporting piece of evidence may be a sentence or passage from an encyclopedia that contains keywords from the question and candidate answer. During the final stage, thousands of answer scorers each produce a numeric score representing the degree to which a piece of evidence supports or refutes a candidate answer. [19] The goal of the final stage is to combine all evidence scores for each candidate answer in order to determine the best candidate answer and its corresponding confidence score. To best combine evidence scores, DeepQA uses machine learning to train over a corpus of previously used questions and their correct answers [19]. DeepQA then produces a model describing how the evidence scorers should be combined (i.e., assigning different weights to different evidence scorers). Sometimes a single evidence scorer or group of evidence scorers are highly indicative of the correctness of an answer, and therefore should be given more importance when aggregating scores.

3 Goal Selection Using Case-Based Reasoning

3.1 Case Representation

We describe a system that takes an approach to goal selection that is inspired by IBM's Watson. As mentioned before, Watson ranks candidate answers according to scores produced by what are called evidence scorers. Evidence scorers are essentially functions that take the question posed to the contestant combined with a candidate answer, and a piece of evidence (i.e. a sentence or paragraph from an encyclopedia) and produce a score of how well that piece of evidence supports the given candidate answer for the given question. This can be represented as a triple: ⟨question, answer, score⟩. All of the scores from each piece of evidence for a candidate answer are aggregated into a single score for that candidate answer. This aggregated score reflects how well the pieces of evidence support the candidate answer. The candidate answers are then ranked based on their aggregated scores and the highest scoring answer is chosen.

Our baseline evidence scorer system takes the same approach, except instead of evidence scoring functions that take a candidate answer and a piece of textual evidence (such as a paragraph), our evidence scoring functions take a goal and features of the current game state in the RTS game Wargus. Analogous to the representation of ⟨question, answer, score⟩ in Watson playing *Jeapordy!*, we use ⟨gamestate features, goal, score⟩ as the representation in our system playing Wargus. In the same way as Watson, we produce aggregated scores for each goal that we may decide to pursue next. After each goal's scores are aggregated, the highest scoring goal is chosen.

Such an approach allows the goal selection component to neither restrict nor conform to models of the game state used in other GDA components. For example, perhaps the planning component of a GDA agent uses a compact state representation (as is the case in LGDA and GRL [1, 15]). The goal selection component is not forced to use that state representation, nor does it impose any restriction on the planning component's use of a compact game state. The goal selection component of the GDA agent may make use of more or all information in the game state at the time of goal selection. This is a benefit of modularity and would allow a component like the one presented in this paper to easily fit into a current GDA system.

3.2 Information Flow

Our main motivation is for situations in which the system neither has access to the internal functioning of the evidence scorers nor to the evidence scorers themselves. For example, the evidence scorers are humans that we observed in past instances solving problems. Ontañón *et al.*, 2007 show how domain experts annotate input traces by the goals they achieve [6]; in our situation, we would ask the experts to also annotate the goals' scores. The primary objective is to create a system by reusing previous instances of these evidence scorers providing scores for specific situations. We present a case-based reasoning system that approximates an evidence scoring system by reusing past instances.

Given a sufficiently large number of past instances ⟨gamestate features, goal, score⟩ from an evidence scoring system, a case-based reasoning component can be constructed. For each instance, it is necessary to have the results from each evidence scorer and features from the game state at the time of the instance.

Figure 1 depicts a high level overview of the Watson-inspired evidence scoring component as well as the information available to the case-based reasoning system. Immediately to the right of the "Watson-inspired Component" are the evidence scorers, denoted as the functions $ES_1(G_i, S), ES_2(G_i, S), \ldots ES_N(G_i, S)$. Each evidence scorer is invoked for every goal, resulting in $N * M$ intermediate scores denoted by $G_{1,E_1} \ldots G_{M,E_1}, G_{1,E_2}, \ldots G_{M,E_2}, G_{1,E_N} \ldots G_{M,E_N}$. These intermediate scores are then aggregated to produce a single score, one for each goal, denoted by the $G_1 \ldots G_M$ scores. The goal with the highest aggregate score is chosen. The evidence scoring functions each take an additional argument, S, representing features from the current game state from Wargus. The case-base of the case-based reasoning component is shown in the lower right. The area within the double line represents all of the information available to be stored in each new case. Our case-based reasoning system records features from the game state, the highest scoring goal, and the score.

4   A CBR System for Goal Scoring

We now present a detailed walk-through of a system that implements the ideas discussed in the previous section. A goal is a task we want to achieve. Akin to [1], in our implementation we assume there is one way to achieve a goal. However, our ideas are amenable to situations in which there is more than one way to achieve a goal. Table 1 shows the goals used in our implementation. These are high level goals that require multiple actions in order to be achieved.

The baseline evidence scoring component that chooses goals within Wargus uses three specific evidence scoring functions, described in Table 2. Each of these evidence scorers produce a score based on specific features of the current game state and a goal. Designing evidence scorers can require significant knowledge

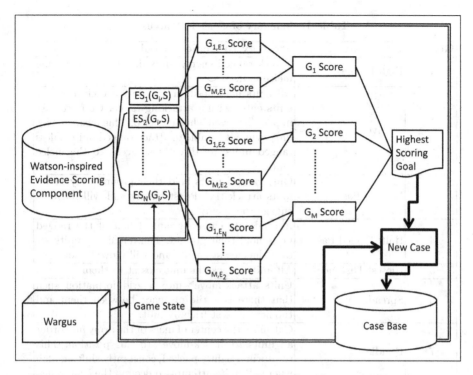

Fig. 1. Cases built from observing Watson-inspired evidence scoring component

engineering. Every instance of the evidence scoring system selecting a goal, a new case ⟨map features,goal,score⟩ is created that is then used in the case-based goal selection component.

We now walk through an example taking place using the scenario in Figure 2. In this scenario, the darker tiles are water and inaccessible by land units, and the lighter tiles are land tiles. In our experiments, goals were selected at the beginning of a scenario and their corresponding strategy was continuously executed. An agent playing full RTS games would encounter many such scenarios. In a GDA agent, components related to acting on discrepancies would determine when new goals were chosen. We take the perspective that new goals would only be chosen at the start of each micro-battle similar to those depicted in these scenarios.

4.1 Evidence Scoring System

When each game starts, the evidence scoring component will first calculate intermediate scores for each goal by invoking each evidence scorer on each goal. (Refer back to Figure 1, the intermediate scores are $G_{1,E_1} \ldots G_{M,E_1}, G_{1,E_2}, \ldots G_{M,E_2}, G_{1,E_N} \ldots G_{M,E_N}$). With $M = 7$ goals and $N = 3$ evidence scorers, $N * M = 21$ intermediate scores produced. It is important that the evidence scorers only score

Table 1. Summary of Goals in Wargus

Goal #	Goal Name	Strategy to Achieve Goal
1	High Range	Attack ranged enemy units first before attacking melee units
2	Passive	All units hold position and only attack if enemy units enter within attacking range. Units remain in position, even when engaged. This means melee units will not be able to attack ranged units unless ranged units are directly adjacent to the melee unit.
3	Ranged Passive	Only ranged units hold their ground, any melee units attack the closest enemy and will move to engage the enemy.
4	Half Ranged Passive	A randomly chosen group of half of the ranged units hold their ground. The remaining units attack the closest enemy and will move to engage.
5	Closest Distance	All units attack the unit closest to them.
6	Spread Out	Units attack-move* into a grid formation such that there is 1 tile of space between them and the next closest friendly unit.
7	Huddle	Calculate the center of mass of the army and order all units to attack-move* to this position. This results in a tightly packed, constantly shifting blob where all units attempt to occupy the center tile.

* attack-move is a Wargus game command that orders units to move to a specific location. Unlike the move command, units commanded to attack-move will pursue and attack any enemy that comes into their line of sight at any time.

goals that have supporting evidence. When no supporting evidence exists, the evidence scorer returns a value of zero. For the sake of space, we only show the intermediate scores that are not zero in Table 3. Observe that the goal Half Ranged Passive was scored by two evidence scorers, but because Huddle was scored so highly by a single evidence scorer, and we do not weight intermediate scores, Huddle obtains the highest aggregated score. In our implementation, evidence scorers produced scores between 0 and 7. Assigning weights to different evidence scorers is another place where knowledge engineering is required (although machine learning can be used to figure out how to best combine intermediate scores - IBM's Watson made heavy use of machine learning to combine the thousands of intermediate scores that were produced by thousands of evidence scorers over hundreds of answers. See [19] for more details.).

The Army Distance evidence scorer calculated that the distance between each army in this scenario was relatively far, and therefore gave a score of 5 to the Closest Distance goal (Table 3). The Army Distance evidence scorer produced zeros for all other goals, indicating it did not think the current game state provided any evidence to choose a goal when taking into account the distance between opposing units.

Table 2. Summary of Evidence Scorers in Wargus

#	Evidence Scorer Name	Function
1	Army Distance	Finds the single minimum distance of all distances between each friendly unit and the closest enemy unit. Goals are scored based on this distance. Distance is calculated between each units location using the formula: $\sqrt{(x_2 - x_1)^2 + (y_2 - y_1)^2}$
2	Ratio Water to Land	Draws a straight line between each of friendly unit and the closest enemy unit, and counts all tiles touching this line, recording if each tile is a water tile or land tile. If the ratio of water to land is high, that indicates there is a greater chance of a choke point between the two armies. Scores goals based on this ratio of water to land tiles between the opposing forces.
3	Ratio Ranged to Melee	Calculates the ratio of ranged units to melee units. Scores goals based on the this ratio.

The Ratio Ranged to Melee evidence scorer gave a low score of 1 to goals Passive, Ranged Passive, and Half Ranged Passive, indicating that given the current number of ranged and melee units, it would be slightly advantageous for ranged units to hold position. When there are no ranged units, this evidence scorer produces a score of all zeros for every goal. This evidence scorer also produced a high score of 7 for Huddle, indicating that whenever there is a low ratio of ranged to melee units, Huddle is a strong goal. The intuition is that by huddling the army, the chances of surrounding ranged units by melee units increases and would result in ranged units having greater chances of survival and increased damage output. In our implementation, Ratio Ranged to Melee chose Huddle over Spread Out when the ratio of melee units to ranged is greater than 1.

The Ratio Water to Land evidence scorer detects a low ratio of water to land tiles in between the enemy armies, suggesting that there is a wide chokepoint, in which case it is somewhat advantageous to keep half of the ranged units holding position. Thus, this evidence scorer produces a score of 4 for the HalfRangedPassive goal. For more narrow chokepoints, it would rate Ranged Passive or Passive higher, with the intuition being that the more narrow the chokepoint, the more ranged units should hold their ground on the opening of the chokepoint.

It is easy to see that the Huddle goal is the highest ranked goal when the intermediate scores for each goal are aggregated. At this point the evidence scoring system executes the Huddle goal and the goal selection process finishes.

4.2 Case-Based Goal Selection Component Example

The case base is populated with each instance of the evidence scoring system selecting a goal. The case-based system performs on each scenario only after a

Table 3. Intermediate Scores for each goal

Evidence Scorer	Goal	Score
Army Distance	ClosestDistance	5
Ratio Ranged to Melee	Passive	1
Ratio Ranged to Melee	RangedPassive	1
Ratio Ranged to Melee	HalfRangedPassive	1
Ratio Ranged to Melee	Huddle	7
Ratio Water to Land	HalfRangedPassive	4

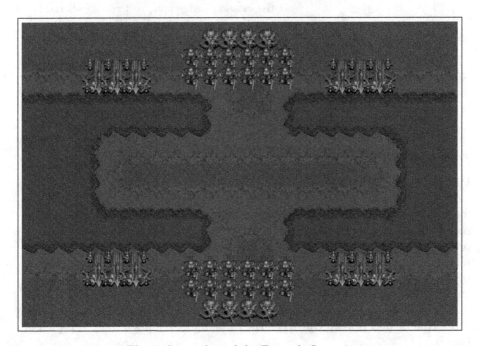

Fig. 2. Screenshot of the Example Scenario

sufficient number of cases are constructed from observing the evidence scoring system in action. At the start of the scenario, the case-based system obtains game state information, specifically the number of water and land tiles and the numbers of each type of unit (however, much more information could be used, such as the locations of each unit). In the scenario in Figure 2, it counted 6, 4, 12 for ballista's, rangers, and footmen respectively and 129 and 895 water and land tiles specifically (Figure 2 shows only the main part of the scenario; some water and land tiles are not shown). The case-based system uses a straightforward similarity function shown in Figure 3 where WT_{c_i} and LT_{c_i} represent the number

$$sim(c_1, c_2) = 1 - \left(\frac{1}{4}\right)\frac{WT_{c_1} - WT_{c_2}}{WT_{max}} - \left(\frac{1}{4}\right)\frac{LT_{c_1} - LT_{c_2}}{LT_{max}} - \left(\frac{1}{2}\right)\sum_{i=1}^{K}\left(\frac{1}{K}\right)\frac{U_{i,c_1} - U_{i,c_2}}{U_{max}}$$

Fig. 3. CBR Similarity Function

of water tiles and land tiles for case c_i, respectively, and U_{i,c_i} represents the number of units of type i found in case c_i. WT_{max} and LT_{max} are the maximum number of water and land tiles of any scenario. K is the number of different types of units in each team.

In one of our experiments, for this scenario, the case-based reasoning system retrieved a very similar case where the number of ballistas and rangers differed by 2 and 8 respectively (the number of land and water tiles remained the same). In the retrieved case, the evidence scoring system had chosen the goal Spread Out instead of Huddle. Surprisingly, Spread Out ended up being a slightly better goal than Huddle. While the case-based reasoning system was incorrect in choosing the same goal as the evidence scoring system, it actually ended up performing better. This is the result of two important considerations. First, the evidence scoring system is not perfect, and would need machine learning techniques such as ensemble methods [21] to learn the appropriate weights to achieve a very high accuracy (as well as a proper set of evidence scorers). One reason DeepQA is attributed to the success of IBM's Watson is the ease in which different answer scorers could be experimented with. Described as a process like a running trial and error, the sets of evidence scorers that performed increasingly well were kept and continuously revised [19]. Both the set of evidence scorers and the corresponding weights play a significant role. Second, the strategies to achieve different goals in Wargus have varied performance and often pursuing multiple goals results in decent performance in some scenarios. This is much different from *Jeopardy!*, where it is rare for more than one answer to be correct.

5   Experiments

Our hypothesis is that a case-based reasoning system that is able to observe an evidence-scoring system can learn to accurately choose the same goals, given a sufficiently large and comprehensive case base. Additionally we hypothesize that the case-based reasoning system can perform close to the performance of the evidence-scoring system. While the case-based reasoning system may choose a different goal than the evidence scoring system, resulting in lower accuracy for the first hypothesis, it may still be a fairly good goal (perhaps even better), and therefore result in relatively good performance to that of the evidence scoring system.

In order to test both hypotheses, we hand crafted 48 unique scenarios for the goal-selection systems. Each scenario was one of 6 unique terrain layouts, 2 of which were pure land maps (no water tiles) and 4 of which had varying amounts of water tiles. For each of the 6 unique terrain maps, 8 different configurations of number of units and unit types were created. The configurations of units was either all melee, melee outnumbering ranged, ranged outnumbering melee, or all ranged. For each of these four relative unit configurations, two different maps were created, differing slightly. For all 8 variations for each unique terrain map, the locations of units were kept approximately the same. Every scenario was symmetrical about the terrain and units, except for 1 scenario in which one team surrounded another. Both goal systems faced the same opponent implemented by the strategy achieving the goal Closest Distance. This is the most general strategy and generally performed well in every scenario. For every match (scenario) the system played on either side and the resulting score is the average difference in scores from both runs. The score in Wargus is calculated by adding the score for every enemy unit you defeat, with different units being worth different point values. For example, rangers are worth 70 points and ballistas are worth 100. So if both teams score 1000, it means they each killed 1000 points worth of units on the opposing team, resulting in a tie.

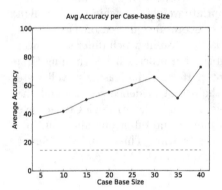

Fig. 4. Accuracy per Case-base Size

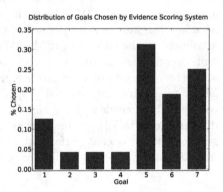

Fig. 5. Goal Selection Distribution

5.1   Results

The first graph, Figure 4, shows the accuracy of the case-based system against the evidence-based system. We ran the evidence scoring system on each of the 48 scenarios, and recorded the goals chosen. Next, we randomly picked X cases (where X varied from 5 to 40 by intervals of 5) from those 48 scenarios to use as cases in the case base. Each case only recorded the game state features and final goal chosen by the evidence scoring system. Each data point is the result of the average of 5 rounds, where each round consisted of randomly picking a case base of size X and recording the accuracy of choosing the same goal as the evidence scoring system on the remaining scenarios. For example, using a case base of size 15, the testing set was of size $48 - 15 = 33$ scenarios.

A random system would choose the correct goal once out of seven times, roughly 14% shown by the straight line in Figure 4. The first case base of size 5 exceeds random goal selection, and this is partly due to obtaining a diverse case base and to the distribution over the goals chosen by the evidence scoring system shown in Figure 5. Because goals 5, 6, and 7 were chosen much more often by the evidence scoring system, if the case base of size 5 contained cases where goals 5 and 6 were chosen, it would likely score highly in the remaining scenarios. Only if the case base consisted of mostly goals 2, 3 or 4 would the performance be near or worse than random. This distribution of goals in Figure 5 is also the reason for the dip in accuracy for case base of size 5. Depending on what scenario's were chosen in the case base, the range of the accuracy is quite large. The important result from Figure 4 is that the case base system becomes more accurate with more cases in the case base, and approaches 70% accuracy.

The following graphs, Figures 6 to 9, show the average performance of the case-based reasoning system compared to the evidence scoring system. Each bar represents the difference in score of the goal selection system against the opponent (the strategy for Closest Distance). These results are broken into 4 different figures to allow for closer inspection. Each graph depicts the difference in scores for 12 scenarios. For each scenario, the first bar is the evidence scoring system, the second is the case-based system with a case base of 5 cases, the third bar is the case based system using a case base of 10 cases, etc until the last bar is the case base containing 40 cases. The angled lines above or below each set of bars for each scenario help display the difference from the evidence scoring system and the last case based system. A angle with an end point higher than the starting point represents the case based system outperforming the evidence based system (example is scenario 31) and vice versa. Whenever the goal selection system tied with the opponent (each army defeated the other) the difference in scores is 0, and no bar is shown. Generally, we expect to see that the evidence-scoring system scores relatively highly (the first bar should be a high positive value), and the case-based systems progressively get closer and closer to the evidence scoring system score. We see this happen approximately in Figure 6 for scenarios 1, 2, 5, 11, and 12, Figure 7 for scenarios 15 and 16, Figure 8 for scenarios 26, 28, 31, 34, and 35 and in Figure 9 scenarios 38, 40, 44, 46, 47, and 48. Even when the evidence scoring system performs poorly (shown by negative bars) the case-based system approximates it. Also, notice that in Figure 7 the last 8 scenario's are almost all blank (the lone vertical bar in them is the case-based system of size 5). This is because the evidence based system and case-based system (except for case base size of 5 in some instances) chose the same goal as the opponent, Closest Distance, and resulted in tied games, and further shows that the case-based system performs approximately as well as the evidence scoring system, despite only achieving a 70% accuracy in pure goal selection. For 70% of the scenarios, when the case base had the most cases, the case-based prediction system had the same or better performance than the evidence scoring system.

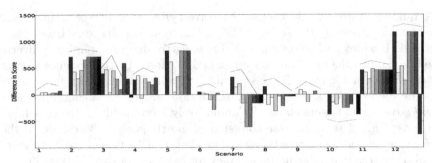

Fig. 6. Average Performance in Each Scenario (1 to 12)

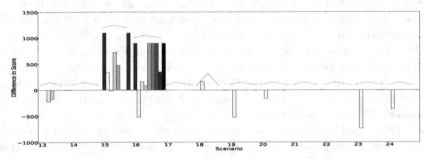

Fig. 7. Average Performance in Each Scenario (13 to 24)

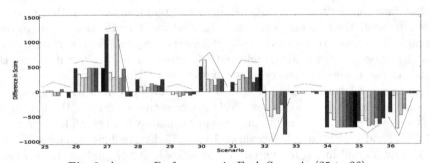

Fig. 8. Average Performance in Each Scenario (25 to 36)

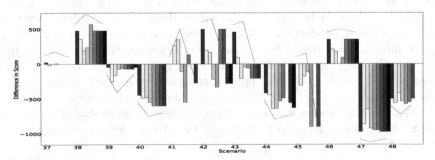

Fig. 9. Average Performance in Each Scenario (37 to 48)

When running these experiments, occasionally matches would fail for seemingly no reason, but in a deterministic manner. We ran 336 unique match ups (goal strategy and unique scenario), and only 9 of these failed before the game finished. However, 3 of these failed with only 1 unit left on one team and at least 4 units alive on the other team. For these 3, we calculated the worth of the single unit and assumed it would have been killed, and manually adjusted the winning team's score appropriately. For the remaining 6 failed matches, we re-ran the match until right before it crashed, recorded the score and the locations of each of the remaining units on each team. We then reconstructed a new match exactly as it was left off, ran it, and added the score to the match before it failed. Because there is no way to start a Wargus match so that units have less than maximum health, there was some information loss, but because this only occurred on 6 out of the 336 matches the overall results were not significantly affected.

6   Related Work

We discussed IBM Watson in Section 2. Here, we discuss other related works. As previously mentioned our objective is to embed these goal selection techniques into GDA agents. GDA agents select their goals based on the explanation of a discrepancy. A number of techniques have been suggested for goal selection in GDA research. These include using rules that map the explanation to the goal to pursue next [11, 14]. A more common technique is to rank the goals according to priority lists (i.e., the goals having higher priorities are more likely to be selected than goals that have a lower priority) (e.g., [10]). We believe that using priority lists is a natural way to integrate our work with GDA; since each goal will be assigned a score, these scores can be used to update the priorities in the list. Mechanisms will be needed to merge the priorities suggested by GDA with the scoring suggested by the CBR system.

Outside of GDA research, goal selection has been a recurrent topic in planning research [8]. Typically, the higher level goals to achieve are fixed and the problem is to select subgoals that achieve those goals. Research has been done to relax the requirement that the goals are fixed; over-subscription planning aims at finding the maximal subset of the goals that can be achieved [12]. In principle, the user could input a large set of possible goals and let the system figure out which subset of these goals can be achieved in the given situation.

Wargus has been extensively used as a testbed in case-based reasoning research. Among many others, Mehta et al. (2009) used case-based learning techniques to learn from failure patterns in a Wargus game trace [9]. It has also been used to retrieve cases aimed to counter an adversary [4] and to evaluate online case-based adaptation algorithms [3]. Outside of CBR, Wargus has been used to demonstrate concurrent reinforcement learning techniques [5] and to acquire playing strategies using evolutionary computation among many others [2]. A common motivation among these works for using Wargus is that it provides a rich environment for decision making. This is precisely the motivation for using

Wargus in our work as we want to test the goal selection mechanism and observe how it affects this environment.

7   Conclusions and Future Work

IBM's Watson demonstrated the effectiveness of using a variety of evidence scorers to rank potential answers. However, one of the biggest issues is the knowledge engineering burden to create the ranking algorithms. In this paper we explored a CBR solution to an instance of this problem in which episodic knowledge of the form ⟨gamestate features, goal, scores⟩ are retained and reused. We tested our ideas in the Wargus RTS game using a hand-crafted evidence scoring system inspired from IBM's Watson as our baseline. We use this baseline system to generate cases that are fed into a CBR system. Our experiments demonstrated that the CBR system can attain comparable performance to the baseline system after sufficient cases have been retained in the case base.

For future research directions we wish to explore a number of ideas. (1) Given the independent nature of evidence scorers, we would like to explore running more complex and potentially computationally expensive evidence scorers in parallel. (2) We will like to use ensemble methods [21] to aggregate the individual information from the scoring algorithms. The tuning of this feature can lead to significant performance improvements. (3) We will like to study the use of an adaptation algorithm, where there can be multiple alternative ways to take the k-retrieved cases and select goals (such as taking into account the intermediate scores from the evidence scorers). These alternative ways could be ranked using techniques such as the ones shown in this paper.

Acknowledgments. We would like to thank David W. Aha (Naval Research Laboratory) for suggesting the idea of using IBM Watson in the context of goal-driven autonomy research. We would also like to thank Ulit Jaidee for his code used to run automated experiments in Wargus. This research is funded in part by NSF 1217888.

References

1. Jaidee, U., Muñoz-Avila, H., Aha, D.W.: Learning and Reusing Goal-Specific Policies for Goal-Driven Autonomy. In: Agudo, B.D., Watson, I. (eds.) ICCBR 2012. LNCS, vol. 7466, pp. 182–195. Springer, Heidelberg (2012)
2. Ponsen, M., Muñoz-Avila, H., Spronck, P., Aha, D.: Automatically Acquiring Domain Knowledge For Adaptive Game AI Using Evolutionary Learning. In: Proceedings of the Seventeenth Innovative Applications of Artificial Intelligence Conference (IAAI 2005). AAAI Press (2005)
3. Sugandh, S., Ontañón, S., Ram, A.: On-Line Case-Based Plan Adaptation for Real-Time Strategy Games. In: Proceedings of the AAAI Conference (AAAI 2008). AAAI Press (2008)

4. Aha, D.W., Molineaux, M., Ponsen, M.: Learning to win: Case-based plan selection in a real-time strategy game. In: Muñoz-Ávila, H., Ricci, F. (eds.) ICCBR 2005. LNCS (LNAI), vol. 3620, pp. 5–20. Springer, Heidelberg (2005)
5. Marthi, B., Russell, S., Latham, D., Guestrin, C.: Concurrent hierarchical reinforcement learning. In: International Joint Conference of Artificial Intelligence (IJCAI 2005). AAAI Press (2005)
6. Ontañón, S.: Case Acquisition Strategies for Case-Based Reasoning in Real-Time Strategy Games. In: FLAIRS 2012. AAAI Press (2012)
7. Ontañón, S., Mishra, K., Sugandh, N., Ram, A.: Case-Based Planning and Execution for Real-Time Strategy Games. In: Weber, R.O., Richter, M.M. (eds.) ICCBR 2007. LNCS (LNAI), vol. 4626, pp. 164–178. Springer, Heidelberg (2007)
8. Ghallab, M., Nau, D.S., Traverso, P.: Automated Planning: Theory and Practice. Morgan Kaufmann (2004)
9. Mehta, M., Ontañón, S., Ram, A.: Using Meta-reasoning to Improve the Performance of Case-Based Planning. In: McGinty, L., Wilson, D.C. (eds.) ICCBR 2009. LNCS, vol. 5650, pp. 210–224. Springer, Heidelberg (2009)
10. Molineaux, M., Klenk, M., Aha, D.W.: Goal-driven autonomy in a Navy strategy simulation. In: Proceedings of the Twenty-Fourth AAAI Conference on Artificial Intelligence. AAAI Press, Atlanta (2010)
11. Cox, M.T.: Perpetual self-aware cognitive agents. AI Magazine 28(1), 23–45 (2007)
12. van den Briel, M., Sanchez Nigenda, R., Do, M.B., Kambhampati, S.: Effective approaches for partial satisfaction (over-subscription) planning. In: Proceedings of the Nineteenth National Conference on Artificial Intelligence, pp. 562–569. AAAI Press, San Jose (2004)
13. Powell, J., Molineaux, M., Aha, D.W.: Active and interactive discovery of goal selection knowledge. In: To Appear in Proceedings of the Twenty-Fourth Conference of the Florida AI Research Society. AAAI Press, West Palm Beach (2011)
14. Muñoz-Avila, H., Jaidee, U., Aha, D.W., Carter, E.: Goal-Driven Autonomy with Case-Based Reasoning. In: Bichindaritz, I., Montani, S. (eds.) ICCBR 2010. LNCS, vol. 6176, pp. 228–241. Springer, Heidelberg (2010)
15. Jaidee, U., Muñoz-Avila, H., Aha, D.W.: Integrated learning for goal-driven autonomy. In: Proceedings of the Twenty-Second International Joint Conference on Artificial Intelligence, vol. 3. AAAI Press (2011)
16. Ferrucci, D.A.: Introduction to This is Watson. IBM Journal of Research and Development 56(3.4), 1 (2012)
17. Ferrucci, D., Brown, E., Chu-Carroll, J., Fan, J., Gondek, D., Kalyanpur, A., Lally, A., Murdock, J.W., Nyberg, E., Prager, J., Schlaefer, N., Welty, C.: BBuilding Watson: An Overview of the DeepQA Project. AI Mag. 31(3), 59–79 (2010)
18. Lally, A., Fodor, P.: Natural Language Processing With Prolog in the IBM Watson System (retrieved June 15, 2011)
19. Gondek, D.C., et al.: A framework for merging and ranking of answers in DeepQA. IBM Journal of Research and Development 56(3-4), 14:1–14:12 (2012)
20. Giunchiglia, F., Walsh, T.: A theory of abstraction. Artificial Intelligence 57(2-3), 323–390 (1992)
21. Dietterich, T.G.: Ensemble Methods in Machine Learning. In: Kittler, J., Roli, F. (eds.) MCS 2000. LNCS, vol. 1857, pp. 1–15. Springer, Heidelberg (2000)
22. Hunter, J.D.: Matplotlib: a 2D graphics environment. Computing in Science & Engineering, 90–95 (2007)

Opinionated Product Recommendation

Ruihai Dong, Markus Schaal, Michael P. O'Mahony, Kevin McCarthy,
and Barry Smyth

CLARITY: Centre for Sensor Web Technologies
School of Computer Science and Informatics
University College Dublin, Ireland

Abstract. In this paper we describe a novel approach to case-based
product recommendation. It is novel because it does not leverage the
usual static, feature-based, purely similarity-driven approaches of traditional case-based recommenders. Instead we harness experiential cases,
which are automatically mined from user generated reviews, and we use
these as the basis for a form of recommendation that emphasises similarity and sentiment. We test our approach in a realistic product recommendation setting by using live-product data and user reviews.

1   Introduction

Recommendation services have long been an important feature of e-commerce
platforms, making automated product suggestions that match the learned preferences of users. Ideas from case-based reasoning (CBR) can be readily found in
many of these services — so-called content-based (or case-based) recommenders
— which rely on the similarity between *product queries* and a database of *product
cases* (the case base). However, the relationship between CBR — which emphasises the reuse of *experiences* — and many of these 'case-based' recommenders
can be tenuous. For example, many case-based recommenders do borrow similarity assessment techniques from CBR, as a basis for query-product similarity,
but the idea that product cases (which are typically static feature-based records)
are experiential is at best a stretch. Does this matter? After all such approaches
have met with considerable success and have proven to be useful in practical
settings. But how might we harness *genuine* experiential knowledge as part of a
case-based product recommender? This is the question that we address in this
paper. We do this by describing and evaluating a novel approach to product
recommendation that relies on product cases that are genuinely experiential in
nature as well as a unique approach to retrieval that is based on the combination
of feature similarity and user sentiment.

Consider the *Fujifilm X100* camera. At the time of writing the *product features* listed by Amazon cover technical details such as *resolution* (12.3 MP),
sensor-type (APS-C), *aperture* (f2), and *price*($1,079.00). These are the type of
features that one might expect to find in a conventional product recommender,
facilitating the recommendation of other products that share similar values for
these same features. The features are clearly few in number: this limits the scope

S.J. Delany and S. Ontañón (Eds.): ICCBR 2013, LNAI 7969, pp. 44–58, 2013.

of assessing inter-product similarity at recommendation time. Moreover, features are often technical in nature, so it can be difficult to judge the importance of feature similarities in any practical sense. Is a 13.3 MP camera more or less similar to the $X100$ than a 11.3 MP alternative? However, the $X100$ has 149 reviews which encode valuable insights into a great many features of the $X100$, from its beautiful design to its quirky interface, and from its great picture quality to the limitations of its idiosyncratic auto-focus or the lack of optical stabilisation. Clearly these features capture far more detail than the handful of technical 'catalog' features. The reviews also encode the *opinions* of users and as such provide a subjective basis for comparison; all other things being equal, for example, the *"beautiful retro design"* of the $X100$ certainly beats another camera suffering from *"terrible design"*. A key idea of this work is that we can mine these opinion-rich features directly from user-generated reviews and use them as detailed experiential product cases to provide a basis for recommendation.

The key contributions of this work are three-fold. Firstly, we describe how product features can be automatically mined from the plentiful user-generated reviews on sites like Amazon.com and TripAdvisor etc. These features are aggregated at the product-level to produce *product cases*. Secondly, we explain how these product features can be associated with sentiment information to reflect the opinions of reviewers, whether positive, negative, or neutral. The resulting product cases are thus genuinely *experiential* in nature, in the sense that they are based wholly on the opinions and experiences of the users of these products. Thirdly, we describe a novel approach to *"more-like-this"* style recommendations that are based on a combination of similarity and sentiment, to prioritise products that are similar to, but better than, a given target (query) product.

2   Related Work

Recent research highlights how online product reviews have a significant influence on the purchasing behavior of users; see [1–3].To cope with growing review volume retailers and researchers have explored different ways to help users find high quality reviews and avoid malicious or biased reviews. This has led to a body of research focused on classifying or predicting review helpfulness. For example [4–7] have all explored different approaches for extracting features from user-generated reviews in order to build classifiers to identify helpful versus unhelpful reviews as the basis for a number of review ranking and filtering strategies.

It is becoming increasingly important to weed out malicious or biased reviews, so-called *review spam*. Such reviews can be well written and so appear to be superficially helpful. However reviews of this nature often adopt a biased perspective that is designed to help or hinder sales of the target product [8]. For example, Li et al. describe an approach to spam detection that is enhanced by information about the identity of the spammer as part of a two-tier, co-learning approach [9]. O'Callaghan et al. use network analysis techniques to identify recurring spam in user generated comments associated with YouTube videos by identifying discriminating comment *motifs* that are indicative of spambots [10].

In this work we are also interested in mining useful information from reviews and employ related feature extraction and opinion mining techniques to the above. However, our aim is to use this information to build novel product case descriptions that can be used for recommendation rather than review filtering or classification. As such our work can be framed in the context of past approaches for case-based product recommendation including *conversational recommenders* [11] and *critiquing-based* techniques [12], for example. For the most part, such past approaches are unified by their use of static case descriptions based around technical features. It is not the type of case representation that is situated in any experiential setting. In contrast the cases that we produce from reviews are experiential: they are formed from the product features that users discuss in their reviews and these features are linked to the opinions of these users. Past approaches also rely (usually exclusively) on query-case similarity as the primary recommendation ranking metric. In this work, while acknowledging that query similarity is an important way to *anchor* recommendations, we argue the importance of looking for cases that also differ from the query case, at least in terms of the opinions of users at the feature level; see also [13]. We recommend cases that are similar to the query but *preferred* by end users.

3   Recommending Experiential Product Cases

A summary of our overall approach is presented in Figure 1. Briefly, a case for a product P is made up of a set of product features and their sentiment scores mined from $Reviews(P)$, the set of reviews written for product P. The sentiment of each feature is evaluated at the review-level first and then aggregated at the case-level as an overall sentiment score for that feature. At recommendation time suitable cases are retrieved and ranked based on their similarity and sentiment with respect to a given query case Q.

3.1   Extracting Review Features

When it comes to extracting features from reviews for a particular product category (for example, *Laptops, Tablets*), we consider two basic types of features — *bi-gram* features and *single-noun* features. We use a combination of shallow NLP and statistical methods, by combining ideas from Hu and Liu [14] and Justeson and Katz [15]. To produce a set of bi-gram features we look for bi-grams in the review cases which conform to one of two basic part-of-speech co-location patterns: (1) an adjective followed by a noun (AN) such as *wide angle*; and (2) a noun followed by a noun (NN) such as *video mode*. These are candidate features but need to be filtered to avoid including AN's that are actually opinionated single-noun features; for example, *great flash* is a single-noun feature (*flash*) and not a bi-gram feature. To do this we exclude bi-grams whose adjective is found to be a sentiment word (for example, *excellent, great, terrible, horrible* etc.) using Hu and Liu's sentiment lexicon [16].

To identify single-noun features we extract a candidate set of nouns from the reviews. Often these candidates will not make for good case features however; for

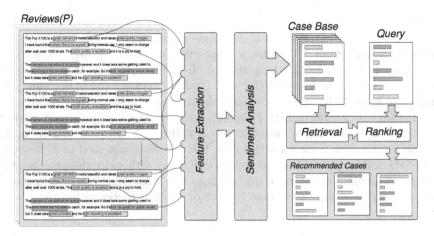

Fig. 1. An overview of how we mine user-generated reviews to create experiential product case bases for sentiment-based recommendation

example, they might include words like *family* or *day* or *vacation* which do not relate to product features. Hu and Liu [16] propose a solution to validate such features by eliminating those that are rarely associated with opinionated words. The intuition is that nouns that frequently co-occur with opinion laden words in reviews are likely to be relevant product features. We calculate how frequently each feature co-occurs with a sentiment word in the same sentence (again, as above, we use Hu and Liu's sentiment lexicon [16]), and retain a single-noun only if its frequency is greater than some threshold (in this case 70%).

This produces a set of bi-gram and single-noun features which we filter based on their frequency of occurrence, keeping only those features that occur in at least k of the s reviews; in this case, for bi-gram features we set $k_{bg} = s/20$ and for single noun features we set $k_{sn} = 10 \times k_{bg}$, where s is the total number of reviews for a category.

3.2 Evaluating Feature Sentiment

For each feature we evaluate its sentiment based on the sentence containing the feature. We use a modified version of the *opinion pattern mining* technique proposed by Moghaddam and Ester [17] for extracting opinions from unstructured product reviews. Once again we use Hu and Liu's sentiment lexicon as the basis for this analysis. For a given feature F_i and corresponding review sentence S_j from review R_k, we determine whether there are any sentiment words in S_j. If there are not then this feature is marked as *neutral*, from a sentiment perspective. If there are sentiment words then we identify the word w_{min} which has the minimum word-distance to F_i.

Next we determine the part-of-speech (POS) tags for w_{min}, F_i and any words that occur between w_{min} and F_i. The POS sequence corresponds to an *opinion pattern*. For example, in the case of the bi-gram feature *noise reduction* and the review sentence, *"...this camera has great noise reduction..."* then w_{min} is the word *"great"* which corresponds to an opinion pattern of *JJ-FEATURE* as per Moghaddam and Ester [17]. After a complete pass of all features through all reviews we can compute the frequency of all opinion patterns that have been recorded. A pattern is deemed to be valid (from the perspective of our ability to assign sentiment) if it occurs more than the average number of times. For valid patterns we assign sentiment to F_i based on the sentiment of w_{min} and subject to whether S_j contains any negation terms within a 4-word-distance of w_{min}. If there are no such negation terms then the sentiment assigned to F_i in S_j is that of the sentiment word in the sentiment lexicon. Otherwise this sentiment is reversed. If an opinion pattern is deemed not to be valid (based on its frequency) then we assign a *neutral* sentiment to each of its occurrences within the review set.

3.3   Generating Experiential Cases

For each review R_i the above methods generate a set of valid features $F_1, ..., F_{m_i}$ and their associated sentiment scores *positive, negative,* or *neutral*. We can now construct experiential product cases in a straightforward fashion, as a set of product features paired with corresponding sentiment scores as per Equation 1.

$$Case(P) = \{(F_j, Sentiment(F_j, P)) : F_j \in Features(P)\} \qquad (1)$$

The case features $(Features(P))$ for a product P are the union of the valid features extracted from its reviews. Each of these features may be present in a number of P's reviews and with different sentiment scores. To assign a sentiment score to a feature at the case-level we aggregate the individual review-based sentiment scores according to Equation 2, where $Pos(F_j, P)$ is the number of positive sentiment instances of F_j among the reviews of product P, and likewise for $Neg(F_j, P)$ and $Neutral(F_j, P)$. Thus, $Sentiment(F_j, P)$ will return a value between -1 (*negative* sentiment) and +1 (*positive* sentiment). For example, one of the features extracted for the $X100$ camera mentioned earlier is its *lens quality* which is invariably mentioned in a positive fashion across many reviews. As such its overall sentiment score is 0.72 (25 positive mentions, 5 neutral mentions, and only 2 negative mentions).

$$Sentiment(F_j, P) = \frac{Pos(F_j, P) - Neg(F_j, P)}{Pos(F_j, P) + Neg(F_j, P) + Neutral(F_j, P)} \qquad (2)$$

3.4   From Case Retrieval to Sentiment-Enhanced Recommendation

Now that we have a case base of experiential cases we can describe our approach to recommendation. First it is worth stressing again that, unlike many more conventional approaches to product recommendation, these experiential cases do not have a fixed set of shared static features. Instead each case is represented by its own (possibly unique) set of features, mined from its own product reviews.

We must ensure some minimal set of shared features between cases to serve as the basis for comparison. First we define *k-comparability* as a boolean property of two cases P_u and P_v which is true if and only if P_u and P_v share at least k features. During retrieval we only consider cases that are at least *k-comparable* (have at least k features in common) with the target query case Q; see Equation 3, where CB denotes the case base of all product cases.

$$Retrieve_k(Q) = \{P \in CB : k - comparable(Q, P)\} \qquad (3)$$

In a conventional product recommender system, we would likely rank these cases in decreasing similarity to the query case, for some suitable similarity metric (for example, Jaccard or Cosine similarity). However, we adopt a very different approach in this work. Remember that the values of our case features are sentiment scores; that is, overall judgements by real users about how good or bad a given feature is. It stands to reason that we would like to rank cases according to how much *better* their respective feature scores are compared to the query case. If the query case has a sentiment score of 0.5 for *lens quality* then we would surely prefer to rank another case with a score of 0.8 for *lens quality* ahead of a case with a *lens quality* of score 0.6, all other things being equal, and even though the latter case has a more similar *lens quality* sentiment score than the former case compared to the query. Thus, cases that have a *better* sentiment score across their shared features $(Features(Q) \cap Features(C))$ should be preferred.

$$better(F, Q, P) = \frac{Sentiment(F, P) - Sentiment(F, Q)}{2} \qquad (4)$$

$$Better(Q, P) = \frac{\sum_{\forall F \in Features(Q) \cap Features(P)} better(F, Q, P)}{|Features(Q) \cap Features(P)|} \qquad (5)$$

We compute a *better* score between the sentiment for a feature F in a query Q and a retrieved case P; see Equation 4. This returns a value from -1 (the sentiment for F in Q is better than in P) to +1 (the sentiment for F in P is better than in Q). Then we calculate a *Better* score at the case-level as the average better scores for the features shared between Q and P; see Equation 5.

Thus, for a given query case we first retrieve a set of *k-comparable* cases for a suitable value of k ($k = 15$ in this work) and then these cases are ranked in terms of the degree to which their sentiment scores are better over the shared features in the query case. We then return the top n ranked products as recommendations.

4   Evaluation

In this section we test how well this experience-based product recommendation works in practice. We do this by using a large corpus of more than 12,000 product reviews for about 1,000 consumer electronic products, ultimately comparing the performance of our sentiment-based recommendation to a more conventional recommendation approach using a reliable and objective ground-truth.

4.1 Data Sets and Setup

The review data for this experiment was extracted from Amazon.com during October 2012. We focused on 3 different product categories: *GPS Devices, Laptops, Tablets*. For the purpose of our experiments, we filtered for products with 10 or more reviews. Table 1 shows the number of products and reviews in the raw data, and the number of products with at least 10 reviews, per product category. Each of these 3 product categories is turned into a product case base using the approach described previously, and Table 1 also shows the mean and standard deviation of the number of features extracted per product type.

Table 1. Evaluation of product categories and case bases

Category	#Reviews	#Prod.	#Prod.(Filtered)	# Features Mean (Std. Dev.)
GPS Devices	12,115	192	119	24.32 (10.82)
Laptops	12,431	785	314	28.60 (15.21)
Tablets	17,936	291	166	26.15 (10.48)

4.2 Feature Sparsity and Case *k-Comparability*

The statistics above suggest that cases are being generated with a rich set of 20-30 features. This bodes well but it is of limited use unless these features are shared among the products in a case base. Small numbers of shared features greatly restrict our ability to compare cases during retrieval and lead to the type of *sparsity problem* that is common in collaborative filtering systems [18].

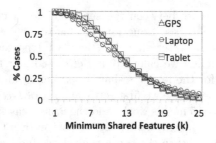

Fig. 2. The average percentage of cases with a minimum of k shared features with the query case

To explore this we can examine the average size of the *k-comparability* sets, as a percentage of case base size, for each product type across different levels of k; see Figure 2. We can see that for $k = 15$, in all three case bases (Laptop, GPS Devices, and Tablets) the average number of *k-comparable* cases is about 35%

of all cases (approximately 109 Laptops, 41 GPS Devices, or 58 Tablets). This is important for a few reasons. It helps to validate the type of features that we are extracting from reviews. The fact that there are so many cases sharing at least k features, even for relatively large values of k, means that we are extracting features that are frequently recurring in product reviews. It also means that there is no significant feature sparsity problem in any of the case bases examined. From a recommendation standpoint this means that we are able to compare product cases based on a rich set of shared features since it is now practical to set k to be between 10 and 20, for example, in these case bases and to still ensure large enough retrieval sets to form the basis for final recommendation.

4.3   Recommendation Quality as Sentiment Improvement

It is good that we are extracting many features from reviews and it is promising that these features tend to recur across many case base products. But are these features, and their associated sentiment scores, effective from a recommendation perspective? Do they facilitate the recommendation of useful products? To test this we consider a number of recommendation strategies as follows:

- *Jaccard* — cases are ranked based on a simple Jaccard metric ($|Features(Q) \cap Features(C)|/|Features(Q) \cup Features(C)|$) over case features; that is, sentiment information is not used and cases are preferred if they share a higher percentage of features with the query case.
- *Cosine* — cases are ranked based on a standard cosine similarity metric calculated from the sentiment scores of shared features.
- *Better* — cases are ranked based on the *Better* metric described previously, which prioritises cases that enjoy improved sentiment scores relative to the query. Again, *Better* scores are calculated from shared features only.

Using a standard *leave-one-out* methodology, each one of these strategies produces a different ranking of retrieved cases. To evaluate the quality of these rankings we consider two different types of ground-truth data: (1) relative sentiment and (2) independent product ratings.

Relative Sentiment. First, we can measure the relative improvement in the sentiment of the recommended cases compared to the query case from the average *Better* scores for the recommended products. Obviously this is biased towards the *Better* technique because it uses this very metric to rank its own recommendations. Nevertheless it provides a useful guide to understanding the extent to which there is room to improve upon alternative strategies. The results are presented in Figure 3 as a graph of the average *Better* scores for recommended cases for increasing large recommendation lists and for each of the recommendation strategies above; we use a *k-comparability* threshold of $k = 15$. We can see that there is a significant uplift in the relative sentiment of the *Better* recommendation lists, which is sustained, albeit decreasing, over all values of n. For instance, for *Laptops* we can see that for recommendation lists of size 3 the *Better* strategy recommends cases that are more positively reviewed on average than the query case ($Better(Q, P) = 0.11$). By comparison the cases recommended by the *Jaccard* or *Cosine* strategies offer little or no sentiment improvement.

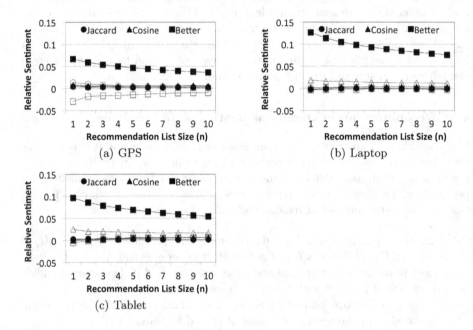

(a) GPS   (b) Laptop

(c) Tablet

Fig. 3. Mean relative sentiment versus recommendation list size ($k = 15$)

Residual Sentiment. As an aside it is also worth considering the relative sentiment of those features in the query case and the recommended cases that are *not* among the k or more shared features. We can estimate a *residual sentiment score* (RSS) for these features by subtracting the average sentiment score for the residual query case features from the average sentiment score of each of the recommended cases' residual features, such that an $RSS > 0$ means that the residual features of the recommended cases tend to be better than those of the query case. These results are also shown in Figure 3 for each of the recommendation strategies but using dashed lines. The RSS values tend to fall close to zero in most scenarios especially for larger values of n, thereby indicating that there is little lost sentiment due to these residual features. Nonetheless there does appear to be some limited sacrifice associated with the *Better* strategy since its RSS scores tend to be slightly less than 0 for the GPS product case base. In general, however, this type of tradeoff seems justified given the much improved sentiment scores of the shared features for the *Better* strategy.

4.4   Recommendation Quality as Relative Ratings Differences

Obviously the above results are somewhat limited by the fact that our measure of recommendation quality (sentiment improvement) is closely coupled to the ranking metric used by the *Better* strategy. As an alternative, in this section we consider our second ground-truth — the average user-provided product ratings — as a truly independent measure of recommendation quality. In other words, we can evaluate recommendation quality in terms of whether or not recommended cases tend to attract higher overall product ratings than the query case. Rather than report changes in product ratings directly we look at the rank improvement of a recommended product relative to the query case with respect to the rating-ordered list of products in the case base. In other words we rank all cases in a case base by overall ratings score and then calculate the quality of a recommended case in terms of its percentage rank difference to the query case. Thus if the query case is ranked 50th in a case base of 100 products by rating and a recommended case is ranked 25th then the relative rank improvement is +25%. We do this because it provides a more consistent basis for comparison across different case bases with different numbers of products and ratings distributions.

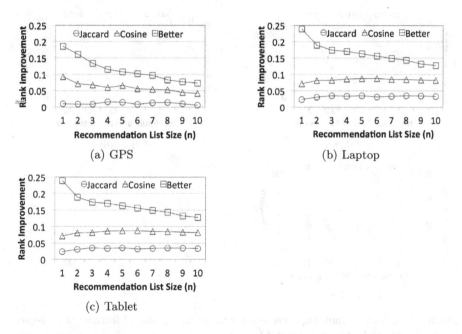

Fig. 4. Mean relative rank improvement versus recommendation list size($k = 15$)

Relative Ratings Ranks. The results are presented in Figure 4 as the relative rank difference (or improvement) versus n (for $k = 15$). These results demonstrate a clear benefit to the *Better* strategy for all values of n and across all product types. For example, for *Laptops* we can see that the *Better* strategy tends to recommend cases with a relative rank improvement of between 24% ($n = 1$) and 13% ($n = 10$); given the 314 laptop cases this means that this strategy is recommending cases that are, on average, up to 75 rank positions better than the query case in terms of overall product rating ($n = 1$). This is a significant improvement compared to the baseline strategies, which achieve a rank improvement of only about 4% or 12-13 rank positions (*Jaccard*) and 9% (or 28 rank positions) for *Cosine*. In other words for each recommendation cycle the *Better* strategy is capable of suggesting new cases that are objectively better than the current query case. It is also worth highlighting that this particular quality measure is obviously considering the 'whole product' quality of recommended cases; we do not need to separately consider the quality of the shared and residual features because user ratings are applied at the product case level and not at the feature level. Thus, even though our recommendations are made on the basis of a set of $k = 15$ or more shared features we can say with some certainty that the products recommended by *Better* are better overall than the query product, and not just with respect to the features that they share with the query case.

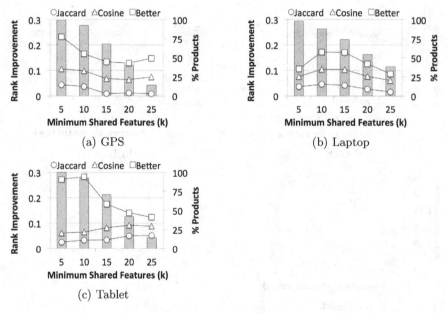

(a) GPS

(b) Laptop

(c) Tablet

Fig. 5. Mean relative rank improvement versus minimum shared features and average percentage of product cases retrieved ($n = 3$)

Relative Rank Versus k. We have yet to consider the impact of k (the minimum number of overlapping features between query and retrieved cases) on recommendation quality. By increasing k we can narrow the scope of comparable products considered for recommendation. This will inevitably limit our ability to recommend products that offer large improvements in quality relative to the query product; it is the nature of product spaces that competition increases as one narrows the product focus, thereby offering less scope for improvement from one product to the next. To test this hypothesis we fix the number of recommended cases ($n = 3$) and compare the relative rank improvement for different values of k, as shown in Figure 5; for reference, we also include a bar chart of the average number of cases retrieved at these k values. Once again we can see a consistent benefit accruing to our *Better* strategy compared to *Cosine* and *Jaccard*. And as predicted, by and large, the relative rank improvement tends to decrease for increasing k. For example, for *Tablets* we see that the relative rank difference of 27% (around 45 rank positions) for $k = 10$ falls to about 12% (or 20 rank positions) for $k = 25$. This compares to rank improvements that are less than 10% for *Cosine* and hardly more than 5% for *Jaccard*, a comparison that is broadly repeated for the *GPS* and *Laptop* case bases too.

4.5 Query Case Similarity

These recommendation quality improvements demonstrate the ability of the *Better* strategy to recommend higher quality products than both alternatives. This is very encouraging but one point that is not clear is the balance between query similarity and this sentiment improvement. If, for example, the cases recommended by *Better* were very different from the query then perhaps these sentiment improvements would be less appealing. To explore this Figure 6 presents the average similarity between the query case and the top n ($n = 10$) recommended cases for the 3 case bases; the standard Jaccard similarity metric is used over features in the query case and the recommended cases.

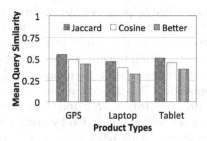

Fig. 6. The average feature similarity between the query case and recommended cases ($n = 10$) for *Jaccard*, *Cosine* and *Better* strategies ($k = 15$)

The results show that there is a reduction in query case similarity for the *Cosine* and *Better* techniques, as expected. However the scale of the reduction is small since it corresponds to only about 2 or 3 features of difference between *Jaccard* and *Better*, for example. And in practice the type of products recommended by *Better* are similar to those recommended by the other techniques.

Moreover, we know from the results above that this relaxation in similarity translates into significant improvements in the quality of recommended cases.

5   Discussion

Our core contributions are: (1) mining product features from user-generated reviews; (2) assigning sentiment to these features to produce experiential product cases; and (3) a novel approach to recommendation that combines product similarity and sentiment to improve recommendation quality.

We believe that the combination of these contributions is important for a number of reasons. Conceptually it keeps the spirit of experience reuse that is the core of CBR, but which is often not deeply ingrained in more traditional case-based recommenders; our product cases are fundamentally *experiential* in the sense that they are based wholly on the experiences of the users of these products. Furthermore, the combination of similarity and sentiment during retrieval facilitates the prioritisation of cases that are not only similar to a query case but also objectively better, at least with respect to the views and usage experiences of product owners. Moreover, the proposed approach is eminently practical: user generated reviews are plentiful even if the type of technical feature specifications used in more traditional product case representations are not. And this means that our approach can be readily deployed in most real-world settings without the need for additional knowledge. Finally, the approach is adaptive and self-regulating. As product reviews accumulate over time the views of users on a particular product or feature may change, and these changes will be reflected in the product cases as they are regularly re-generated from the evolving review-set.

Our results show this approach to be practical and it delivers strong product recommendations that are objectively better than the query, instead of just similar to it. This is important when helping users to *explore* a product space during the early stages of their pre-purchase research. These users are unlikely to have a clear picture of the product they want. The role of the recommender is to help them to explore the trade-offs within the product-space but without prematurely narrowing their search. Critiquing-based recommenders [12, 19] and other forms of conversational recommenders [11, 20] do support this type of discovery, but are based on fixed feature similarities. Our approach combines similarity and sentiment and allows to guide recommendation by quality rather than just similarity. Thus, a user who starts with a point-and-shoot camera might be guided towards more flexible and powerful DSLR models based on superior picture quality, flexibility, and general price-performance features. These products might not be considered in more traditional case-based recommenders due to the lack of similarity between point-and-shoot and DSLR market segments.

In a practical setting for the product types considered in this paper a k comparability score of 15 to 25 provides the right balance of similarity (between the query and retrieved cases) and opportunity for recommendation improvement. These levels of minimum feature overlap provide a suitable basis for case comparison. They constrain the type of cases that are retrieved to be more or less related to the query case and at the same time include cases that are likely to offer improved quality. Smaller values of k provide even greater opportunities for higher quality recommendations but run the risk that the retrieved cases will no longer be sufficiently similar to the query case.

6 Conclusions

In this paper we have described a novel approach to case-based product recommendation. Experiential cases are automatically mined from plentiful user-generated product reviews as the basis for a novel sentiment-based product recommendation strategy. We have demonstrated the benefits of this approach across a number of product domains, in a realistic recommendation setting, and using objective real-user judgements as an objective ground-truth. The results are very promising:

- The generated cases are feature-rich, in the sense that typical cases include 25-30 distinct features and corresponding sentiment scores;
- There is a reasonably dense pattern of overlapping features between cases, thus providing a strong basis for comparison and recommendation;
- It is possible to make recommendations that represent significant improvements in quality with respect to the query case.

In closing our aim has been to describe and demonstrate the viability of a novel approach to case-based product recommendation. But in doing so we have only taken the first step in what has the potential to be a powerful general approach to recommendation on the *experience web*. There is much potential to improve and extend this work by exploring different techniques for topic mining and feature extraction, for example, or alternative ways to evaluate and aggregate sentiment. And of course there are many opportunities to further improve case retrieval, for instance by exploring the use of different feature weighting models. These and other matters reflect our current priorities for future research.

Acknowledgments. This work is supported by Science Foundation Ireland under grant 07/CE/I1147.

References

1. Zhu, F., Zhang, X.M.: Impact of online consumer reviews on sales: The moderating role of product and consumer characteristics. Journal of Marketing 74(2), 133–148 (2010)
2. Dhar, V., Chang, E.A.: Does chatter matter? the impact of user-generated content on music sales. Journal of Interactive Marketing 23(4), 300–307 (2009)
3. Dellarocas, C., Zhang, M., Awad, N.F.: Exploring the value of online product reviews in forecasting sales: The case of motion pictures. Journal of Interactive Marketing 21, 23–45 (2007)

4. Kim, S.-M., Pantel, P., Chklovski, T., Pennacchiotti, M.: Automatically assessing review helpfulness. In: Proceedings of the Conference on Empirical Methods in Natural Language Processing, Sydney, Australia, July 22-23, pp. 423–430 (2006)
5. Baccianella, S., Esuli, A., Sebastiani, F.: Multi-facet rating of product reviews. In: Boughanem, M., Berrut, C., Mothe, J., Soule-Dupuy, C. (eds.) ECIR 2009. LNCS, vol. 5478, pp. 461–472. Springer, Heidelberg (2009)
6. Hsu, C.-F., Khabiri, E., Caverlee, J.: Ranking comments on the social web. In: International Conference on Computational Science and Engineering, CSE 2009, vol. 4, pp. 90–97. IEEE (2009)
7. O'Mahony, M.P., Smyth, B.: Learning to recommend helpful hotel reviews. In: Proceedings of the 3rd ACM Conference on Recommender Systems (RecSys 2009), New York, NY, USA, October 22-25 (2009)
8. Lim, E.-P., Nguyen, V.-A., Jindal, N., Liu, B., Lauw, H.W.: Detecting product review spammers using rating behaviors. In: Proceedings of the 19th ACM International Conference on Information and Knowledge Management, CIKM 2010, pp. 939–948. ACM, New York (2010)
9. Li, F., Huang, M., Yang, Y., Zhu, X.: Learning to identify review spam. In: Proceedings of the Twenty-Second international Joint Conference on Artificial Intelligence, IJCAI 2011, vol. 3, pp. 2488–2493. AAAI Press (2011)
10. O'Callaghan, D., Harrigan, M., Carthy, J., Cunningham, P.: Network analysis of recurring Youtube spam campaigns. In: ICWSM (2012)
11. Bridge, D., Göker, M.H., McGinty, L., Smyth, B.: Case-based recommender systems. The Knowledge Engineering Review 20(03), 315–320 (2005)
12. Reilly, J., McCarthy, K., McGinty, L., Smyth, B.: Dynamic critiquing. In: Funk, P., González Calero, P.A. (eds.) ECCBR 2004. LNCS (LNAI), vol. 3155, pp. 763–777. Springer, Heidelberg (2004)
13. Smyth, B., McClave, P.: Similarity vs. diversity. In: Aha, D.W., Watson, I. (eds.) ICCBR 2001. LNCS (LNAI), vol. 2080, pp. 347–361. Springer, Heidelberg (2001)
14. Hu, M., Liu, B.: Mining and summarizing customer reviews. In: Proceedings of the tenth ACM SIGKDD International Conference on Knowledge Discovery and Data Mining, KDD 2004, pp. 168–177. ACM, New York (2004)
15. Justeson, J., Katz, S.: Technical terminology: Some linguistic properties and an algorithm for identification in text. In: Natural Language Engineering, pp. 9–27 (1995)
16. Hu, M., Liu, B.: Mining opinion features in customer reviews. Science 4, 755–760 (2004)
17. Moghaddam, S., Ester, M.: Opinion digger: An unsupervised opinion miner from unstructured product reviews. In: Proceedings of the 19th ACM International Conference on Information and Knowledge Management, CIKM 2010, pp. 1825–1828. ACM, New York (2010)
18. Sarwar, B., Karypis, G., Konstan, J., Riedl, J.: Item-based collaborative filtering recommendation algorithms. In: Proceedings of the 10th International Conference on World Wide Web, pp. 285–295. ACM (2001)
19. Burke, R.D., Hammond, K.J., Yound, B.: The findme approach to assisted browsing. IEEE Expert. 12(4), 32–40 (1997)
20. Thompson, C.A., Goeker, M.H., Langley, P.: A personalized system for conversational recommendations. J. Artif. Intell. Res. 21, 393–428 (2004)

Mining Features and Sentiment
from Review Experiences

Ruihai Dong, Markus Schaal, Michael P. O'Mahony, Kevin McCarthy,
and Barry Smyth

CLARITY: Centre for Sensor Web Technologies
School of Computer Science and Informatics
University College Dublin, Ireland
http://www.clarity-centre.org

Abstract. Supplementing product information with user-generated content such as ratings and reviews can help to convert browsers into buyers. As a result this type of content is now front and centre for many major e-commerce sites such as Amazon. We believe that this type of content can provide a rich source of valuable information that is useful for a variety of purposes. In this work we are interested in harnessing past reviews to support the writing of new useful reviews, especially for novice contributors. We describe how automatic topic extraction and sentiment analysis can be used to mine valuable information from user-generated reviews, to make useful suggestions to users at review writing time about features that they may wish to cover in their own reviews. We describe the results of a live-user trial to show how the resulting system is capable of delivering high quality reviews that are comparable to the best that sites like Amazon have to offer in terms of information content and helpfulness.

1   Introduction

User-generated product reviews are now a familiar part of most e-commerce (and related) sites. They are a central feature of sites like Amazon[1], for example, featuring prominently alongside other product information. User-generated reviews are important because they help users to make more informed decisions and ultimately, improve the conversion rate of browsers into buyers [13].

However, familiar issues are starting to emerge in relation to the quantity and quality of user-generated reviews. Many popular products quickly become overloaded with reviews and ratings, not all of which are reliable or of a high quality [6, 9]. As a result some researchers have started to look at ways to measure review quality (by using information such as reviewer reputation, review coverage, readability, etc.) in order to recommend high quality reviews to users [8, 10, 12]. Alternatively, others have focused on supporting users during the review-writing phase [1–3], the intent being to encourage the creation of high quality, more informative reviews from the outset. For example, the work of

---

[1] http://www.amazon.co.uk

S.J. Delany and S. Ontañón (Eds.): ICCBR 2013, LNAI 7969, pp. 59–73, 2013.

Healy and Bridge [3] proposed an approach to suggest noun phrases, which were extracted from past product reviews that were similar to the review the user was currently writing; see also the work of Dong et al. [1] for a comparison of related approaches. More recently, Dong et al. described a related approach that focused on recommending product topics or features, rather than simple nouns or noun phrases, to users, based on a hand-coded topic ontology [2].

In this work, we focus on supporting the user at the review-writing stage. We describe a browser-based application called the Reviewer's Assistant (RA) that works in concert with Amazon to proactively recommend product features to users that they might wish to write about. These recommendations correspond to product features which are extracted from past review cases; for example, a user reviewing a digital camera might be suggested a feature such as *"image quality"* or *"battery life"*. This paper extends our previous work [2] in two ways. First, unlike our previous work [2], which relied on hand-coded product features/topics, this paper will describe an approach to automatic feature extraction that does not rely on any hand-coded ontological knowledge. Second, in addition to mining topical features we also evaluate the sentiment of these features, as expressed by the reviewer, to capture whether specific product features have been discussed in a positive, negative or controversial sense. For example, a reviewer might be told that *"image quality"* has been previously reviewed positively while *"battery life"* has largely received negative reviews. We demonstrate how these extensions can be added to the RA system and compare different versions, with and without sentiment information, to examine the quality of the reviews produced.

2   Mining Product Review Experiences

This work is informed by our perspective that user-generated product reviews are an important class of *experiential knowledge* and that, by adopting a case-based reasoning perspective, we can better understand the value of these experiences as they are reused and adapted in different ways to good effect. For example, O'Mahony et al. described how past review cases can be used to train a classifier that is capable of predicting review quality [12]. In this paper, we adopt a different challenge. We are interested in supporting the review writing process and we describe how we can do this by reusing similar past review experiences as the basis for recommending topics to a reviewer for consideration.

The summary RA system architecture is presented in Figure 1. Briefly, the starting point for this work is the availability of a case-base of user-generated product review cases $\{R_1, ..., R_n\}$ for a given class of products such as Digital Cameras, for example. These cases are simply composed of the product id, the text of the review, an overall product rating, and a helpfulness score (based on user feedback). The RA system extends these review cases by augmenting them with a set of *review features* $\{F_1, ..., F_m\}$ and corresponding *sentiment scores*, which correspond to the features covered in the review text. These features and scores are automatically mined from the review case-base, mapped back to the relevant review text, and then used as the basis for recommendation during review writing as described below.

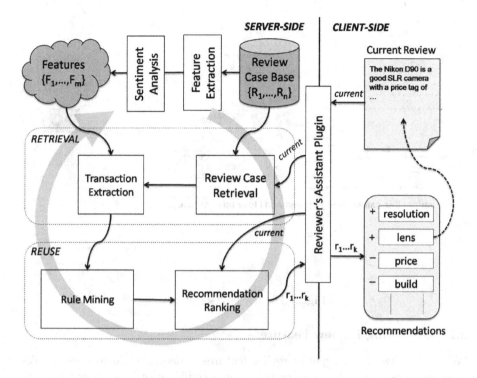

Fig. 1. System architecture

The client-side component of the RA system is designed as a browser plugin that is 'sensitive' to Amazon's review component, which is to say that it becomes activated when the user lands on a review page. When activated it overlays a set of recommendations $r_1, \ldots, r_k$, marked as the suggestion box in Figure 2. These recommendations are essentially sets of product features that have been automatically mined from past reviews for this product and, by default, they are ranked based on the review text at a particular point in time. In this example, the recommendations are enhanced with additional sentiment information, which has also been mined from past reviews by aggregating the sentiment predictions for different review sentences mentioning the feature in question. The colour of the recommendation indicates the relative sentiment label, whether positive (green), negative (red), controversial (yellow), or without sentiment (blue); controversial features are those which divide reviewer opinions. In addition each feature is annotated with a sentiment bar to visualise the number of positive, negative, and neutral instances for the feature in question. For example, the *battery* feature is marked as negative (red) and the sentiment bar shows that the vast majority of users have reviewed the *battery* of this camera as either negative or neutral, with very few positive opinions expressed.

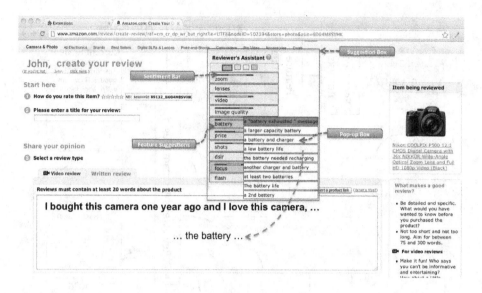

Fig. 2. The RA browser plugin

2.1   Extracting Review Features

We consider two basic types of review features — *bi-gram* features and *single-noun* features — which are extracted using a combination of shallow NLP and statistical methods, by combining ideas from related research [4, 7]. Briefly, to produce a set of bi-gram features we look for bi-grams in the review cases which conform to one of two basic part-of-speech co-location patterns: (1) an adjective followed by a noun (AN) such as *wide angle*; and (2) a noun followed by a noun (NN) such as *video mode*. These are candidate features but need to be filtered to avoid including AN's that are actually opinionated single-noun features; for example, *great flash* is a single-noun feature (*flash*) and not a bi-gram feature. To do this we exclude bi-grams whose adjective is found to be a sentiment word (e.g. *excellent, good, great, lovely, terrible, horrible,* etc.) using Hu and Liu's sentiment lexicon [5].

To identify the single-noun topics we extract a candidate set of (non stop-word) nouns from the review cases. Often these single-noun candidates will not make for good case features however; for example, they might include words such as *family* or *day* or *vacation*. The work of Hu and Liu [5] proposes a solution for validating such features by eliminating those that are rarely associated with opinionated words. The intuition is that nouns that frequently occur in reviews and that are often associated with opinion laden words are likely to be popular product features. We calculate how frequently each feature co-occurs with a sentiment word in the same sentence (again, as above, we use Hu and Liu's sentiment lexicon [5]), and retain the single-noun only if its frequency is greater than some threshold (in this case 70%).

This produces a set of bi-gram and single-noun features which we further filter based on their frequency of occurrence in the review cases, keeping only those features ($\{F_1, \ldots, F_m\}$) that occur in at least k reviews out of the total number of n reviews; in this case, for bi-gram features we set $k_{bg} = n/20$ and for single noun topics we set $k_{sn} = 10 \times k_{bg}$ via manual testing. The result is a master list of features for a product case-base and each individual case can then be associated with the set of features that occur within its review text.

2.2 Evaluating Feature Sentiment

Next for each case feature we can evaluate it's sentiment based on the review text that covers the feature. To do this we use a modified version of the *opinion pattern mining* technique proposed by Moghaddam and Ester [11] for extracting opinions from unstructured product reviews. Once again we use the sentiment lexicon from Hu and Liu [5] as the basis for this analysis. For a given feature, F_i, and corresponding review sentence, S_j, from review case C_k (that is the sentence in C_k that mentions F_i), we determine whether there are any sentiment words in S_j. If there are not then this feature is marked as *neutral*, from a sentiment perspective. If there are sentiment words ($w_1, w_2, \ldots$) then we identify that word (w_{min}) which has the minimum word-distance to F_i.

Next we determine the part-of-speech (POS) tags for w_{min}, F_i and any words that occur between w_{min} and F_i. The POS sequence corresponds to an opinion pattern. For example, in the case of the bi-gram topic *noise reduction* and the review sentence, *"...this camera has great noise reduction..."* then w_{min} is the word *"great"* which corresponds to an opinion pattern of *JJ-TOPIC* as per [11].

Once an entire pass of all features has been completed we can compute the frequency of all opinion patterns that have been recorded. A pattern is deemed to be valid (from the perspective of our ability to assign sentiment) if it occurs more than some minimum number of cases (we use a threshold of 2). For valid patterns we assign sentiment based on the sentiment of w_{min} and subject to whether S_j contains any negation terms within a 4-word-distance either side of of w_{min}. If there are no such negation terms then the sentiment assigned to F_i in S_j is that of the sentiment word in the sentiment lexicon. If there is a negation word then this sentiment is reversed. If an opinion pattern is deemed not to be valid (based on its frequency) then we assign a *neutral* sentiment to each of its occurrences within the review set.

As a result our review cases now include not only the product features identified in their text but also the sentiment associated with these features (positive, neutral, negative). Each of these features is also linked to the relevant fragment of text in the review.

2.3 Reusing Review Cases for Feature Recommendation

For the RA system the primary purpose of review cases is to provide product insights to reviewers for consideration as they write new reviews. This means recommending product features, from relevant past reviews, which fit the context of the current review. This is triggered as the user is writing their review: whenever the user has written a couple of words, or completed a sentence, for example, the recommender returns a new (or updated) set of recommendations.

The recommendations are ranked by default according to a *relevance* metric based on an association rule mining technique which orders features based on their frequency of occurrence in a subset of the *most similar* reviews to the target review so far. This approach is based on the technique described in Dong et al. [2] and is summarised as follows. The relevance ranking process includes the following key steps: (1) review case retrieval; (2) rule mining; (3) transaction extraction; and (4) recommendation generation.

Review Case Retrieval. The current review text is used as a textual query against a relevant set of review cases for the same product to retrieve a set of similar reviews. In the current implementation we rely on a simple term-based Jaccard similarity metric to retrieve a set of review cases that are most similar to the query.

Transaction Extraction. Each of these review cases is converted into a set of sentence-level transactions and review-level transactions. Briefly, each sentence is converted into the set of features it mentions. If, for example, the review is *"The camera takes good pictures. A flash is needed in poor light."*, then we would have sentence transactions {*camera, pictures*} and {*flash, light*}. And the review level transaction corresponds to the set of features mentioned in the review; if in the above example the review was made up just of these two sentences then the review-level transaction would be {*camera, pictures, flash, light*}.

Rule Mining. We apply standard association rule mining techniques across all transactions from the k similar cases to produce a set of feature-based association rules, ranked in descending order of their confidence. For example, we may identify a rule *weight → batterylife* to indicate that when reviews mention camera weight they tend to also discuss battery-life.

Recommendation Ranking. To generate a set of ranked recommendations we apply each of the extracted rules, in order of confidence, to the features of the current review text. If the current review text triggers a rule of the form $F_x \rightarrow F_y$, that is because it mentions feature F_x, then the feature F_y is added to the recommendation list. This process terminates when a set of k recommendations has been generated.

2.4 Discussion

This completes our overview of the RA system. Its aim is to provide users with targeted product feature suggestions based on their review to date and the features discussed in similar reviews that have proven to be helpful in the past.

Ultimately our objective with this work is to make a *systems* contribution. That is to say our aim is to develop a novel system and evaluate it in the context of a realistic application setting. Specifically, the primary contribution of this work is to describe the RA as a system that combines automatic feature extraction and sentiment analysis techniques as part of a recommendation system that is designed to support users during the product review process. This builds on previous work by Dong et al. [2] but distinguishes itself in two important ways: (1) by the use of automatic techniques for feature extraction, versus hand-crafted topics; and (2) by exploring the utility of sentiment as part of the recommendation interface.

3   Evaluation

How well does the RA system perform? Does it facilitate the generation of high quality reviews? How do these reviews compare with the best of what a site like Amazon has to offer? What is the impact of including sentiment information as part of the recommendations made to reviewers? These are some of the questions that we will seek to answer in this section via an initial live-user trial of the new RA system.

3.1   Setup

This evaluation is based on an authentic digital camera product review set containing 9,355 user-generated reviews for 116 distinct camera products mined from Amazon.com during October 2012. We implemented two versions of the RA system: (1) *RA*, which uses automatic feature extraction but does not use sentiment information; (2) *RA + S*, which uses automatic feature extraction and uses sentiment information to distinguish between, for example, positive and negative features as part of the RA recommendation interface.

For the purpose of this evaluation we recruited 33 participants (mainly college students and staff with ages between 17 and 50). These trial participants were mostly novice or infrequent review writers. When asked, 48% (16 out of 33) said they had never submitted an online product review and of those who had, 65% (11 out of 17) of them had written less than 5 product reviews. Each participant was randomly assigned to one of the versions of the RA system; 17 participants were assigned to *RA* and 16 were assigned to *RA + S*. Each participant was asked to produce a review of a digital camera that was familiar to them and the text of their review was stored for later analysis.

As a competitive baseline for review quality we also extracted 16 high-quality camera reviews from the Amazon data-set; we will refer to these as the *Amazon*(+) review set. In order to ensure comparability, we chose these reviews of be of similar lengths as the ones created manually with the help of *RA* and *RA + S*. These 16 reviews were chosen from the subset of the most helpful Amazon reviews by only selecting reviews with a helpfulness score of greater than 0.7. As a result the average helpfulness score of these *Amazon*(+) reviews was 0.86, meaning that 86% of users found them to be helpful. These are clearly among the best of the user-generated reviews found on Amazon for digital cameras. Therefore this constitutes a genuinely challenging baseline review-set against which to judge the quality of the reviews produced by the trial participants.

3.2 Depth, Breadth and Redundancy

We describe a quantitative analysis of the three sets of reviews (RA, $RA+S$ and $Amazon(+)$) by adopting the approach taken by Dong et al. [2]. For each review we note its length and compute its *breadth*, *depth* and *redundancy*. Briefly, the *breadth* of a review is the number of product features covered by the review. The *depth* of a review is the number of words per feature; that is the word-count of the sentences referring to a given feature. And finally, the *redundancy* of a review is the word-count of the sentences that are not associated with any particular feature.

Table 1. A quantitative analysis of review depth, breadth and redundancy; * indicates pairwise significant difference between Sentiment(RA+S)/ Non-Sentiment(RA) and Amazon+ only, at the 0.05 level; ** indicates significant difference between all pairs at the 0.1 level (using two-tailed t-test)

	$RA+S$	RA	$Amazon(+)$
Breadth*	8.44	7.53	3.63
Depth*	9.41	9.01	17.23
Redundancy**	3.75	10.24	23.63
Length	81.88	81.94	81.50

The result of this analysis, for the three sets of reviews, are presented in Table 1 as averages for review breadth, depth, redundancy and length. We can see that both RA systems (RA and $RA+S$) deliver reviews that are broader (greater feature coverage) than the high-quality Amazon reviews, and with less redundancy. For example, RA and $RA+S$ both lead to reviews that cover more than twice as many product features as the $Amazon(+)$ reviews with less than half of the redundancy. The best performing $RA+S$ condition produces reviews that cover 8.44 product features on average compared to less than 4 product features per review for $Amazon(+)$. Moreover, the $RA+S$ reviews display very low levels of redundancy (3.75 words per review on average) compared to more than 10 and 23 redundant words per review for RA and $Amazon(+)$, respectively. However the reviews produced by RA and $RA+S$ offer less depth of feature coverage than $Amazon(+)$, so although RA and $RA+S$ participants are writing about more features, they are not writing as much about each individual feature.

In relation to the breadth differences, our view is that the RA system helps take some of the "guess work" out of the review-writing process. Reviewers have instant access to a list of meaningful product features (and examples of what other reviewers have written about these features). This reduces some of the friction that is inherent in the review-writing process since the users are no longer solely responsible for prioritising a set of features to write about. Thus users find it easier to identify a set of features to write about and they are naturally inclined to discuss more of these features.

Concerning the difference in depth between the sets of reviews, it is reasonable to take review length as a proxy for the amount of time that users spend writing a review. All three sets of reviews are similar in this regard. Then, per unit time spent writing a review, it is perhaps not surprising that the $Amazon(+)$ reviews enjoy improved depth of feature coverage when compared to RA and $RA + S$; if all 3 sets of users are spending the same time on reviews and $Amazon(+)$ reviewers are covering fewer features, then either they are covering these features in greater depth or they are including more redundant sentences in their reviews. As it turns out both effects are evident: there is a greater depth of coverage for the $Amazon(+)$ reviews but there is also a significant amount of additional redundancy.

There is less of a difference between the RA and $RA + S$ conditions. The additional depth and breadth values for $RA + S$ compared with RA are not statistically significant in this trial. It is worth noting, however, that $RA + S$ does enjoy significantly less levels of redundancy than the RA reviews (an average of 3.75 versus 10.24 redundant words per review). Given that RA and $RA + S$ reviews are similar in terms of depth and breadth, then perhaps there are other metrics that might help us to understand other meaningful differences between these review sets — we consider such metrics in the following sections.

Finally, we appreciate that our measurement of breadth, depth and redundancy depends on the performance of our feature extraction method and so we examined its accuracy against the Amazon data-set. We randomly selected 200 sentences from the more than 99,000 review sentences contained in the 9,355 reviews. From each of these sentences we manually identified a set of features (typically a word or pair of words) and manually judged their sentiment as positive, negative or neutral. This manual annotation process was conducted by 4 independent 'experts' and serves as our ground-truth. We compared our predicted features (sentence by sentence) to the ground-truth for the corresponding sentences and found a precision of 63% and a recall of 67%. The overall accuracy of sentiment prediction is 71%. While these results indicate that there is scope to improve our feature extraction method, it is important to note that the results correspond to a strict matching criterion, i.e. a predicted feature *lens* would not match a ground-truth feature *lens quality*. Given this approach and the large (and statistically significant) differences in breadth, depth and redundancy between the $RA + S/RA$ and $Amazon(+)$ reviews, we believe that the findings as reported above reflect true differences in performance.

3.3   Sentiment Density

Clearly the process by which $RA + S$ reviews are produced is different in one important way from the process that produces RA reviews. The former is informed by indicators of sentiment attached to recommended features. Do these labels influence the actual reviews that are produced? Are users more likely to express opinions on sentiment-laden features?

One way to explore this is to look at what we call the *sentiment density* of a review, by which we mean the percentage of sentences that discuss features in an opinionated manner. The intuition here is that reviews that contain content that is neutral is likely to be less useful, when it comes to making a decision. Sentiment density can be calculated in a straightforward fashion by counting the number of review features with positive or negative sentiment as a fraction of the total number of features in reviews.

Table 2. The sentiment density of RA, $RA + S$ and $Amazon(+)$ reviews; * indicates significant difference between Sentiment and Non-Sentiment at the 0.05 level; ** indicates significant difference between $RA + S$ and Amazon(+) at the 0.1 level (using two-tailed t-test)

	$RA+S$	RA	$Amazon(+)$
Density*/**	65%	48%	49%

Table 2 presents the sentiment density results for our three sets of reviews and clearly points to a significant benefit for those produced using the $RA + S$ condition. The sentiment density of the $RA + S$ reviews is 65% compared to 48% and 49% for the non-sentiment RA and $Amazon(+)$ conditions. In other words, almost two thirds of the features discussed in $RA + S$ reviews are discussed in an opinionated manner; i.e. the reviewer expresses a clear positive or negative viewpoint. By comparison a little less that half of the features mentioned in the RA and $Amazon(+)$ reviews are discussed in an opinionated manner.

As a result, one might expect there to be some benefit in the utility of the $RA + S$ reviews, at least in so far as they contain opinions or viewpoints that are more likely to influence buyers. Clearly the sentiment information that is presented alongside the feature recommendations is influencing users to express stronger (more polarised) opinions for those features that they choose to write about. One caveat here is whether or not the sentiment information is *biasing* what the reviewers write? For example, if they see that *image quality* has been previously reviewed in a positive manner for a particular product, then is the user more likely to write positively about this feature? Obviously this would not be desirable and we will return to this point later.

3.4 Review Quality

Clearly there is a difference between the type of reviews produced with recommendation support (whether with or without sentiment) when compared to the Amazon(+) reviews: both RA and $RA+S$ reviews tend to cover more topics but in less detail than the $Amazon(+)$ reviews; the RA and $RA+S$ reviews contain less redundancy; and the $RA+S$ reviews tend to contain more opinionated content. But how does this translate into the perceived utility of these reviews from a user perspective? The $Amazon(+)$ reviews have been selected from among the most helpful of Amazon's reviews. How will the reviews produced by the less experienced reviewers using RA and $RA+S$ compare?

To answer this question we recruited a set of 12 people to perform a blind evaluation of the three sets of reviews. Each evaluator was asked to rate the *helpfulness, completeness* and *readability* of the reviews on a 5-point scale (with a rating of 1 indicating 'poor' and a rating of 5 indicating 'excellent'). Every review was evaluated by 3 of the 12 participants and their ratings were averaged to calculate mean helpfulness, completeness and readability scores for each set of reviews.

Table 3. A qualitative analysis of review quality showing mean (median) ratings

	$RA+S$	RA	$Amazon(+)$
Helpfulness	3.42 (4)	3.33 (3)	3.23 (3)
Completeness	3.06 (3)	3.08 (3)	2.71 (3)
Readability	3.60 (4)	3.51 (4)	3.69 (4)

The results are presented in Table 3 as mean and median (bracketed) ratings. As expected the $Amazon(+)$ reviews are rated highly, they are after all among the best reviews that Amazon has to offer. Importantly, we can see however that the reviews produced using the RA and $RA+S$ conditions perform equally well and, in fact, marginally better in terms of review helpfulness and completeness. Although these findings are not definitive — the differences were not found to be statistically significant, not surprising given the scale of the trial — the data bodes well for the approach we are taking. At the very least the additional breadth of coverage offered by RA and $RA+S$ reviews is found to be just as helpful as the best Amazon reviews, for example.

3.5 System Usability and Influence

At the end of the trial each participant was also asked to rate the RA system on a 3-point scale (*agree, neutral, disagree*) under the following criteria:

1. *User Statisfaction* – Were you satisfied with the overall user experience?
2. *Helpfulness* – Did the RA help you in writing a review?

3. *Relevance* – Were the specific recommendations relevant to the review you were writing?
4. *Comprehensiveness* – Did the recommendations comprehensively cover the product being reviewed?

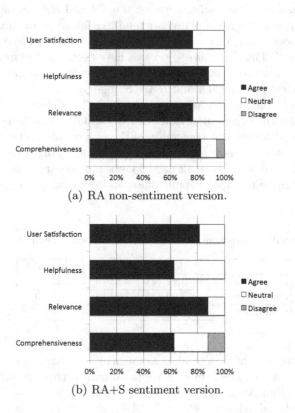

(a) RA non-sentiment version.

(b) RA+S sentiment version.

Fig. 3. User feedback

The results of this feedback for RA and $RA + S$ are presented in Figures 3(a) and 3(b). Broadly speaking users were very satisfied with the RA variations; about 78% of RA users and 82% of $RA + S$ users found the system to be satisfactory and none of the users reported being unhappy with the overall experience. Users also found the reviews to be relevant and mostly helpful, although the $RA + S$ suggestions were judged to be less helpful (62%) that those for the RA system (86%). Interestingly a similar difference is noted with respect to how comprehensive the $RA + S$ suggestions were in comparison to those provided by RA.

Remember that the difference between the RA and $RA + S$ systems is the absence or presence of sentiment information. The above differences would seem to be a result of this interface difference. It is a matter of future work to further explore this by testing different interface choices and different ways to display sentiment information.

Finally,we mentioned earlier the possibility that by displaying sentiment information to users at review time we may lead to biased reviews. As part of the the post-trial feedback (for $RA + S$ participants only) we also asked them to comment on this aspect of the trial as follows:

1. *Influence* – Do you think that the sentiment information influences your own judgement?
2. *Encouragement* – Does the additional sentiment information encourage you to write about your own judgement?
3. *Interruption* – Do you think the additional sentiment information interrupted the review writing process?

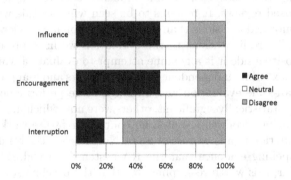

Fig. 4. User feedback on influence, encouragement and interruption – RA+S version

The results are presented in Figure 4. On the positive side, the participants agreed strongly that the recommendations did not interrupt the review writing process. This finding is not surprising since, as above, participants found the recommendations to be mostly helpful and relevant. A majority of $RA + S$ participants (58%) felt that the availability of sentiment information actually encouraged them to write about features, with less than 20% disagreeing with this proposition. Again this is not surprising given that the $RA + S$ reviews benefit from improved breadth characteristics in particular.

However, a small majority of participants (58%) also felt that the availability of sentiment information was likely to influence the reviews they wrote. This may be an issue and certainly raises the need for additional work to explore this particular aspect of the $RA+S$ system, especially if it turns out to be responsible for reviews that are biased with respect to the sentiment of the recommended features.

3.6 Discussion

The primary objective of this work has been to explore the role of the RA system when it comes to helping users to write high quality reviews based on the recommendation of mined features and sentiment information. The evidence suggests that there are good reasons to be optimistic about this approach. For example, the overall review quality, completeness, and readability of reviews produced using RA and $RA + S$ is at least equivalent to the best of Amazon's reviews even though they were produced by more novice reviewers. The reviews produced with support from RA and $RA + S$ tend to offer broader coverage of product features with less redundancy and so, perhaps, provide a useful counterpoint to the more in-depth Amazon reviews that tend to focus on a narrower set of product features.

There are a number of questions that remain to be answered. For example, there is evidence, as discussed above, that the display of sentiment information at review writing time may exert undue influence over reviewers, which may lead to more biased reviews. It remains to be seen whether this will help users to make more informed decisions than with less opinionated reviews.

Of course there are limitations to the evaluation we have presented in this work. On the positive side it is a genuine attempt to evaluate a working system in a realistic context using independent trial participants and real products. However, it is a small-scale evaluation and although some performance differences were found to be statistically significant, others were not, which ultimately limits what we can conclude from the results. Of course our future work will seek to expand this evaluation to a larger set of users. Nevertheless the results presented do provide compelling evidence that the RA system is providing a useful service. In particular, it is worth re-emphasising that the baseline Amazon reviews chosen as a benchmark were selected among the best quality Amazon reviews available, and so represent a particularly high benchmark for our evaluation.

4 Conclusions

This paper describes an experience-based recommender system that is designed to help users to write better product reviews by passively making suggestions to reviewers as they write. It extends the work of Dong et al. [2] in two important ways. First it is based on a fully automatic approach to review feature extraction without the need for hand-crafted topics or ontologies as in [2]. Secondly, it explores the use of feature sentiment during recommendation and presentation. We have described the results of a detailed live-user trial to consider review quality in terms of metrics, such as feature depth, breadth and sentiment density, demonstrating the quality of RA reviews compared to the best that sites like Amazon has to offer.

Acknowledgments. This work is supported by Science Foundation Ireland under grant 07/CE/I1147.

References

1. Dong, R., McCarthy, K., O'Mahony, M.P., Schaal, M., Smyth, B.: Towards an intelligent reviewer's assistant: Recommending topics to help users to write better product reviews. In: Procs. of IUI: 17th International Conference on Intelligent User Interfaces, Lisbon, Portugal, February 14-17, pp. 159–168 (2012)
2. Dong, R., Schaal, M., O'Mahony, M.P., McCarthy, K., Smyth, B.: Harnessing the experience web to support user-generated product reviews. In: Agudo, B.D., Watson, I. (eds.) ICCBR 2012. LNCS, vol. 7466, pp. 62–76. Springer, Heidelberg (2012)
3. Healy, P., Bridge, D.: The GhostWriter-2.0 system: Creating a virtuous circle in web 2.0 product reviewing. In: Bridge, D., Delany, S.J., Plaza, E., Smyth, B., Wiratunga, N. (eds.) Procs. of WebCBR: The Workshop on Reasoning from Experiences on the Web (Workshop Programme of the Eighteenth International Conference on Case-Based Reasoning), pp. 121–130 (2010)
4. Hu, M., Liu, B.: Mining and summarizing customer reviews. In: Proceedings of the Tenth ACM SIGKDD International Conference on Knowledge Discovery and Data Mining, KDD 2004, pp. 168–177. ACM, New York (2004)
5. Hu, M., Liu, B.: Mining opinion features in customer reviews. In: AAAI 2004, vol. 4, pp. 755–760 (2004)
6. Jindal, N., Liu, B.: Review spam detection. In: Proceedings of the 16th International Conference on World Wide Web, WWW 2007, pp. 1189–1190. ACM, New York (2007)
7. Justeson, J., Katz, S.: Technical terminology: Some linguistic properties and an algorithm for identification in text. In: Natural Language Engineering, pp. 9–27 (1995)
8. Kim, S.-M., Pantel, P., Chklovski, T., Pennacchiotti, M.: Automatically assessing review helpfulness. In: Proceedings of the Conference on Empirical Methods in Natural Language Processing (EMNLP 2006), Sydney, Australia, July 22-23, pp. 423–430 (2006)
9. Lappas, T.: Fake reviews: The malicious perspective. In: Bouma, G., Ittoo, A., Métais, E., Wortmann, H. (eds.) NLDB 2012. LNCS, vol. 7337, pp. 23–34. Springer, Heidelberg (2012)
10. Liu, Y., Huang, X., An, A., Yu, X.: Modeling and predicting the helpfulness of online reviews. In: Proceedings of the 2008 Eighth IEEE International Conference on Data Mining (ICDM 2008), Pisa, Italy, pp. 443–452. IEEE Computer Society (2008)
11. Moghaddam, S., Ester, M.: Opinion digger: An unsupervised opinion miner from unstructured product reviews. In: Proceedings of the 19th ACM International Conference on Information and Knowledge Management, CIKM 2010, pp. 1825–1828. ACM, New York (2010)
12. O'Mahony, M.P., Smyth, B.: Learning to recommend helpful hotel reviews. In: Proceedings of the 3rd ACM Conference on Recommender Systems (RecSys 2009), New York, NY, USA, October 22-25 (2009)
13. Zhu, F., Zhang, X(M.): Impact of online consumer reviews on sales: The moderating role of product and consumer characteristics. Journal of Marketing 74(2), 133–148 (2010)

Multi-Agent, Multi-Case-Based Reasoning

Susan L. Epstein[1,2], Xi Yun[1], and Lei Xie[1,2]

[1] Department of Computer Science,
The Graduate Center of The City University of New York, New York, NY 10016, USA
[2] Department of Computer Science,
Hunter College of The City University of New York, New York, NY 10065, USA
{susan.epstein,lei.xie}@hunter.cuny.edu, xyun@gc.cuny.edu

Abstract. A new paradigm for case-based reasoning described here assembles a set of cases similar to a new case, solicits the opinions of multiple agents on them, and then combines their output to predict for a new case. We describe the general approach, along with lessons learned and issues identified. One application of the paradigm schedules constraint satisfaction solvers for parallel processing, based on their previous performance in competition, and produces schedules with performance close to that of an oracle. A second application predicts protein-ligand binding, based on an extensive chemical knowledge base and three sophisticated predictors. Despite noisy, biased biological data, the paradigm outperforms its constituent agents on benchmark protein-ligand data, and thereby promises faster, less costly drug discovery.

Keywords: multiple cases, multiple agents, confidence-based reasoning.

1   Introduction

As the problems presented to computers become increasingly difficult, the techniques researchers develop to address them become increasingly sophisticated and complex. Although these programs may perform unevenly, ensembles of them often smooth performance [1]. At the same time, data pertinent to difficult problems has burgeoned, even though it is often noisy and incomplete. Rather than trust the evidence of a single data point, it may be more informative, to consider several. The case-based reasoning paradigm described here, *MAMC* (Multi-Agent Multi-Case-based reasoning), takes both routes: it consults multiple agents and it uses multiple cases. We report here on two MAMC applications: construction of a parallel schedule for constraint satisfaction search, and prediction about the binding energy between two proteins, a key to rational drug design. Although many such *agents* (here, solvers or predictors) exist, none consistently outperforms all the others on a large, diverse set of benchmark examples. The thesis of this paper is that the effectiveness of a set of agents on a set of similar cases supports reasoning about the agents' performance on a new case. Given a new case, MAMC selects from its knowledge base the cases most similar to it, and examines the accuracy of a set of agents on those cases. The principal result reported here is that MAMC improves predictive accuracy in both applications.

S.J. Delany and S. Ontañón (Eds.): ICCBR 2013, LNAI 7969, pp. 74–88, 2013.

Input: new case *e*, case base *C*, agents *A*, pairwise similarity metric *s*,
     number of reference cases *q*
Output: prediction or recommendation for *e*

Select a subset *L* of *q* cases in *C* most similar to *e* as measured by *s*
Predict or recommend on each case in *L* with A_j for all $A_j \in A$
Combine output from all $A_j \in A$ by their performance on *L* as output for *e*

Fig. 1. High-level pseudocode for MAMC

MAMC is outlined in Figure 1. For a particular domain, MAMC's pre-existing agents *A* are assumed to be the result of extensive research and development, and generally regarded by experts as among the best. The corresponding case base *C*, shared by all the agents, consists of published results for those agents on their common task. Finally, the features on which similarity is gauged to select the reference cases *L* are assumed available from other experts' work in the domain of interest, but deliberately differ from those of any individual agent for their prediction.

Unlike earlier work with multiple cases, which drew from case bases for related tasks [2] or focused on distributed resources across multiple machines [3], MAMC reasons over multiple cases resident on the same computer and for the same task. Rather than use portions of multiple cases or multiple agents to produce a solution, as in [4], MAMC uses multiple cases to select one agent most appropriate for a new case. MAMC can also estimate the reliability of its output, an essential but rarely available property in bioinformatics. MAMC's confidence is based not on the quality of its cases, as in [5], but on the degree of similarity it detects between the new case *e* and the reference cases *L*, along with the performance of the selected agent on those cases.

The applications described here face similar challenges: development of an extensive case base with an incisive feature-based index, a pairwise similarity metric for cases, and a way to combine the agents' output to make a decision. Each of the next two sections details an application, with relevant background; related work; the origin of *C*, *A*, and *s;* and empirical design and results. The discussion mines this experience to establish commonalities, issues, and promise for future MAMC applications.

2   Parallel Portfolio Construction for Constraint Satisfaction

Constraint satisfaction is a powerful representation for many real-world problems, but search for a solution to a constraint satisfaction problem (*CSP*) is in general NP-hard. Many solvers succeed on quite difficult problems, but unevenly and unpredictably. Our goal is to schedule solvers on multiple processors to solve one problem.

Here, a CSP <*X, D, R*> is a set *X* of variables, a set *D* of finite domains associated with those variables, and a set *R* of constraints that restrict how values from their respective domains may be assigned to variables simultaneously. A *solution* assigns a value to each variable and satisfies all of *R*. A *solvable* CSP has at least one solution. The solvers used here assign a value to one variable at a time, and temporarily remove inconsistent values from the domains of as-yet-unassigned variables. If a domain becomes empty, the solver backtracks to the most recent alternative and chooses a new value. Search returns the first solution found, or halts when it shows the problem is *unsolvable* (i.e., the domain of the variable at the top of the search tree becomes empty). A *variable-ordering* heuristic determines the order in which the solver addresses variables, and a *restart policy* begins search on the problem again, probably with a different root variable. This remainder of this section summarizes work that appeared in the optimization literature, and detailed its rationale and development saga more theoretically [6].

2.1   The Task for MAMC

In this application, the agents A are CSP solvers, C is a set of CSPs, and T is a set of consecutive, unit time intervals. A *simple schedule* σ for a problem on one processor specifies at most one agent to address the problem in each time interval, that is, $\sigma: T \to A$. A *schedule* for k processors is a set of k simple schedules, one for each processor. On any one processor, at most T time can be allotted to any solver on any problem. We represent the performance of A on C in a $|C| \times |A|$ *performance matrix* τ, where entry $\tau_{ij} \in \{1,2,..., T\}$ means that the jth solver solves the ith problem in time τ_{ij}; otherwise the problem goes unsolved in time T. The solvers used here are *deterministic*, that is, each τ_{ij} is a fixed, positive integer.

A CSP *portfolio* is a combination of solvers intended to outperform each of its constituents [7-12]. A *solver portfolio* $<A, k, S, T>$ proposes a set S of k simple schedules that deploy agents A on k processors to solve a problem in time T. For $k = 1$, a solver portfolio is *simple;* for $k > 1$, it is *parallel*. If it is deterministic, neither a simple nor a parallel solver portfolio can exceed the performance of an *oracle*, which always selects the single fastest solver. We focus here on *offline* solver portfolio constructors, which observe the performance of A on C and then build a portfolio to optimize performance on new cases [8, 9, 11]. We consider only *switching* schedules, which preserve each solver's intermediate search state when its time elapses, for reuse only by that solver if time is allocated to it on the same processor later.

Given cases C, a new problem e, similarity metric s, and performance matrix τ for solvers A on C, MAMC's task is to find the best parallel schedule σ that uses A to solve e on k processors. The portfolio constructors most relevant here, CPHYDRA [8] and GASS [13], were both intended for a single processor. CPHYDRA is case-based; it selects a small set of problems in C similar to e, and searches all possible schedules, in time $O(2^{|A|})$. It weights problems in C by their Euclidean distance from e, and seeks an optimal schedule, one that maximizes the number of problems solved within T. In contrast, GASS is greedy, and its performance depends on $|C|$. At each step, GASS selects the agent that maximizes the number of problems solved per unit of time, and counts only problems solved for the first time during the current time step. GASS creates schedules in time $O(|C| \cdot |A| \log |C| \cdot \min\{|C|, T \cdot |A|\})$ that are at most four times worse than optimal; any better approximation was shown to be NP-hard [13].

2.2   Cases, Similarity, and Combination

The case base was developed from the 3307 problems in 5 categories at the Third International CSP solver competition (*CPAI'08*), where problems represent a wide

Table 1. CPAI'08 problems by category

Applicable solvers	Category	Competition problems	Experiment problems	Solvable experiment problems
17	GLOBAL	556	493	256
22	k-ARY-INT ($k\geq2$)	1412	1303	739
23	2-ARY-EXT	635	620	301
24	N-ARY-EXT ($N>2$)	704	449	156

Table 2. Problems in Table 1 solved by non-parallel solver portfolios in 1800 seconds

Solver	Oracle	GASS	CPHYDRA10	CPHYDRA40
Number solved	2865	2773	2577	2573
% solved	100%	96.79%	89.95%	89.81%

variety of challenges and are intended to be difficult [14]. Cases were represented by the same 36 numeric features (e.g., number of variables, maximum domain size) used by CPHYDRA. To extract feature values, we ran the solver Mistral 1.550 on an 8 GB Mac Pro with a 2.93 GHz Quad-Core Intel Xeon processor. We excluded any problem whose full set of features was not calculated within 1 second, and any problem never solved by any solver at CPAI'08. Table 1 summarizes the remaining 2865 problems.

Agents in A are the 24 solvers in CPAI'08. They include CPHYDRA10 and CPHYDRA40, versions of CPHYDRA that used the same 3 solvers, but with 10 or 40 cases respectively (CPHYDRA won all but one category at CPAI'08.) The CPAI'08 results provide the performance matrix τ. For *neighbor set ratio r* ($0 \leq r \leq 1$), the *neighbor set L* of any case e is the $r \cdot |C|$ problems in C with feature vectors most similar to e.

Portfolio construction experiments performed 10-fold cross-validation. Each iteration partitioned the 2685 into a set of testing problems and a set of training problems (i.e., the case base C). Stratified partitioning maintained the proportions of problems from different categories in each subset. Table 2 reports the performance of an oracle and three non-parallel solver portfolio constructors, given 1800 seconds per problem.

RSR-WG (Retain, Spread and Return with a Weighted Greedy approach) is the MAMC implementation for this task. As in Figure 2, it formulates a parallel schedule for problem e based on τ, A, C, and L, the set of cases most similar to e. RSR-WG tries to build a schedule that solves as many cases as it can from L, under the assumption that the same schedule will then do well on e. RSR-WG tries, heuristically and greedily, to schedule L, and measures the similarity of case $c_i \in L$ to e based on its Euclidean distance $d(c_i, e)$ from e (here, $\varepsilon = 0.001$):

$$s_i = 1 - \frac{(1-\varepsilon)\left[d(c_i,e) - d_{min}\right]}{d_{max} - d_{min}} \tag{1}$$

where $d_{max} = \max(\{d(c_i, e) \mid c_i \in L\})$ and $d_{min} = \min(\{d(c_i, e) \mid c_i \in L\})$. Given execution time t for A_j, RSR-WG counts (from performance matrix τ) and weights (with corresponding similarity) how many problems A_j could solve from L in time t:

$$N_j^z(t) = \sum_{x_i \in L} s_i \zeta_{ij}(t) \text{ where } \zeta_{ij}(t) = \begin{cases} 1 & \text{if } \tau_{ij} \leq t \\ 0 & \text{otherwise} \end{cases} \tag{2}$$

Then, at time t, RSR-WG greedily maximizes (2) per unit of time expended over all solvers and their possible execution duration Δ_z, that is, it calculates:

$$\underset{A_j, \Delta_z}{\text{argmax}} \frac{N_j^z(t + \Delta_z)}{\Delta_z} \tag{3}$$

and removes those now-solved similar problems from L. *Retain* (line 6) places solver A_j on processor π_u if that maximizes equation (2) per unit of expended time and π_u still has time available ($T^u < T$). Among such processors, Retain prefers one that has hosted A_j earlier ($t_{uj} \neq 0$); otherwise it selects one that has thus far been used the least (i.e., has minimum T^u). If a parallel schedule σ solves all of L without making full use of all the processors, *Spread* (line 11) places the solver A_j that solves the most problems in L but does not appear in σ on a processor that was idle throughout σ (if such an A_j exists), breaking ties at random. (The rationale here is that A_j may be generally effective but not outstanding on a particular e.) Finally, if a processor is not fully used in σ (i.e., $T^u < T$), *Return* (line 14) places the first solver it executed there on that processor until the time limit. Obviously, RSR-WG achieves the performance of an oracle when $k = |A|$, but it is also effective when k is relatively small compared to $|A|$.

Input: case base C, solvers A, time limit T, testing problem e, distance function d,
 similarity function s, neighbor set ratio r, processors $\{\pi_1, \pi_2, ..., \pi_k\}$
Output: schedule $\sigma = \{\sigma_1, \sigma_2, ..., \sigma_k\}$ for a parallel switching portfolio

1   Compute distance $d(c_i, e)$ for all c_i in C
2    $L \leftarrow \{100r\%$ of problems in C closest to $y\}$
3   Compute similarity s_i for each c_i in L with equation (1)
4   Initialize time step $z \leftarrow 1$, overall time $T^u \leftarrow 0$ on processor π_u,
      time $t_{uj} \leftarrow 0$ for A_j on π_u
5   While $L \neq \emptyset$ and $T^u < T$ for at least one u
6   Select A_j on π_u with time Δ_z to maximize equation (3)   ** *Retain* **
7   Remove from L all problems solved by A_j during step z
8   Schedule A_j with execution time Δ_z on π_u
9   Update times: $t_{uj} \leftarrow t_{uj} + \Delta_z$, $T^u \leftarrow T^u + \Delta_z$, and $z \leftarrow z + 1$
10   For each π_u where $T^u < T$
11   If $T^u = 0$ ** *Spread* **
12   then assign A_j to π_u for T, where A_j solves the most problems in L and $A_j \notin \sigma$
13   update times: $t_{uj} \leftarrow T, T^u \leftarrow T$, and $z \leftarrow z + 1$
14   else π_u executes the first solver placed on π_u until T ** *Return* **
15   update times: $t_{uj} \leftarrow t_{uj} + (T - T^u), T^u \leftarrow T$, and $z \leftarrow z + 1$
16   Return σ

Fig. 2. High-level pseudocode for RSR-WG

2.3   Experimental Design and Results

The parallel constructors tested here are RSR-WG, PGASS (a naïve parallel version of GASS with uniform weights $s_i = 1$), and PCPHYDRA (a naïve parallel version of CPHYDRA that randomly partitions L into k subsets and then uses CPHYDRA on each subset to construct a schedule for each processor). PCPHYDRA selects $|L| = 10k$ neighbors, randomly distributes them to k processors, and executes a complete search for the optimal schedule on each processor. If it does not produce an optimal schedule within 180 seconds, it takes the best schedule it has found so far. For RSR-WG, we simulated all 24 solvers from the original competition [15].

All portfolio experiments ran on a Dell PowerEdge 1850 cluster with 696 Intel 2.80 GHz Woodcrest processors. We gauged performance as in recent competitions, on the number of problems solved with a fixed, per-problem time limit, with ties broken on average solution time across solved problems [15, 16]. Time for RSR-WG included both portfolio construction (i.e., scheduling) and search, but time for PGASS and PCPHYDRA excluded portfolio construction, which gave them a slight advantage.

As Table 3 shows, for $k > 1$, RSR-WG consistently solved more problems than PGASS or PCPHYDRA. For RSR-WG only, we also tested $k = 16$ processors, which produced near-oracle performance. Although 2 of the 10 runs for $k = 16$ were perfect, this becomes very nearly a race among solvers that did well on the cases in L. The near-optimal performance of $k = 8$, or even $k = 4$, along with the fact that only RSR-WG was charged for scheduling time, is more noteworthy.

Figure 3 compares an oracle's runtime to that of RSR-WG with $r = 0.005$. Each circle represents one of the 2865 problems. Those at the far right are problems unsolved by RSR-WG in 1800 seconds; those on the diagonal were solved by RSR-WG as quickly as an oracle. Clearly, more processors solved more problems (from 2769 to 2859 in this particular run) and solved more problems as quickly as an oracle.

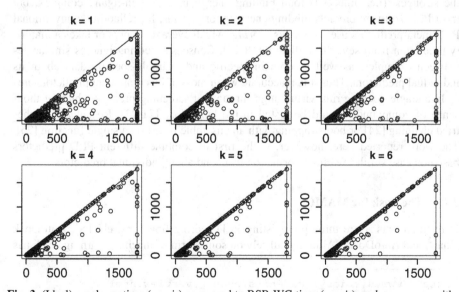

Fig. 3. (Ideal) oracle runtime (*y-axis*) compared to RSR-WG time (*x-axis*) on k processors with neighbor set ratio $r = 0.005$. Each circle is a result on one of the 2865 problems.

Table 3. Mean performance of 3 constructors on 2865 problems over 10 runs, with the (significantly) best value for k processors in boldface. * denotes RSR-WG outperformed PGASS; † denotes RSR-WG outperformed PCPHYDRA ($p < 0.005$). Neighbor set ratio was $r = 0.005$.

	$k = 1$	$k = 2$	$k = 3$	$k = 4$	$k = 5$	$k = 6$	$k = 7$	$k = 8$
PCPHYDRA	2779	2807	2817	2827	2830	2831	2834	2834
PGASS	2771	2801	2808	2810	2817	2821	2823	2825
RSR-WG	2773	**2826***†	**2841***†	**2850***†	**2855***†	**2857***†	**2858***†	**2859***†

MAMC's computational cost for parallel schedule construction is worthwhile, compared to that of other schedulers. Recall that the time allotted to each problem is 1800 seconds. RSR-WG constructs its schedules quickly. For example, it averages less than 15 seconds ($\sigma \approx 6$) with $r = 0.16$ and 8 processors, and is faster for smaller r and k. Recall, however that PCPHYDRA sometimes fails to compute its (optimal) schedule within 180 seconds. Indeed, given its $O(2^{|A|})$ complexity, it produced no schedule at all on 4.81% of the cases when $k = 1$, and 14.39% of the cases when $k = 8$. Because it learns on all the cases, however, GASS was even slower; its single entry in Table 2 required more than 5 days to compute.

3   Protein-Ligand Docking

The central topic of rational drug design is protein-ligand interaction, where a small molecule (a *ligand*) binds to a specific position (e.g., an open cavity) in a protein [17]. Protein-ligand docking (*PLD*) evaluates the ligand's orientations and conformations (three-dimensional coordinates) when bound to a receptor. PLD seeks ligands with the strongest (i.e., minimal) total binding energy in a protein-ligand complex, but most PLD software predicts binding energy poorly. Thus, for reliability, conventional PLD meta-predictors use *consensus scoring*, which averages scores or takes a majority prediction from several predictors [18-20]. Consensus scoring ignores similarities between examples, as well as domain-specific and example-specific data about its individual predictors. Thus it is inaccurate when most of its component predictors are.

In a single docking run, virtual high-throughput screening predicts which of thousands of compounds should be tested in the laboratory [21-23]. Recent approaches tried chaining [24] or bootstrapping with an ensemble based on a single function [25]. The work reported here, however, is the first to combine different PLD predictors based on case similarity plus information from and about individual predictors.

3.1   The Task for MAMC

Here the agents A are three pre-existing PLD scoring functions: eHiTS[1], AutoDock Vina[2], and AutoDock[3]. Although all rely on some form of machine learning, each has

---

[1] http://www.simbiosys.ca/ehits/ehits_overview.html

[2] http://vina.scripps.edu/

[3] http://autodock.scripps.edu/

its own conformational sampling, scoring, and feature-based representation. They often perform dramatically differently on the same data, with no consistent winner.

A case is the binding energy measured in the laboratory between a given *receptor* (a target in a protein) and a chemical compound. Each compound is a potential ligand, represented by a feature vector that reports chemical properties (e.g., whether it is a hydrogen-bond donor, or whether its topological distance between two atoms lies in some range). These are different features from those used by the agents; the agents consider three-dimensional chemical conformation, while the cases describe physio-chemical and topological properties derived from two-dimensional chemical structure. In a case base C, all cases address the same receptor. To describe a case, values for its standard chemical *footprint* of 1024 boolean features were calculated offline with programs such as openbabel[4]. Given a case base C, a new chemical e, chemical-similarity metric s, and performance matrix τ for the agents A on C, MAMC's task is to predict the binding energy between C's receptor and e.

3.2 Cases, Similarity, and Combination

We tested MAMC on datasets from *DUD* (Directory of Useful Decoys), a set of benchmarks for virtual screening [26]. A *decoy* is a molecule similar to its ligand in its physical properties but dissimilar in its topology. DUD has multiple ligands for each receptor, and 36 decoys for each of its ligands. We considered two receptors from DUD: gpb and pdg. All three agents perform relatively poorly on them. More-over, eHiTS, is the worst of the three on gpb but the best on pdg. Together these receptors challenge MAMC to choose the most accurate predictor for each chemical.

The similarity metric s on example e and case $c_i \in C$ is defined by the *Tanimoto coefficient*, the ratio of the number of features present in their intersection to the number of features present in their union, where $N(c)$ is the number of 1's in c:

$$s(e, c_i) = N(e \ \& \ c_i) \ / \ N(e \mid c_i) \tag{4}$$

Each predictor A_j was asked to calculate a score for each ligand and decoy in the data-set. We eliminated the very few chemicals that did not receive three scores; this left 1901 chemicals for gpb, and a separate set of 5760 for pdg. The agents' incompara-ble scales, however, required a simple but robust *rank-regression scoring* mechanism to map raw scores uniformly to a normalized rank score that reflects only the prefe-rence of an agent for one case over another. For each agent $A_j \in A$, MAMC sorts the raw scores from A_j for all $c_i \in C$ in ascending order, replaces each score with its rank, and normalizes the ranks in [0,1]. The normalized rank, denoted by $p(c_i, A_j)$, predicts the score of A_j on c_i. Higher-ranked cases thereby receive lower scores, in line with the premise that lower binding energy is better.

We again represent the performance of A on C in a $|C| \times |A|$ performance matrix τ. To evaluate the performance $\tau(e, A_j)$ of $A_j \in A$ on e, we use the set of cases L similar to e, but weight more heavily those more similar to it:

---

[4] http://openbabel.org/wiki/Main_Page

$$\tau(e, A_j) = \sum_{c_i \in L} s(e, c_i)\tau_{ij} \qquad (5)$$

Then, to predict on e with all of A, we take the agent A_j with the highest $\tau(e, A_j)$ and combine its predicted scores on L, again weighting more similar cases more heavily:

$$p(e) = \sum_{c_i \in L} s(e, c_i)p(c_i, A_j) \qquad (6)$$

Intuitively, a scoring function that accurately distinguishes ligand set G from decoy set Y (where $Y \cup G = C$) should predict lower scores for ligands and higher scores for decoys. In other words, agent A_j is more accurate on ligand g only if its prediction for g is generally lower than its predictions for Y, and it is more accurate on decoy y only if its prediction for y is generally higher than its predictions for G. The *performance score* of agent A_j on case c is thus

$$\tau(c, A_j) = \begin{cases} \dfrac{\left|\{y \in Y | p(y, A_j) > p(c, A_j)\}\right|}{\left|\{y \in Y\}\right|} & \text{if } c \in G \\[4mm] \dfrac{\left|\{g \in G | p(g, A_j) < p(c, A_j)\}\right|}{\left|\{g \in G\}\right|} & \text{if } c \in Y \end{cases} \qquad (7)$$

Scores in equation (7) lie in [0,1], where a higher value indicates better performance.

3.3   Experimental Design and Results

Each of these experiments predicts the binding energy of a chemical e to a receptor. We examine the accuracy of five predictors: the three individual agents (eHiTS, AutoDock Vina, AutoDock) and two meta-predictors: MAMC and *RankSum*, a typical bioinformatics consensus-scoring meta-predictor. To predict the score on example e, RankSum adds the rank-regression scores from the three predictors, where a lower sum is better. In advance, for MAMC, we computed the similarities between all $_nC_2$ pairs of chemicals (about 1.8 million for gpb and 16.6 million for pdg) with equation (4), and recorded the five chemicals most similar to each chemical, along with their similarity scores. Experiments ran on an 8 GB Mac Pro with a 2.93 GHz Quad-Core Intel Xeon processor, and analysis used the R package ROCR.

First, we evaluated the three individual predictors with leave-one-out validation: in turn, each of the n chemicals for a receptor served as the testing example e, while the other n-1 served as C. MAMC extracted the $|L|$ cases in C most similar to e, and then used equation (5) to gauge the accuracy of each individual predictor across all the cases in L. MAMC then chose the individual predictor with the best predictive accuracy on L and reported as a score the rank-regression score on e from that best individual predictor as in (6).

We compare predictors' performance by their hit ratio across C. ROC (Receiver Operating Characteristic) curves illustrate the tradeoff between true positive and false positive rates, an important factor in the decision to test a likely ligand in the laboratory. (Classification accuracy alone would be less helpful, because the prevalence of so many decoys heavily biases the data sets. Simple prediction of every chemical as a decoy would be highly accurate but target no chemicals for investigation as likely ligands.) A predictor p on any $c \in C$ produces true positives $C_1 = \{g \in G \mid p(g) \leq p(c)\}$ and false positives $C_2 = \{y \in Y \mid p(y) \leq p(c)\}$. Thus the true positive rate for c is $|C_1|/|G|$ and the false positive rate is $|C_2|/|Y|$.

We report first on $|L| = 1$, using the single case c_i most similar to e. (For $|L| > 1$, see the next section.) In this case, MAMC need only reference τ_{ij} for each $A_j \in A$. The ROC curves in Figure 4 compare the performance of all five predictors on receptors gpb and pdg for $|L| = 1$, based on the predictors' scores and DUD's class labels.

MAMC clearly outperforms the other predictors on both gpb and pdg. In particular, MAMC outperforms the best individual predictor eHiTS on pdg, even though the majority of its individual predictors perform poorly. In contrast, the performance of the consensus scorer RankSum on pdg was considerably worse than MAMC; it requires accurate rankings from most of its constituent predictors for satisfactory performance, rankings the individual predictors could not provide.

4   Discussion

We remind the reader that each of the applications described here was developed in part because carefully honed individual agents produced after many millions of hours of development had proved unsatisfactory. Not only is there no reliable way to predict the difficulty of a CSP, but also the solvers' performances vary from one problem to the next. A similar situation exists with predictors for PLD binding energy: their performance varies unpredictably. Both are hard problems on which MAMC has made some progress. Some choices, however, require further examination.

MAMC assumes that an agent's accuracy on similar cases will also be similar, but the number of those cases (i.e., the size of L) is an important decision. For portfolio construction, Table 3 reports on $r = 0.005$ which, given 10-fold cross validation, selects $|L| = 13$ cases from among the 2578 eligible ones. This enabled RSR-WG to outperform its competitors for $k > 1$ ($p < 0.005$). Table 4 explores values of r that enlarge L to as many as 412 cases. The data there suggest that, while the smallest r is reliable, occasionally a larger neighbor set pays off, particularly for the (non-parallel) $k = 1$.

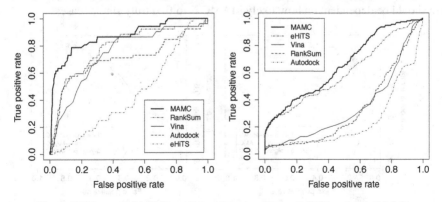

Fig. 4. ROC curves for PLD predictors on receptors gpb (left) and pdg (right)

Table 4. Mean and standard deviation for the number of problems solved by RSR-WG out of 2865 over 10 runs with k processors. Best value for k processors is in boldface, $p <0.005$.

k	\multicolumn{2}{c}{0.005}										

k	\multicolumn{12}{c}{Neighbor set ratio r}
	0.005		0.01		0.02		0.04		0.08		0.16	
1	2773	3.65	2779	3.20	2786	2.30	**2789**	3.17	2788	3.09	**2789**	2.51
2	**2826**	3.51	2821	2.49	2823	3.16	2816	2.97	2810	2.99	2809	2.87
3	**2841**	2.12	2836	1.93	2839	2.56	2832	2.07	2827	2.27	2819	2.07
4	**2850**	2.15	2847	1.57	2847	2.63	2843	2.06	2838	2.22	2832	2.50
5	**2855**	1.37	2851	2.35	2852	0.88	2850	1.78	2845	2.72	2843	3.26
6	**2857**	0.95	2855	1.07	2856	1.26	2853	1.64	2851	1.03	2850	1.07
7	**2858**	0.79	**2858**	0.57	2857	0.82	2855	1.83	2854	2.35	2854	1.14
8	2859	1.18	**2860**	1.34	2858	1.06	2858	1.18	2856	0.74	2855	1.43
16	**2864**	0.42	**2864**	0.00	**2864**	0.00	2863	0.00	2861	0.42	2861	0.47

Next we consider the impact of larger $|L|$ on PLD prediction. Again, each of the three scoring functions predicts for e, and then MAMC evaluates its performance on L with equation (5). MAMC then combines the predicted scores from all three predictors with equation (6), which allows it to consider the overall weighted performance of each predictor on a set of similar cases, and then takes the weighted prediction of the agent with the best overall performance on those similar cases. Although Figure 5 shows a clear performance improvement for $|L| = 2$ on both receptors, the improvement of $|L| = 3$ over $|L| = 2$ is only marginal, despite the fact that under leave-one-out validation, $|C| = 1900$ for gpb and 5759 for pdg.

The nature of the data, we believe, accounts for the difference in the appropriate choice for $|L|$. The PLD data is inherently noisy and incomplete; the cases in C are only those that have been tested by a laboratory and made publicly available. The dismal performance of a case, for example, may have dissuaded further testing of similar ones. For such a case there would be very few similar cases, so a larger L would provide little benefit. In contrast, competition CSPs are typically submitted by researchers who expect their own solvers to have an advantage on those problems. A *class* of CSPs consists of problems that may vary somewhat in their size or anticipated difficulty but have some structural or modeling commonality. Each CPAI'08 problem class typically had dozens of problems, so that, even under 10-fold cross-validation, MAMC is likely to find more than a few similar cases. Thus neighbor set size should be dependent on how likely MAMC is to find truly similar cases.

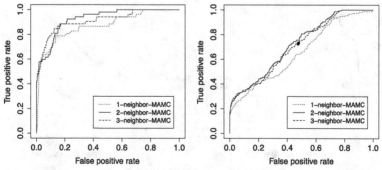

Fig. 5. ROC curves for MAMC with different $|L|$. on gpb (left) and pdg (right)

Given a fixed size for L, MAMC's confidence about its results is still likely to vary from one case to the next. For example, as noted above, a particular case may have an L whose members are only slightly similar to it. Moreover, individual agents may perform poorly on some members of L. In both situations, MAMC should be less confident about its prediction on the original case. Intuitively, if MAMC could categorize individual cases by confidence level, it might improve its performance on the cases where its confidence level is high.

Our confidence analysis considers three kinds of predictions, demonstrated here on protein-ligand docking where confidence before real-world laboratory testing is particularly important. Two cases c_i and c_j are said to be *similar* if and only if $s(c_i, c_j) > t_1$ (here, 0.8), and *dissimilar* otherwise. A *reliable* predictor is one whose performance, as calculated by equation (7), is greater than t_2 (here, 0.9); otherwise it is *unreliable*. Together t_1 and t_2 define three categories of agent A_j's ability to decide on example e. A prediction has *high confidence* if e's closest neighbor c is similar to e and A_j is reliable on c. A prediction has *low confidence* if c is dissimilar to e and A_j is unreliable on c. In all other situations, a prediction has *normal confidence*.

Figure 6 isolates the performance of MAMC at these three confidence levels for gpb and pdg. For gpb, 31.77%, 49.50%, and 18.73% of the chemicals had high, normal, and low confidence, respectively. For pdg, these percentages were 19.77%, 60.63%, and 19.60%. As expected, MAMC performed far better on the high-confidence chemicals for both receptors than it did on the full set. The benefit introduced by the confidence-based classification for pdg is particularly promising: although most candidate scoring functions had unreliable performance, confidence-based MAMC achieved almost perfect prediction on the high-confidence chemicals.

Might s alone have accurately predicted whether a chemical was a ligand? To investigate, we ranked by similarity all pairs of cases that included at least one ligand: each pair is either a *match* (two ligands) or a *non-match* (a ligand and a decoy). Ideally, match pairs should have higher similarity scores than non-match pairs. In Figure 7, the ROC curve for each receptor is based only on s and whether or not a pair was a match. Although chemical similarity alone clearly distinguishes ligands from decoys in the DUD benchmark data set, it provides fewer likely ligands than MAMC, whose predictive performance is considerably better, especially when its confidence is high.

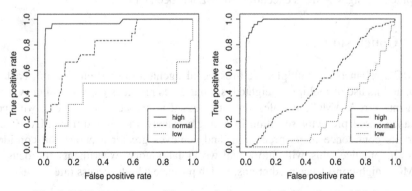

Fig. 6. ROC curves for confidence analysis on gpb (left) and pdg (right)

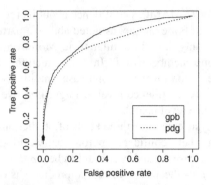

Fig. 7. ROC curves for gpb and pdg based on computed similarity and match/non-match labels of chemical pairs. The marks at lower left correspond to the predictions based only on the minimum chemical similarity score 0.8.

Diversity is essential in MAMC. For parallel scheduling we used 24 agents, 36 features, and 2578 cases per run under 10-fold cross validation. For PLD, we used 3 agents, 1024 features, and 1710 or 5183 cases per run. We believe that MAMC succeeds in part because it can draw upon such diversity. For PLD, there may be only three agents, but they are quite diverse both in their knowledge base and in their approach: AutoDock and AutoDock Vina use a genetic algorithm to search the ligand conformational space, while eHiTS uses systematic search in the conformational space of ligands. They variously consider force fields, energy terms (e.g., Van der Waals, electrostatic), empirical binding affinities, and knowledge-based scores trained from known protein-ligand complexes. Moreover, the different features in the PLD case representation provided an entirely different perspective on the cases. Although the construction of each case base required machine-months of computation, we view the increased availability of data not as a burden but as an opportunity.

Many sophisticated agents now offer the ability to set parameters. Future work will investigate the use of copies of one agent with different parameter settings or an element of randomization, as well as information flow algorithms to improve the case-similarity metric s. Moreover, although here τ was known in advance, it could be computed during MAMC's execution, after L has been chosen.

5   Conclusion

MAMC's reliance on multiple, well-respected agents draws strength from each of them, and its use of multiple, weighted, similar cases provides greater resiliency when its agents err. MAMC integrates similarity-based reference-case selection with performance-based predictor selection in a single framework. In addition, MAMC can report its confidence in its prediction, and achieves greater accuracy on confident cases. In practice, this will allow real-world laboratory experiments to focus on MAMC's high-confidence predictions, which promise a high success rate.

MAMC's portfolios of deterministic constraint solvers outperform thóse from naïve parallel versions of popular portfolio constructors. With only a few additional processors MAMC's schedules are competitive with an oracle solver on one processor. MAMC's improved predictions on compound virtual screening for protein-ligand docking suggests that PLD could support real drug-discovery. Moreover, with careful formulation of C, s, and A, MAMC should readily apply to other challenging bioinformatics and chemo-informatics tasks, including prediction of two- and three-dimensional protein structures, protein-protein interaction, protein-nucleotide interaction, disease-causing mutation, and the functional roles of non-coding DNA. Highly-confident predictions from MAMC there should be worthy of particular attention.

Acknowledgements. This work was supported in part by the National Science Foundation under grants IIS-1242451, IIS-0811437, CNS-0958379 and CNS-0855217, and the City University of New York High Performance Computing Center. Weiwei Han prepared the cases for PLD.

References

1. Dietterich, T.G.: Ensemble methods in machine learning. In: Kittler, J., Roli, F. (eds.) MCS 2000. LNCS, vol. 1857, pp. 1–15. Springer, Heidelberg (2000)
2. Leake, D.B., Sooriamurthi, R.: Automatically Selecting Strategies for Multi-Case-Base Reasoning. In: Craw, S., Preece, A.D. (eds.) ECCBR 2002. LNCS (LNAI), vol. 2416, pp. 204–233. Springer, Heidelberg (2002)
3. Plaza, E., McGinty, L.: Distributed Case-Based Reasoning. The Knowledge Engineering Review 20(3), 261–265 (2005)
4. Redmond, M.: Distributed Cases for Case-Based Reasoning: Facilitating Use of Multiple Cases. In: Proceedings of the Eighth National Conference on Artificial Intelligence (AAAI 1990), pp. 304–309 (1990)
5. Kar, D., Chakraborti, S., Ravindran, B.: Feature Weighting and Confidence Based Prediction for Case Based Reasoning Systems. In: Agudo, B.D., Watson, I. (eds.) ICCBR 2012. LNCS, vol. 7466, pp. 211–225. Springer, Heidelberg (2012)
6. Yun, X., Epstein, S.L.: Learning Algorithm Portfolios for Parallel Execution. In: Hamadi, Y., Schoenauer, M. (eds.) LION 2012. LNCS, vol. 7219, pp. 323–338. Springer, Heidelberg (2012)
7. Guerri, A., Milano, M.: Learning Techniques for Automatic Algorithm Portfolio Selection. In: Proceedings of the Sixteenth European Conference on Artificial Intelligence, pp. 475–479 (2004)
8. O'Mahony, E., Hebrard, E., Holland, A., Nugent, C., O'Sullivan, B.: Using Case-Based Reasoning in an Algorithm Portfolio for Constraint Solving. In: Proceedings of the Nineteenth Irish Conference on Artificial Intelligence and Cognitive Science (2008)
9. Silverthorn, B., Miikkulainen, R.: Latent Class Models for Algorithm Portfolio Methods. In: Proceedings of the Twenty-Fourth AAAI Conference on Artificial Intelligence, pp. 167–172 (2010)
10. Xu, L., Hoos, H.H., Leyton-Brown, K.: Hydra: Automatically Configuring Algorithms for Portfolio-Based Selection. In: Proceedings of the Twenty-Fourth AAAI Conference on Artificial Intelligence, pp. 210–216 (2010)

11. Xu, L., Hutter, F., Hoos, H.H., Leyton-Brown, K.: SATzilla: Portfolio-Based Algorithm Selection for SAT. Journal of Artificial Intelligence Research 32, 565–606 (2008)
12. Horvitz, E., Ruan, Y., Gomes, C.P., Kautz, H.A., Selman, B., Chickering, D.M.: A Baye- sian Approach to Tackling Hard Computational Problems. In: Proceedings of the Seven- teenth Conference in Uncertainty in Artificial Intelligence, pp. 235–244. Morgan Kauf- mann Publishers Inc. (2001)
13. Streeter, M., Golovin, D., Smith, S.F.: Combing Multiple Heuristics Online. In: Proceed- ings of the Twenty-Second National Conference on Artificial Intelligence, pp. 1197–1203 (2007)
14. Mistral, http://4c.ucc.ie/~ehebrard/Software.html
15. Third International CSP Solver Competition (CPAI 2008), http://www.cril.univ-artois.fr/CPAI08/
16. Fourth International CSP Solver Competition (CSC 2009), http://www.cril.univ-artois.fr/CSC09/
17. Huang, S.-Y., Zou, X.: Advances and Challenges in Protein-Ligand Docking. International Journal of Molecular Science 11, 3016–3034 (2010)
18. Charifson, P.S., Corkery, J.J., Murcko, M.A., Walters, W.P.: Consensus Scoring: A Me- thod for Obtaining Improved Hit Rates from Docking Databases of Three-Dimensional Structures into Proteins. Journal of Medicinal Chemistry 42, 5100–5109 (1999)
19. Clark, R.D., Strizhev, A., Leonard, J.M., Blake, J.F., Matthew, J.B.: Consensus Scoring for Ligand/Protein Interactions. Journal of Molecular Graphics Modelling 20, 281–295 (2002)
20. Wang, R., Wang, S.: How Does Consensus Scoring Work for Virtual Library Screening? An Idealized Computer Experiment. Journal of Chemical Information and Computer Sciences 41, 1422–1426 (2001)
21. Zsoldos, Z., Reid, D., Simon, A., Sadjad, B.S., Johnson, P.A.: Ehits: An Innovative Ap- proach to the Docking and Scoring Function Problems. Current Protein and Peptide Science 7, 421–435 (2006)
22. Trott, O., Olson, A.J.: Autodock Vina: Improving the Speed and Accuracy of Docking with a New Scoring Function, Efficient Optimization and Multithreading. Journal of Com- putational Chemistry 31, 455–461 (2010)
23. Morris, G.M., Goodsell, D.S., Halliday, R.S., Huey, R., Hart, W.E., Belew, R.K., Olson, A.J.: Automated Docking Using a Lamarckian Genetic Algorithm and Empirical Binding Free Energy Function. Journal of Computational Chemistry 19, 1639–1662 (1998)
24. Miteva, M.A., Lee, W.H., Montes, M.O., Villoutreix, B.O.: Fast Structure-Based Virtual Ligand Screening Combining Fred, Dock, and Surflex. Journal of Medicinal Chemistry 48, 6012–6022 (2005)
25. Fukunishi, H., Teramoto, R., Takada, T., Shimada, J.: Bootstrap-Based Consensus Scoring Method for Protein-Ligand Docking. Journal of Chemical Information and Modeling 48, 988–996 (2008)
26. Huang, N., Shoichet, B.K., Irwin, J.J.: Benchmarking Sets for Molecular Docking. Journal of Medicinal Chemistry 49, 6789–6801 (2006)

Case-Based Learning of Applicability Conditions for Stochastic Explanations

Giulio Finestrali and Héctor Muñoz-Avila

Department of Computer Science & Engineering, Lehigh University, Bethlehem, PA 18015
gif311@lehigh.edu, munoz@cse.lehigh.edu

Abstract. This paper studies the problem of explaining events in stochastic environments. We explore three ideas to address this problem: (1) Using the notion of *Stochastic Explanation*, which associates with any event a probability distribution over possible plausible explanations for the event. (2) Retaining as cases *(event, stochastic explanation)* pairs when unprecedented events occur. (3) Learning the probability distribution in the stochastic explanation as cases are reused. We claim that a system using stochastic explanations reacts faster to abrupt changes in the environment than a system using deterministic explanations. We demonstrate this claim in a CBR system, incorporating the 3 ideas above, while playing a real-time strategy game. We observe how the CBR system when using *stochastic* explanations reacts faster to abrupt changes in the environment than when using *deterministic* explanations.

Keywords: case-based learning, explanations, stochastic explanations.

1 Introduction

A good way to evaluate an agent's intelligence is by measuring its ability to react to unprecedented events. An agent interacts with the environment by performing actions, and expecting these actions to have a certain effect. In a stochastic domain (i.e., agent's actions can have multiple outcomes and events happen independently of the agent's actions), though, unexpected and often unprecedented events can always happen. Whenever the results of an agent's action on the environment are different than what the agent expected them to be, this situation is called a *discrepancy* (Cox, 2007). Whenever an unexpected event occurs, we have two choices: relying on the previously collected knowledge to *react* to the event, or trying to *explain* the event and then reacting accordingly to the explanation. The first method can be effective but if the unexpected event is completely new, the knowledge acquired in the past might not be sufficient to react effectively to it.

Explaining unexpected events is a much more complicated task, but offers a higher degree of flexibility than merely relying on previously acquired knowledge to react to discrepancies. The process of explaining an event can be divided in two tasks: given an event, provide a set of possible explanations for it; after that, pick the best explanation among the possible ones. In this paper, we present a novel method to perform the second task efficiently using Case-Based Reasoning (CBR).

At the center of our approach are three ideas: (1) Using the notion of *Stochastic Explanations*, which associates with any event a probability distribution over possible

S.J. Delany and S. Ontañón (Eds.): ICCBR 2013, LNAI 7969, pp. 89–103, 2013.

plausible explanations for the event. Stochastic explanations are necessary when operating in a stochastic domain, because in such domains it is implausible to have a set of fixed cause-effect rules for unexpected events. (2) Retaining as cases *(event, stochastic explanation)* pairs when such unexpected events occur. (3) Learning the probability distribution in the stochastic explanation as the cases are retrieved in subsequent episodes.

In stochastic environments, abrupt changes can make explanations that were previously true suddenly become false. We claim that a system using stochastic explanations reacts faster to such changes in the environment than a system using deterministic explanations. We demonstrate this claim with a CBR system, implementing the three ideas above, while playing the real-time strategy Wargus game. This game is an example of a stochastic domain. We observe how the CBR system when using *stochastic* explanations reacts faster to abrupt changes in the environment than when using *deterministic* explanations.

2   Related Work

There have been many studies about explanations. A thorough discussion on the types of reasoning failure can be found in Cox (1996). Using Cox's categorization, in our work we deal with *contradiction*: a failure condition characterized by an agent's expectations not matching the resulting state; we refer to this as *unexpected failure*. Whenever an agent performs an action, it expects the environment to be modified in a certain way as a result of the action performed. When the actual state of the environment does not match these expectations, we have a contradiction.

Cox (2007) provides an interesting insight on the difference between Learning Goals and Achievement Goals. Whenever an anomaly happens, we try to generate an explanation for it; this process involves finding the culprit of the anomaly. An intelligent agent must be able to blame itself and its own cognitive process in order to improve it: this can be done by generating a Learning Goal. Achievement Goals are the regular goals the agent pursues to accomplish its task. In our work we use stochastic explanations to determine the learning goal.

Existing work on generating explanations to reasoning failures use deterministic explanations (e.g. (Molineaux *et al.*, 2011; 2012)). We label the kind of explanations generated by these systems as *deterministic* because when provided with the same discrepancy, the system will output the same explanation. DiscoverHistory relies on a deterministic Hierarchical Task Network planner. The planner cannot account for every plausible event in the domain; it makes assumptions that the conditions of the environment are static to generate its plan. The system then uses the DiscoverHistory algorithm whenever one of these assumptions fails. The way DiscoverHistory works is by modifying the generated plan, introducing new elements that solve discrepancies, or by modifying the initial assumptions. Often, this process introduces new contradictions, which are then recursively solved until the whole plan is consistent, or until a maximum number of modifications have been made. This process is deterministic: if we provide to the algorithm the same anomaly twice, given the same state of knowledge, the system will produce the same explanation.

More recently, Klenk et al. (2012) presented ARTUE, which uses a direct application of explanations in a strategy simulation. ARTUE is based on a modified version of the SHOP2 planner, and it tries to explain discrepancies similarly to

DiscoverHistory. When ARTUE cannot create an explanation, it discards the discrepancy entirely. ARTUE's approach is deterministic and does not deal with anomalies that should trigger the generation of Learning Goals. This means that ARTUE is not able to blame itself to explain the discrepancies that arise.

Explaining discrepancies is a hard problem, and research efforts to solve it are a recurrent research topic. DiscoverHistory and ARTUE generate explanations by blaming the environment for any given discrepancy, and by then finding a set of modifications to the previous plan that resolve the inconsistencies. This solution is greedy and deterministic, and does not change over time. Stochastic explanations represent a novel approach to the problem: instead of blaming the environment, we blame the agent itself. When the culprit of the discrepancy is the environment, blaming the agent will make the agent learn that it is operating in an environment where this kind of discrepancies can happen, and should account for them in the future. Not only does our system reacts to discrepancies by selecting an explanation, but the explanation can change over time, adapting to changes in the environment. Once an explanation is selected, the agent follows the results of the reaction, increasing the likelihood of selecting the same explanation again in the future in case of success, and decreasing it otherwise. Obviously, a stochastic approach requires many iterations to perform well, because it learns by failure. Nonetheless, we believe that the flexibility and dynamicity of this approach justify the learning curve.

3   Stochastic Explanation

We begin by defining explanations in general, and then we will focus on stochastic explanations and how they differ from deterministic explanations. We will show a motivating example for stochastic explanations in the RTS game Wargus.

3.1   Definitions

The task of generating an explanation for a given discrepancy can be divided in two parts: (1) generating a set of possible explanations and (2) choosing among the explanations in the set the one we think is correct. In this work we are tackling the second part.

In our representation, an explanation is merely a label, which can be more or less descriptive of the discrepancy it is trying to explain. More importantly, each explanation is associated with a *reaction*: a goal that the agent will try to pursue after the explanation is picked in attempt to solve a discrepancy. Learning the reaction associated with each explanation is still an open problem.

Let us now discuss what a stochastic explanation is and how it differs from deterministic explanations. Intuitively, stochastic explanations have a probability value associated to them, which expresses the likelihood of an explanation to be correct. Formally, let E be the set of all possible explanations. For any set A, the notation 2^A is the power set of A; the collection of all subsets of A. A stochastic explanation is an element ε in $2^{E \times \{0,1\}}$ such that ε is a probability distribution. That is, $\varepsilon = \{(e_1, p_1), ..., (e_n, p_n)\}$ such that each $e_k \in E$ and $\Sigma_k p_k = 1$ hold.

A *discrepancy* occurs whenever the state s' resulting after taking an action in a state s doesn't match an expectation X. Frequently, the expectation X is defined as the expected state. In this situation a discrepancy occurs if $X \neq s'$ holds (e.g., (Jaidee et al., 2012)). In our work a discrepancy happens when the attempt to accomplish a goal fails. This is because the system keeps a positive outlook on the goal it generates. Since we generate only one kind of goal for now, we also have one possible kind of discrepancy. There are many possible explanations for the same discrepancy. Whenever a discrepancy happens, the system classifies it (see Section 4), and then provides a stochastic explanation ε. The probabilities in ε are constantly updated during the execution of the system. The system will then pick an explanation from ε, according to the probability distribution. If every time we picked the most likely explanation, the one having the highest probability, our system would behave like existing explanation-generating systems. For a given stochastic explanation ε, we refer to the explanation in ε that has the highest probability as the ***greedy explanation*** and denote it by greedy(ε): $greedy(\varepsilon) = max\ arg\ _p\ \{(e_1,p_1),...,(e_n,p_n)\}$.

The probability distribution of a stochastic explanation changes during execution. More concretely, the system learns these probabilities from experience by using reinforcement learning (RL). Once the system picks an explanation for a discrepancy, the reaction associated to the explanation is executed, and has two possible outcomes: success or failure. In case of success, we consider the explanation to be correct, and we increase its probability. In case of failure, we consider the explanation as incorrect and we decrease its probability. After several iterations, RL guarantees that the correct probability distribution of each stochastic explanation is learned.

3.2   Motivating Example

To better understand the shortcomings of greedy explanations, we will now consider a motivating example. First, let us describe in more detail the Wargus domain. We focus on combat tasks in our experiments. Any scenario starts with a predetermined number of units both for the player and the adversary; these units battle and as soon as one player has no units left, the scenario is concluded. The units are balanced with a typical rock-paper-scissors mechanism: every unit has other units that it's strong against, weak against, and fairly balanced against. On top of this, some units can be "upgraded" to become more powerful units. For example, an *archer* can be upgraded to a *ranger*, improving its health points and other stats. Finally, the units are divided in three categories: land units, air units, and sea units. Units usually have constraints on the kinds of units they can attack; for example, knights cannot attack flying units. Our agent starts its execution having none of these notions: it does not know which units every unit can attack, or which units are strong against a specific type of unit. The only information the agent knows at the beginning of the system's execution is how to perform the basic policy *attack unit,* needed to achieve the only goal generated by the system *kill unit A with unit B,* where A and B are picked randomly between the enemy units and our units respectively.

The agent always keeps an optimistic outlook on the generated goal. In case the attack fails, we have a discrepancy. This triggers the *Explanation Generator,* assigning it the task of explaining the discrepancy. As previously explained, the set of possible explanation is predetermined. In our example, we have three possible explanations to why the attack failed: *(e₁) unit A was not upgraded; (e₂) unit A is a bad choice for attacking unit B; (e₃) unit B has an environmental advantage.* The latter explanation needs some clarification. We consider an environmental advantage an obstacle on the map that prevents unit A to attack unit B. For example, unit B could be surrounded by trees, preventing close-ranged land units to attack it. The reactions for these explanations are as follows: (1) *attack unit B with the upgraded version of unit A;* (2) *attack unit B with a better unit;* (3) *attack with a ranged or a flying unit.* As we said before, the agent has no knowledge of which unit is strong against unit B, therefore to perform Reaction 2, the system keeps track of success rates for every attack it performs, learning with time the rock-paper-scissors mechanism of the game.

Our motivating example is composed of two scenarios. We play the first scenario for 50 iterations, and then the second scenario for 50 iterations. The knowledge acquired during the execution of the first scenario, is kept when switching to the second scenario. The agent has no perception about the change of scenario: we treat this as an unexpected event, very similar to the kinds of unpredictable events that can always happen in a stochastic domain. In the first scenario (See Figure 1), the agent has several *knight* units, some *paladin* units (which is the upgraded version of a knight), and also several *griffin* units, which are powerful flying units that can attack land units. The enemy AI, which is a passive script that only attacks back when attacked, only controls one *paladin*, surrounded by trees (environmental advantage). During the execution of the first scenario, whenever the agent attacks the *paladin* with a land unit, it fails. With time, it learns that the right explanation for this fact is that the enemy has an environmental advantage, and therefore the right thing to do is attack with a *griffin*. After many iterations the system will learn a stochastic explanation such as $\varepsilon_1 = \{(e_1,p_1), (e_2,p2), (e_3,p_3)\}$ where $p_3 > p_1$ and $p_3 > p_2$. Hence $greedy(\varepsilon_1) = e_3$ holds.

In the second scenario (see Figure 2), we have basically the same situation, but our agent does not have any *griffin* units, and there are no more trees surrounding the enemy *paladin;* this can happen in a regular RTS game, because the *peasant* units can chop trees down. Now, when attacking the *paladin* with a *knight*, the system fails but the correct explanation changed: the *knight* is not upgraded (i.e., explanation e_1), and we should have attacked with a *paladin*. At first, when we switch to the second scenario, when we fail to kill the enemy *paladin*, the system will have a high probability associated with the explanation *environmental advantage* (i.e., explanation e_3). This means that the greedy explanation will *always* pick *environmental advantage*, as long as the probability is higher than the others. Therefore, a greedy approach would have to wait until the probability of *environmental advantage* falls below the probability of the other explanations, making a lot of mistakes until then. Stochastic explanations, instead, will pick a different explanation sooner, and it will actually realize much faster that the correct explanation changed.

This is the reason that motivates the experiments we present in Section 5. If an intelligent agent operates in a stochastic domain, then greedy explanations will have a hard time reacting effectively and in a timely manner to the sudden changes that happen in such domain. If the agent still had griffin units in the second scenario, then both explanations e_3 and e_1 are plausible, since both their reactions would provide a positive outcome. In this case, stochastic explanations would learn a probability distribution that represents these conditions. Greedy would keep using explanation e_3 instead, since the outcome received from it remained positive. While the performance of the two approaches would be similar in this case, the stochastic explanation's representation is more representative of the situation. If, eventually, the agent won't be able to use griffins anymore, an agent using stochastic explanations will then be able to recover rapidly, while an agent Greedy Explanations will suffer greatly, taking a much longer time to re-establish good performance.

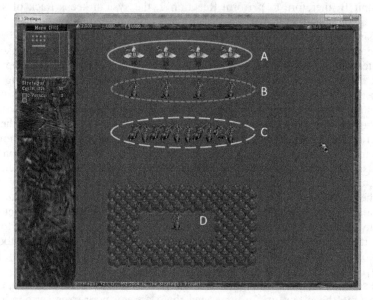

Fig. 1. Scenario 1. A are the griffins, B are the paladins and C are the knights. D is the enemy paladin surrounded by trees and hence inaccessible for land units.

4 Case-Based Learning of Applicability Conditions for Explanations

We use case-based learning techniques to acquire knowledge when generating and testing an explanation for a given discrepancy and use case retrieval to dynamically recognize the applicability of explanations captured in a previous episode.

Whenever a discrepancy happens, the system takes a snapshot of the state of the environment. This snapshot contains a set of feature-value pairs that are domain-dependent. For our experiments in Wargus we use five features: the Euclidean distance between the attacker and the target, the difference in health points between the attacker in the retained case and the query and the same difference for the target in the retained case and the query, and the types of both attacker and target.

Fig. 2. Scenario 2. A are the knights, B are the paladins. C is the enemy paladin, now accessible from land units.

The system then checks if a similar discrepancy ever happened in the past by measuring the similarity of the current discrepancy with the previous ones that are stored in the system's case base. To do so, the system has to compute the local similarity $sim_i(x_i, y_i)$ between pairs of features x_i, and y_i (Ricci and Avesani, 1995). The global similarity between two discrepancies is then computed with an aggregated similarity metric SIM_{agg} is defined as: $SIM_{agg}(X, Y) = \Sigma_{i=1,n} \ \alpha_i \cdot sim_i(x_i, y_i)$. This similarity and its weights α_i are also domain-dependent. If the similarity of the nearest neighbor for the current situation is above a user-defined threshold, then the system considers the unexpected event as "already seen" in the past, and will utilize the learned explanations for it. Otherwise, the current situation is considered as "unprecedented", and the case *(features, equiprobable stochastic explanation)* is retained in the case-base. Since we don't know anything about unprecedented situations, we associate the equiprobable stochastic explanation (i.e., every known explanation has the same probability). These probabilities are refined over time.

The following is a pseudo-code presenting how the system handles the task of retrieving and generating explanations for any given discrepancy. The algorithm requires five elements in input: the case-base CB which contains the discrepancies previously encountered (CB will be empty at the first iteration of the system), the policies Π to execute the goals as defined by their reactions (e.g., pre-coded instructions of what to do next), the pre-determined set E of (explanations, reaction) pairs $<e, r>$, the similarity threshold σ, and the maximum number of goals θ.

EXP_GEN(CB, Π, E, σ, θ)

```
 1:   GoalsInProgress ← ∅; ExplanationsInProgress ← ∅, FailedGoals ← ∅
 2:   while true
 3:
 4:        s ← GetState() // Get the current state from the environment
 5:        if ScenarioOver
 6:            return CB
 7:
 8:        for-each g ∈ FailedGoals  // Generate Explanations for discrepancies
 9:            c ← NewCase(g,s) //Create query case
10:            c' ← GetNearestNeighbor(CB, c)//Retrieve NN(c)
11:            if (Similarity(c',c) ≥ σ)
12:                    // Consider the query case as the same case of c'
13:                    e ← PickStochasticExplanation(c'.Explanations)
14:            else
15:                    // Consider case c as unprecedented
16:                CB.retain(c)
17:                for-each <e,r> in E
18:                    //initialize every explanation as equiprobable
19:                    c.addExplanation() ← (<e,r>, 1/|E|)
20:                e ← PickStochasticExplanation (c.Explanations)
21:            ReactToExplanation(e, Πₑ)
22:            ExplanationsInProgress.add(e)
23:
24:        for-each e ∈ ExplanationsInProgress
25:            UpdateExplanationState(e)
26:            if (e.state == success)
27:                RaiseProbability(e.p)
28:            if (e.state == failed)
29:                LowerProbability(e.p)
30:            if (e.state != in_progress)
31:                ExplanationsInProgress = ExplanationsInProgress – {e}
32:
33:        if GoalsInProgress < θ & ExplanationsInProgress =∅
34:            g ← GenerateGoal()
35:            PerformGoal(g, Πg)
36:            GoalsInProgress.add(g)
37:
38:        for-each g ∈ GoalsInProgress
39:            UpdateGoalState(g)
40:            if (g.state == failed)
41:                FailedGoals.add(g)
42:            if (g.state != in_progress)
43:                GoalsInProgress = GoalsInProgress – {g}
44:   //end while
```

After resetting Goals and Explanations in progress in Line 1, the algorithm enters a loop, which will be terminated when the scenario ends (Line 5). At Line 4, we update the agent's state of the environment representation; if the scenario is over, the algorithm returns the updated case base.

At Line 8, the algorithm loops through the failed goals (if any), generating a new case for each one. Then, at Line 10, the algorithm looks for the most similar case to c in the case base. If the similarity of c' and c is greater than σ (Line 11), the system picks the explanation from c' at Line 13, otherwise the case c is considered unprecedented. It is then added to the case base (Line 16), its explanations are initialized with the given input set E all having the same probability $1/|E|$, and an explanation is then picked at Line 20. The function PickStochasticExplanation takes a stochastic explanations $c'.Explanations$ as input. It picks an explanation $<e,p>$ according to the probability distribution in $c'.Explanations$

At Line 21 we call the function ReactToExplanation, which takes in input the explanation e and the reaction associated to it in order to perform the reaction. Then, at Line 22 we add this explanation to the *ExplanationInProgress* collection. Lines 24-31 handle the update of the state of each explanation in progress. In case the reaction succeeds, which means the explanation was correct, the probability of the explanation e is raised. If the executing the reaction fails to achieve the goal, the probability of e is reduced. Finally, unless the reaction associated with explanation e is still in progress, we remove it from the *ExplanationInProgress* collection at Line 31.

The probability value of an explanation is the sum of the ratio of successes over the total number of trials, and an additive term called *booster* value.

$$P(explanation) = \frac{Times\ Succeeded}{Times\ Tested + 1} + multiplier \cdot boosterValue$$

There are two kinds of booster values: a *reward booster* value and a *punishment* one. The former is positive; the latter is zero or negative. If in the future the explanation is tested again, and the outcome is again positive, the multiplier will be now 2 and therefore the probability will be raised more. The multiplier will increase by one for each consecutive success/failure and will be reset to 0 (no booster) whenever a success-failure or failure-success pattern occurs. The two booster values are user-defined parameters and they can impact on the reaction time of the EXP_GEN agent. In our experiments, we use a value of .12 for rewarding correct explanations and a value of .01 to punish incorrect ones. When applying the formula above, values fall off the [0,1] range: in this case we re-scale them so they will fall into [0,1].

Goal Generation in EXP_GEN. We generate a new goal only if there are no explanations in progress (Line 33) and only if there are no more than θ goals already in progress; in our experiments, we set $\theta = 1$. This makes the system pursue one goal at a time, eliminating possible noise from the performance analysis we will present in Section 5. The GenerateGoal function generates only one kind of goal: *kill unit A with unit B*, where A and B are randomly chosen (so there are many goals that can be pursued, one for every pair *(A,B)*. This goal then gets started by the PerformGoal function, which takes the goal and its associated reaction as input (Line 35). The goal is then added to the *GoalsInProgress* collection (Line 36). At Line 38 we handle the update of the state of the current goal. For each goal g, if g failed then we add it to the *GoalsFailed* collection. And finally, unless the goal is still in progress, we remove it from the *GoalsInProgress* collection (Line 43).

GENERATE_GOAL+. The EXP_GEN pseudo-code learns over time the applicability conditions and probability distribution of the stochastic explanations. We can exploit this learned knowledge to prevent discrepancies from happening in the future. The GENERATE_GOAL+ function uses the case base before selecting the next goal to pursue. In case the goal that was picked originally generated discrepancies in the past (we can compute this with a relaxed similarity metric over the case base), then it has a high chance of failure and the goal should be changed. This will not prevent discrepancies from happening, because we cannot know the future events that might occur in a stochastic domain, but we expect that it will nonetheless dramatically decrease the number of times that our agent will find its expectations inconsistent with the state of the world. The GENERATE_GOAL+ function is represented in the following pseudo-code.

GENERATE_GOAL+(*CB*, τ)
1:     $g \leftarrow$ CreateGoal()
2:     **if** FindDiscrepancies(g, *CB*) $> \tau$
3:         PickBestGoal(g)
4:     **return** g

At Line 1, the function CreateGoal will generate a domain-specific goal. As explained before, the goal generated will always be *KillUnit(A,B)* where A and B can be any unit. At Line 2 the function FindDiscrepancies will scan the case base CB, looking for previous failures that involved the same kinds of unit that were picked for g. This is a retrieval operation from the case base that involves a similar, but more relaxed, similarity metric. Here we consider only the unit types as features to compute the similarity.

If FindDiscrepancies can find more than τ previous failures, where τ is a user-defined threshold, then the algorithm calls PickBestGoal at Line 3. This function will look for the best unit to attack the target among the agent's units. To do this, we rely on knowledge acquired during the execution of the system. Recall that the system starts its execution knowing nothing about the rock-paper-scissors mechanism of the game, but improves its knowledge during its execution. Therefore, this algorithm will improve its performance over time, effectively learning from past mistakes.

5   Empirical Study

We claim that an agent using Stochastic Explanations will better react to sudden changes in the environment than an agent using Deterministic Explanations. Also, we will show how an agent that learns from past mistakes greatly outperforms an agent that does not perform this process.

5.1   Experimental Setup

In our scenarios, we used a similarity threshold $\sigma = 0.7$, and the explanation for each discrepancy is chosen from a predetermined set of three possible explanations: *Unit not upgraded, Wrong Unit,* and *Environmental advantage.* The first represents the situation where the target should be attacked with an upgraded version of the unit that was used originally (e.g. *ranger* instead of *archer*). The second is probably the most intuitive explanation, and means that the attacker should have been a completely different unit (e.g. *knight* instead of *peasant*). Finally, the latter explanation, Environmental Advantage, represents the situation where the enemy has some kind of obstacle protecting it from being attacked by closed-range land units and should therefore be attacked with flying or ranged units.

In the description below we refer to **team A** to the one controlled by our AI and **team B** to the opponent.

As in other works on explanations (Molineaux et al. 2012), our scenarios are hand-crafted to reduce noise and present more accurately the effectiveness of our theories. We will present the results from three scenarios, each composed by two parts. In the first part, the system acquires knowledge about the environment, the discrepancies that can happen within it, and their corresponding explanations. In the second part the scenario is slightly modified, so that the previously acquired knowledge is now flawed and not effective for the new conditions. We will show how Stochastic Explanations react to these events in comparison with Greedy Explanations and Random Explanations.

In the first scenario, the first part presents team A consisting of *archers* and *rangers* attacking team B which consisted of *rangers* only. When archers are selected to accomplish the goal, they will fail because rangers are stronger. The correct explanation for this kind of discrepancy is *Upgraded Unit*. In the second part, we also have *griffins* available, and the enemy *rangers* are now surrounded by trees. When the rangers attack they will sometimes fail. The correct explanation becomes *Environmental Advantage*; it is best to attack with the griffins.

The second scenario is similar to the first one, except that now Team A have *knights* and *paladins* but neither *archers* nor *rangers*; team B controls only *paladins*. The main difference between the first scenario and the second one is that *paladins* cannot attack *griffins*, while *rangers* can. Therefore in the first scenario the *Environmental Advantage* explanation has a small chance to fail, while in the second it can only fail in case of a *timeout,* which happens when the *griffin* does not kill the *paladin* in a reasonable amount of time.

Finally, the third scenario is the same as the second one, except that we switch the order of the first and second part.

Stochastic Explanations behave as explained in Section 3 and 4; Greedy Explanations utilize the same mechanics, except for the explanation decision part: here, the system will always pick the explanation having the highest probability. Random Explanations, intuitively, will just pick a random explanation among the three possible ones for each discrepancy that occurs during the system's execution. We expect Greedy to be a very effective heuristic during the first part of each experiment because we repeatedly try the same scenario, but then to perform much worse than Stochastic in the second part, having a much harder time to adapt to changes in the environment. We also expect Random not to present behavioral differences between part one and two, and to be worse than both Stochastic and Greedy Explanations.

In the charts that we will show, the x-axis will represent the number of iterations: for our experiments, we ran part one of each experiment for 50 iterations before switching to part two, which was then ran for 50 iterations, totaling 100 iterations per-experiment. Each experiment was also run 5 times, and the results were then averaged. The y-axis shows the number of failed explanations – that is, explanations generated by the system that failed when tested, providing a negative feedback.

5.2   Results

Figures 3, 4, and 5 show the results for first, second and third scenario respectively. The results meet the expectations: while in part 1 Greedy Explanations outperform Stochastic Explanations, in part 2 Stochastic outperforms Greedy for all the three scenarios. Random Explanations, as expected, have a consistent behavior in all the scenarios, and are consistently outperformed both in part 1 and 2 and in all three scenarios by both Greedy and Stochastic Explanations.

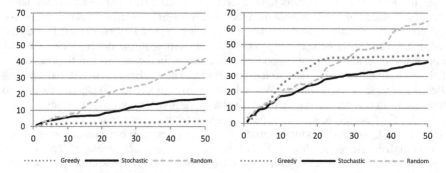

Fig. 3. Results for the first scenario for parts 1 (left) and 2 (right)

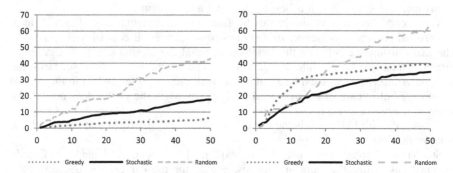

Fig. 4. Results for the second scenario for parts 1 (left) and 2 (right)

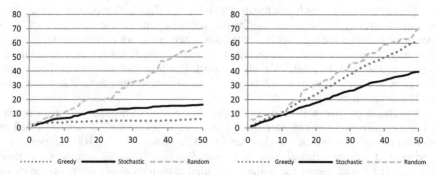

Fig. 5. Results for the third scenario for parts 1 (left) and 2 (right)

The implications of these results are deeper and more subtle that one might think. A stochastic explanation-generating agent, as opposed to a deterministic one, will always question its own experience and expectations. While this line of thought will penalize its performance in deterministic scenarios (represented by the first part of each experiment), it will be extremely helpful in case of sudden environment changes (the second part of our experiments). A deterministic agent cannot react this quickly to unseen events, because for functioning correctly such an agent must entirely rely on its own previous experiences. A stochastic agent, instead, constantly keeps questioning itself and its past experiences, which might be influenced not only by the nature of the environment but also by the limitedness of the agent's perception of the environment. Therefore, depending on the nature of the domain we operate in, one or the other approach can be better. In the case of a stochastic domain, Stochastic Explanations is a more suitable choice. If anomalies are extremely rare in the domain we operate in, then using greedy explanations would be better; if, instead, anomalies are common, then using Stochastic Explanations is preferred.

We did a second experiment showing how GENERATE_GOAL+ (see Section 4) exploits knowledge learned from past mistakes. Learning from past mistakes reduces substantially the number of discrepancies occurring. For the same 3 scenarios, Figures 6, 7, and 8 show a substantial reduction in the number of discrepancies generated with *error learning* (i.e., using GENERATE_GOAL+) compared to *normal* (i.e., using the goal generation code in EXP_GEN). In these experiments, we use Stochastic Explanations in both parts and the settings are the same as the previous experiment.

Using GENERATE_GOAL instead of the goal generation code in EXP_GEN shows the potential of taking advantage of the knowledge acquired by explaining discrepancies. This is even more pronounced in the second part of our scenarios, where the agent is presented with an unexpected event; if an agent is able to reason on its own past mistakes and on the explanation that it gave to those mistakes, then it is able to greatly outperform an agent that does not perform this kind of analysis.

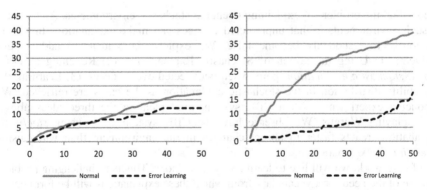

Fig. 6. Results with learning for the first scenario for parts 1 (left) and 2 (right)

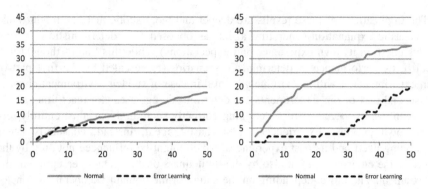

Fig. 7. Results with learning for the second scenario for parts 1 (left) and 2 (right)

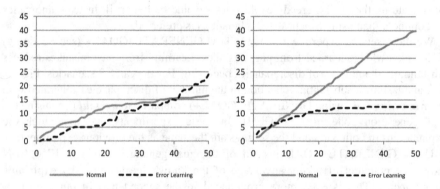

Fig. 8. Results with learning for the third scenario for parts 1 (left) and 2 (right)

6   Conclusions and Future Work

We studied the problem of explaining events in stochastic environments. A challenge in such environments is that unpredicted changes can make explanations that were previously true suddenly become false. We explore three ideas to address this problem: (1) Using the notion of Stochastic Explanations, (2) Retaining as cases *(event, stochastic explanation)* pairs when unexpected events occur. (3) Learning the probability distribution in the stochastic explanation as the cases are retrieved. We conducted experiments with a CBR system, implementing the three ideas above, while playing Wargus. We observe how the CBR system when using *stochastic* explanations reacts faster to abrupt changes in the environment than when using *deterministic* explanations.

In future work, we will like to learn new explanations. This is a challenging problem because of the need for a vocabulary from which these explanations will be formulated. Furthermore, this will also involve learning the reaction for the explanation.

Acknowledgements. This work was supported in part by NSF grant 1217888.

References

Cox, M.T.: Introspective Multistrategy Learning: Constructing a Learning Strategy under Reasoning Failure. Doctoral dissertation, Georgia Institute of Technology (1996)

Cox, M.T.: Perpetual self-aware cognitive agents. AI Magazine 28(1), 32 (2007)

Jaidee, U., Muñoz-Avila, H., Aha, D.W.: Learning and Reusing Goal-Specific Policies for Goal-Driven Autonomy. In: Agudo, B.D., Watson, I. (eds.) ICCBR 2012. LNCS, vol. 7466, pp. 182–195. Springer, Heidelberg (2012)

Molineaux, M., Aha, D.W., Kuter, U.: Learning Event Models that Explain Anomalies. In: Roth-Berghofer, T., Tintarev, N., Leake, D.B. (eds.) Explanation-Aware Computing: Papers from the IJCAI Workshop, Barcelona, Spain (2011)

Molineaux, M., Kuter, U., Klenk, M.: Proceedings of the 11th International Conference on Autonomous Agents and Multiagent Systems (2012)

Klenk, M., Molineaux, M., Aha, D.W.: Goal-Driven Autonomy For Responding To Unexpected Event. Strategy Simulations. In: Computational Intelligence (2012)

Ricci, F., Avesani, P.: Learning a local similarity metric for case-based reasoning. In: Aamodt, A., Veloso, M.M. (eds.) ICCBR 1995. LNCS, vol. 1010, pp. 301–312. Springer, Heidelberg (1995)

Case-Based Reasoning
on E-Community Knowledge

Emmanuelle Gaillard[1,2,3], Jean Lieber[1,2,3],
Yannick Naudet[4], and Emmanuel Nauer[1,2,3]

[1] Université de Lorraine, LORIA 54506 Vandœuvre-lès-Nancy, France
[2] CNRS 54506 Vandœuvre-lès-Nancy, France
[3] Inria 54602 Villers-lès-Nancy, France
firstname.lastname@loria.fr
[4] CRP Henri Tudor, Luxembourg
firstname.lastname@tudor.lu

Abstract. This paper presents MKM, a meta-knowledge model to manage knowledge reliability, in order to extend a CBR system so that it can reason on partially reliable, non expert, knowledge from the Web. Knowledge reliability is considered from the point of view of the decision maker using the CBR system. It is captured by the MKM model including notions such as belief, trust, reputation and quality, as well as their relationships and rules to evaluate knowledge reliability. We detail both the model and the associated approach to extend CBR. Given a problem to solve for a specific user, reliability estimation is used to filter knowledge with high reliability as well as to rank the results produced by the CBR system, ensuring the quality of results.

Keywords: case-based reasoning, meta-knowledge, reliability, filtering, ranking, personalization.

1 Introduction

The social Web is known to generate an enormous amount of information, constituting virtually a big knowledge base about almost any kind of subject. This knowledge base is mainly created by e-communities, consisting of people sharing common ideas, goals or interests, communicating over the Internet or any other technological communication network. The knowledge asset built by e-communities on the Web has a high exploitation potential ranging from knowledge management to, e.g., data-mining applications for detecting interesting trends or tracking users for personalization purposes. Such knowledge is influenced by different human factors such as belief, confidence, or trust, which influence its reliability, not only from a human perspective but also for automated reasoning and decision making [1].

In this context, the work presented here focuses on the exploitation of partially reliable e-community knowledge for Case-Based Reasoning (CBR), in contrast to classical approaches that rely on consensual and validated expert knowledge.

S.J. Delany and S. Ontañón (Eds.): ICCBR 2013, LNAI 7969, pp. 104–118, 2013.
© Springer-Verlag Berlin Heidelberg 2013

In order to preserve the quality of reasoning, we have designed a meta-knowledge model, called MKM, to manage knowledge coming from (non-expert) users. MKM represents knowledge reliability using meta-knowledge such as belief, trust, reputation, quality, so that the CBR engine can take it into account in its reasoning process.

This paper is organized as follows. Section 2 gives the motivations for this research. We present the state of the art related to the use of meta-knowledge to characterize knowledge reliability and especially in CBR systems in Section 3. Section 4 details the global process for meta-knowledge management, and Section 5 shows how to extend a classical CBR system to handle partially reliable knowledge. Section 6 presents a use-case in the cooking domain and Section 7 concludes this paper.

2   Motivation

The research work presented here is motivated by the will to extend the traditional application field of CBR systems, and especially adapt the latter for processing web-originated data. Indeed, traditional CBR systems are used in a closed world where the manipulated knowledge base is fed with expert data, consensual and validated. Because of this, CBR systems can hardly be used when expert data is sparse and especially on the web, where knowledge reliability is not guaranteed and difficult to measure.

Knowing this, an important question is to evaluate if reasoning on not fully reliable knowledge coming from e-communities can be exploited in a CBR approach and if at least results of a similar quality than when reasoning on expert knowledge can be obtained.

Classical CBR systems are usually composed of four knowledge containers: the cases, the domain knowledge (i.e., ontology of the domain), the adaptation knowledge, and the similarity (i.e., retrieval) knowledge. Knowledge is validated by experts; this entails to reason with a limited amount of knowledge. Some studies, like the Kolflow project (http://kolflow.univ-nantes.fr/), have investigated ways to manage knowledge provided by an e-community. The Kolflow approach consists in improving the man-machine collaboration to ensure a collaborative knowledge construction. It allows to collect big amount of knowledge from users, and uses non-regression tests to ensure that new incoming knowledge does not affect the results of the reasoning process that exploits the knowledge [2]. More particularly in the CBR domain, Richards [3] has investigated decision support from a knowledge base constituted of experts and informal knowledge gathered from collective tools. She has proposed a collaborative knowledge engineering approach mixing CBR and rules, named Collaborative Multiple Classification Ripple Down Rules (C-MCRDR). C-MCRDR proved being efficient to build collaboratively a knowledge base. However, partial reliability of knowledge is handled implicitly in a manual human-driven review and negotiation process. Knowledge reliability related to truth, belief, trust or reputation, is not considered and not directly exploited in the reasoning process.

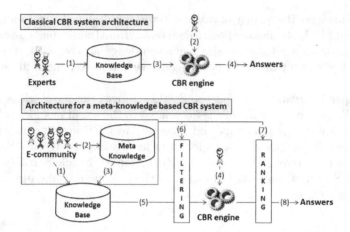

Fig. 1. CBR system classical architecture vs. architecture using meta-knowledge

We propose to associate meta-knowledge about reliability to each knowledge unit (KU) in order to allow inferences on e-community originated knowledge, while ensuring the quality of results. Our solution consists in establishing a meta-knowledge model to describe e-communities knowledge. A new container, the meta-knowledge base, is thus added to the classical CBR architecture.

Fig. 1 shows the difference of architecture between a *classical* CBR system and a CBR system based on meta-knowledge. In the classical CBR system architecture, (1) the knowledge base is produced by experts of the domain. (2) A user queries the CBR engine which (3) uses the expert validated knowledge for (4) computing its answers. In a meta-knowledge based CBR system, (1) the knowledge base is produced by the e-community and (2) users and (3) knowledge are linked to meta-knowledge. When (4) a user queries the CBR engine, the latter (5) uses the e-community knowledge, (6) filtered by the meta-knowledge model. The final answers of the CBR engine (8) may be ranked (7) using also the meta-knowledge. Moreover, if the user who queries the system is a member of the e-community, the filtering and ranking operations produce personalized results since some meta-knowledge are user-specific.

3   State of the Arts

3.1   Meta-knowledge About Reliability

While philosophical studies (e.g, [4]) associate knowledge reliability to safety and robustness, using reliable knowledge element in a knowledge-based system allows to infer knowledge with an acceptable level of trustworthiness. Knowledge reliability is influenced by several factors, sometimes interrelated, as discussed in [1], where a generic model for representing knowledge generated by e-communities is proposed. In an effort to provide a basis for exploiting partially reliable knowledge in a reasoning process, this work identifies the following dimensions for knowledge reliability: origin, context, truth, belief, value, quality, and trust. Origin (or provenance) is the source of a KU; context concerns validity conditions associated to knowledge use; truth is the knowledge validity in the considered world, while belief is the knowledge truth from the author, provider or community perspective; value stands for the importance of the KU for its consumer; quality accounts for characteristics of the KU (e.g., precision, completeness) and finally trust, impacting many relationships in the model and more particularly between knowledge, author/provider and community.

In order to assess at best the impact of these parameters on knowledge reliability, it is necessary to first work with a reduced set of meta-knowledge. In this work we focus on knowledge quality, belief, trust and reputation. Context, as a complex notion, is not considered for the moment, as well as value, since it can be considered that a knowledge of high quality is valuable for the community. Finally, truth is highly correlated with belief notion, and they can be considered as synonyms in a first approach.

Quality is considered by researchers as a complex notion, and it is perceived as a multi-dimensional concept [5]. The nature of data and measurement of its quality have an important impact on the success of decision making. So it is important to evaluate the quality of knowledge, when creating a knowledge base. Several criteria of quality are listed in [5], such as completeness, consistency, freshness, and accuracy. Quality assessment is obtained by using a set of formulas aggregating the different criteria. [6] proposes a data quality measurement framework that takes into account the context of use and the utility of data. The quality measurement in MKM focuses on the satisfaction a user (measured thanks to a 5-star rating system) will have when using a KU.

Trust, belief and reputation are closely related terms. Trust is largely studied in literature and multiple viewpoints exist, because trust is both a component of our everyday life and of each application domain [7]. Trust is generally defined as a ternary relation, valid in a given context, between a truster, a trustee and an object, as originally proposed by Cook et al. [8]. As for Grandison [9], the trusted object is often related to an action performed by the trustee or its ability to do it. In other words, independently of whether the trustee is a person or not, the trustee is viewed as some entity that will actually perform the action the truster expects him to do. Trust is a social process, and evolves dynamically following the history of the relationship. In the human computer interaction domain, Golbeck [10] asserts that "A trusts B if A commits to an action based on the belief that B's future actions will lead to a good outcome." She used this definition in her recommending system for movies, where users can rate both movies and other users. For a given user, movie recommendation scores are computed by taking into account the community opinion: scores depends on movies ratings weighted by user reputation.

Several models have been proposed, both for conceptualizing and evaluating trust. The most common systems exploiting trust are based on reputation [11,12,13,14] and some of them take also belief into account. For example, the model proposed in [15] shows the relation between trust, belief and reputation in a social network. This model relates trust to a set of relevant beliefs on the evaluated user, for example if a user is an expert, or if a user is honest.

In [16], a movie is recommended to a user u if this movie has been well rated by users belonging to the trust network of u. Trust is computed in [16] from the proportion of common ratings users have put on movies they evaluated. The cold start problem is handled by finding in the available users one that is similar to a newcomer, where similarity is computed from demographic data. In our work, the MKM solution relies on predefined default values.

Trust and reputation management has been addressed in multiple domains like in multi-agent systems (MASs) (see, e.g., [17]), network security and Service Oriented Architectures [18]. Regarding e-communities, trust is used in social networks applications and addressed in the Web of Trust initiative.

3.2 Meta-knowledge in CBR Systems

Usually, user feedback on proposed answers in CBR systems allows to judge the quality of the new cases and to repair a failed adaptation [19]. However, some authors think that the feedback approach is insufficient (by observing missing or delayed feedback) and that meta-knowledge is more relevant to improve the reasoning results [20]. In [20], the authors propose to integrate the provenance of a case, as a meta-knowledge, in a CBR system to guide the case base maintenance and increase confidence in future results. For example, a repair is propagated through generated cases from the initial case and the quality of a case is measured by the length of the adaptation path. However, quality of initial cases is set to a same maximum value. The quality does not depend on external factors and additional meta-knowledge like, e.g., provenance of initial cases is not taken into account.

In [21], authors integrate trust in addition to provenance in a CBR approach to propose a model of collaborative web search. During a user search, web pages are filtered and ranked using their relevance to the query and the reputation of users having already selected the pages. Reasoning is guided by preferences of users and not by reliability of cases. Besides, the indicator of quality acts only on the case base containers and on cases which have already been found.

In conclusion, the state of the art shows that meta-knowledge is introduced to provide explanations on CBR results or recommendations. To the best of our knowledge, during the reasoning of a CBR system, meta-knowledge like trust, belief and reputation are only used for cases created by the adaptation process or cases which have already been retrieved during a previous reasoning. Meta-knowledge representing reliability of cases are not yet used for representing new external cases. Besides, reliability of knowledge of the other containers is not represented. The novelty of MKM is to represent reliability of KUs of all the containers: cases (not only learned cases), domain, adaptation and similarity KUs. Reliability is then used to filter KUs and to rank personalized answers returned by the CBR system.

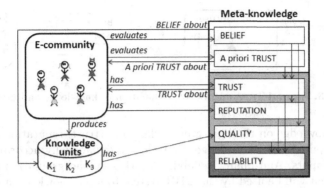

Fig. 2. Dependencies between users, knowledge units and meta-knowledge

4   Meta-knowledge Management for a CBR System

This section presents MKM, the meta-knowledge model, used to represent and to compute the reliability of knowledge coming from an e-community. The objective is to compute the reliability of each KU (of the four containers) for a given user, in order to improve the CBR system reasoning and personalize the answers returned to a user. The reliability is represented by a score that depends on several meta-knowledge elements presented in the following. Some meta-knowledge is inferred by the system while some other is entered in the system by the users who evaluate KUs or by other community members. These evaluations are the foundations of MKM. Fig. 2 introduces the links between users, KUs and meta-knowledge:

- A user u may evaluate a KU ku of a knowledge container by a belief score which represents the belief u has in ku.
- A user u may evaluate another user v by an a priori trust score which represents the trust u has towards v.
- A trust score from a user u towards a user v, which represents how u trusts v, is inferred from the a priori trust score that u has assigned to v and from the belief scores that u has assigned to KUs produced by v.
- A reputation score of a user u, which represents the reputation of u in the e-community, is inferred from all the trust scores about u.
- A quality score of a KU ku, which represents the global quality of ku for the e-community, is inferred from all the belief scores about ku.
- Finally, a reliability score of a KU ku for a user u, which represents the personalized reliability of ku for u, is inferred from quality, reputation and trust scores.

To summarize, the meta-knowledge represented on a white background in Fig. 2 (belief and a priori trust) are initially entered by the users. Sayya et al. [22] show that collecting such items works well when there is a small number of users who rate frequently, leading other users of the community to give feedback.

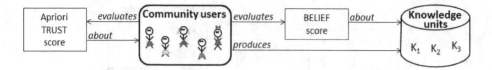

Fig. 3. A priori Trust and Belief: Evaluation of knowledge and users

The meta-knowledge on the light-grey background (trust, reputation and quality) are computed meta-knowledge, that change dynamically according to the community inputs. And finally, reliability (dark-grey background) is the meta-knowledge that will be used by the CBR system to filter knowledge and to rank answers with respect to the users trusts and beliefs. in the following, we detail all these elements and the way they are computed.

User interactions and evaluations. Fig. 3 illustrates how users may interact with the system, by editing some KUs, by evaluating some KUs, and by evaluating other users. In the model, let User be the set of all the users of the system and let KU be the set of all KUs (of all the containers) used in the system.

The ku_from function returns the KUs edited by a user, for $u, v \in$ User, $\text{ku\_from}(u) \cap \text{ku\_from}(v) = \emptyset$ (a $ku \in$ KU has only one producer):

$$\text{ku\_from} : \text{KU} \to 2^{\text{KU}}$$
$$u \mapsto \text{ku\_from}(u)$$

Users may evaluate KUs or other users with, e.g., a numerical scale or a star system. In MKM, these evaluations are normalized in $[0, 1]$.

– When a user u evaluates a $ku \in$ KU, u assigns a belief score to the KU ku. This belief score represents the degree of acceptance, for u, that ku may be true, according to his/her own knowledge.

$$\text{belief} : \text{User} \times \text{KU} \to [0, 1] \cup \{?\}$$
$$(u, ku) \mapsto \text{belief}(u, ku)$$

where ? stands for the unknown value ($\text{belief}(u, ku) = ?$ means that u has not evaluated ku).

– When a user u evaluates another user v, u assigns an a priori trust score for v. This a priori trust score represents the degree of acceptance that v is a trustworthy user for u according to subjective information independently from the KUs produced by v. The higher this a priori trust score is, the higher u expects that a KU ku produced by v is true.

$$\text{a\_priori\_trust} : \text{User} \times \text{User} \to [0, 1] \cup \{?\}$$
$$(u, v) \mapsto \text{a\_priori\_trust}(u, v)$$

Belief scores and a priori trust scores are directly given by users when evaluating KUs or other users. The other meta-knowledge is inferred from these two scores.

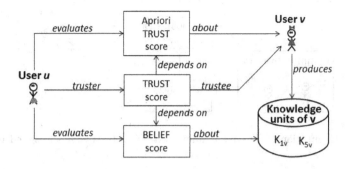

Fig. 4. Trust of a user u towards a user v, computed from the a priori trust score of u towards v, and belief scores of u towards KUs produced by v

Trust. The trust of user u (the truster) regarding a user v (the trustee) represents the degree of expectation that the knowledge brought by v to the community is true for u. The trust score depends on the a priori trust score assigned by u to v, and on all the belief scores assigned by u to the knowledge produced by v (see Fig. 4).

The multi-set of belief scores that u assigns to the knowledge produced by v, is returned by the function $\mathtt{user\_belief\_scores}(u, v)$. This multi-set is inferred thanks to the function $\mathtt{ku\_from}(v)$ and $\mathtt{belief}(u, ku)$ where ku is the KU that v has produced:

$$\mathtt{user\_belief\_scores} : \mathtt{User} \times \mathtt{User} \to 2^{[0,1]}$$
$$(u, v) \mapsto \{\mathtt{belief}(u, ku) \mid ku \in \mathtt{ku\_from}(v)\} \setminus \{?\}$$

$\mathtt{trust}(u, v)$ is a measure of the trust the user u has towards the user v.

$$\mathtt{trust} : \mathtt{User} \times \mathtt{User} \to [0, 1] \cup \{?\}$$

Trust is computed as follows (for $u, v \in \mathtt{User}$):

- If u has never evaluated any KUs produced by v
  (i.e., $\mathtt{user\_belief\_scores}(u, v) = \emptyset$)
  then, $\mathtt{trust}(u, v) = \mathtt{a\_priori\_trust}(u, v) \in [0, 1] \cup \{?\}$
- Else, let $n = |\mathtt{user\_belief\_scores}(u, v)|$, $n \neq 0$.
  - If u has not assigned an a priori trust score to v
    (i.e., $\mathtt{a\_priori\_trust}(u, v) = ?$)
    then, trust is the average of the belief scores that u assigned to KUs produced by v (i.e., $\mathtt{user\_belief\_scores}(u, v)$).

$$\mathtt{trust}(u, v) = \frac{1}{n} \sum_{s \in \mathtt{user\_belief\_scores}(u, v)} s \qquad (1)$$

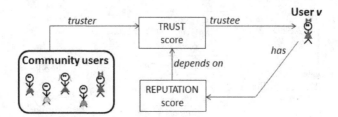

Fig. 5. Reputation of a user v is inferred from all trust scores assigned to v

- Else, the trust is computed by a combination of the a priori trust and the belief scores given by u:

$$\text{trust}(u,v) = \alpha_n \text{ a\_priori\_trust}(u,v)$$
$$+ (1 - \alpha_n) \frac{1}{n} \sum_{s \in \text{user\_belief\_scores}(u,v)} s \qquad (2)$$

where $\alpha_n = \dfrac{1}{n+1}$

The more scores have been assigned by u to the KUs provided by v, the less is the influence of the a priori trust score (since $\lim_{n \to \infty} \alpha_n = 0$): asymptotically, the expressions (1) and (2) are equivalent.

Reputation. The reputation of a user v is the perception all the users in the community have of v, based on their previous experience with v. Reputation provides an indicator of the truth of knowledge produced by v. It depends on the inferred trust scores of the community towards v (see Fig. 5).

The multi-set of trust scores inferred for the community towards v, is returned by the function `community_trust_scores(v)`:

$$\text{community\_trust\_scores} : \text{User} \to 2^{[0,1]}$$
$$v \mapsto \{\text{trust}(u,v) \mid u \in \text{User}\} \setminus \{?\}$$

`reputation(v)` is an estimation of the measure of the trust that has all the community towards v.

$$\text{reputation} : u \to [0,1] \cup \{?\}$$

For $v \in \text{User}$, reputation is computed as follows:

- If $|\text{community\_trust\_scores}(v)| < \tau$
  then $\text{reputation}(v) = \text{default\_reputation}$,
- Else $\text{reputation}(v) = \dfrac{\sum_{u \in \text{User}, u \neq v} \text{trust}(u,v)}{|\text{community\_trust\_scores}(v)|}$.

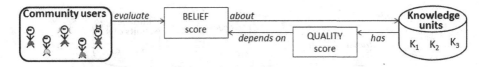

Fig. 6. Quality of a knowledge computed from its belief scores

The reputation score of v is the average of the set of $\text{trust}(u, v)$. For users that have obtained a number of evaluation less than a given threshold τ, the reputation score is assigned to a default value between 0 and 1 (for example, 0.5) and is denoted by $\text{default\_reputation}$.

Quality. The quality score of a KU ku is the computed community quality of ku. This score is independent from the KU's producer and represents the estimation of the degree of satisfaction of users after the use of ku. The quality score of ku depends on all the belief scores assigned to ku (see Fig. 6).

The multi-set of belief scores of a KU ku, denoted by $\text{ku\_belief\_scores}(ku)$, represents all the evaluations the community has assigned to ku:

$$\text{ku\_belief\_scores} : \text{KU} \rightarrow 2^{[0,1]}$$
$$ku \mapsto \{\text{belief}(u, ku) \mid u \in \text{User}\} \setminus \{?\}$$

The quality knowledge of ku, denoted by $\text{community\_quality}(ku)$, is the average of $\text{ku\_belief\_scores}(ku)$:

$$\text{community\_quality} : \text{KU} \rightarrow [0, 1] \cup ?$$
$$ku \mapsto \frac{\sum_{s \in \text{ku\_belief\_scores}(ku)} s}{|\text{ku\_belief\_scores}(ku)|}$$

Reliability. The reliability corresponds to how much a user u can rely on a KU ku, or how ku is actually useful for him/her. This reliability score will be used by the CBR system for filtering knowledge, as well as for ranking.

For a user u (see Fig. 7), the reliability of a KU ku produced by a user v, depends of the reputation of v, the trust score of u towards v (if it exists), and the quality of ku (if it exists). Because reliability is based on the trust score of u towards v, the reliability of knowledge is a personalized score (the trust score varies from one user to another).

$$\text{reliability} : \text{User} \times \text{KU} \rightarrow [0, 1]$$
$$(u, ku) \mapsto w_{reputation}\text{reputation}(v) + w_{trust}\text{trust}(u, v)$$
$$+ w_{quality}\text{community\_quality}(ku)$$
$$\text{where } w_{reputation} + w_{trust} + w_{quality} = 1$$

This function assumes that $\text{trust}(u, v) \neq ?$ and $\text{community\_quality}(ku) \neq ?$. If $\text{trust}(u, v) = ?$ and/or $\text{community\_quality}(ku) = ?$ then they are not taken into account.

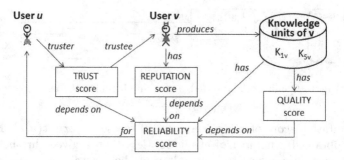

Fig. 7. Reliability of a KU ku for a user u is computed from the quality score of ku, the reputation score of v, the producer of ku, and the trust of u towards v

5   Plugging the Meta-model into a CBR System

This section presents how a classical CBR system can be modified in order to take into account the reliability of knowledge (see Fig. 1). A filter function is used to select the *more reliable* set of knowledge according to the query and to the user who queries the system. A ranking function is used to order the set of answers according to the meta-knowledge associated to the KUs involved in the computation of the results.

5.1   Filtering

According to MKM, the knowledge which is not sufficiently reliable for a user is filtered according to the reliability score. The filtering function, `to_be_filtered`(u, ku), depends on `reliability`(u, ku). It returns true iff the reliability score ku for u is higher than a given threshold $\beta \in [0, 1]$.

$$\texttt{to\_be\_filtered} : \text{User} \times \text{KU} \rightarrow \{\texttt{true}, \texttt{false}\}$$
$$(u, ku) \mapsto (\texttt{reliability}(u, ku) \geq \beta)$$

If `to_be_filtered`(u, ku), then ku is used in inferences by the CBR engine, else it is not considered.

5.2   Ranking

The answers computed by the CBR engine will be ranked according to MKM. The idea is to associate to a user u and an inference, a score, called the *inferred reliability*. Let $\{ku_1, \ldots, ku_n\} \vdash ku$ denote an inference performed by the CBR engine: the ku_i's are the KUs taken from the knowledge containers and ku is an inferred knowledge unit. For example, if a case c is adapted thanks to an adaptation rule a into a case c', $\{c, a\} \vdash c'$ denotes this adaptation inference. The inference reliability is computed thanks to an aggregation function ⊛ applied to the reliability of the ku_i's:

$$\texttt{inferred\_reliability}(u, \{ku_1, \ldots, ku_n\} \vdash ku) \mapsto \underset{1 \leq i \leq n}{\circledast} \texttt{reliability}(u, ku_i)$$

There are several possible aggregation functions ⊛: it can be an average aggregation function (corresponding to a probabilistic approach) or a minimum (corresponding to a necessity measure of a possibilistic approach).

By abuse of notation, $\texttt{inferred\_reliability}(u, \{ku_1, \ldots, ku_n\} \vdash ku)$ is denoted by $\texttt{inferred\_reliability}(u, ku)$. Thus, given several inferred knowledge units $ku^1, \ldots, ku^p$ and a user u, the ku^j's can be ranked according to decreasing $\texttt{inferred\_reliability}(u, ku^j)$.

5.3 Prerequisites

To apply the filter and ranking functions, a classical CBR system that would be adapted to reason on an e-community knowledge, must:

- provide the possibility for users to manage KUs of the different containers in order to assign them a belief score;
- be able to filter KUs it will use according to their reliability;
- return the set of KUs that are involved in the computation of each answers, in order to rank these answers.

6   Use Case: Adapting Cooking Recipes

This section presents a use case, in the framework of TAAABLE (http://taaable.fr/). TAAABLE is a CBR system which retrieves and creates cooking recipes by adaptation [23].

TAAABLE is based on the four classical knowledge containers. The domain knowledge container contains an ontology of the cooking domain that is used to retrieve the source cases that are the most similar to a target case (i.e. the query). This ontology includes several hierarchies (about food, dish types, etc.). A generalization cost is associated to each edge in the hierarchy, and is used to compute similarity between cases and the query. For example, cost associated on the edge connecting *GreenOnion* and *Onion* is 0.3.

The case base is composed of recipes. Each recipe R is transformed into $idx(R)$, the index of the recipe R which is a conjunction of concepts of the domain ontology. Four recipes are given in example in Table 1. $idx(R_1)$ is a formal and abstracted representation of the recipe R_1 which is a gratin dish and whose ingredients are green onion, leek, béchamel sauce, and ham (and nothing else).

According to a query entered in the system by a user, and also represented by a conjunction of concepts, the system searches in the case base for some cases (recipes) satisfying the query. For example, $Q = GratinDish \wedge Leak \wedge Ham \wedge$

Table 1. Four examples of recipes and their indexes

Id	Title	$idx(R_i)$
R_1	Leeks gratin	$GratinDish \wedge Leek \wedge Potato \wedge Bechamel \wedge Ham$
R_2	Carrots gratin	$GratinDish \wedge GreenOnion \wedge Carrot \wedge Bechamel \wedge Ham$
R_3	Endives gratin	$GratinDish \wedge Endive \wedge Lemon \wedge Bechamel \wedge Ham$
R_4	Salmon paste	$GratinDish \wedge Pasta \wedge Salmone \wedge Bechamel \wedge Ham$

Table 2. Reliability inferred for Tom, for recipes and adaptations

Id	Title	Reliability
R_1	Leeks gratin	0.4
R_2	Carrots gratin	0.6
R_3	Endives gratin	0.9
R_4	Salmon paste	0.8

Id	σ	Reliability
A_1	$GreenOnion \rightsquigarrow Leek$	0.5
A_2	$Endive \rightsquigarrow Leek$	0.6
A_3	$Salmon \rightsquigarrow Leek$	0.8

Bechamel represents the query "I would like a recipe of gratin dish with leak, ham and béchamel sauce." Recipes matching exactly the query, if they exist, are returned to the user and are ranked at the first place (for example R_1 matches exactly Q). The system searches also similar recipes, using the hierarchies to generalize the user query. Recipes that have a similarity measure higher than a given threshold are retrieved and ranked by similarity. Adaptation consists in substituting some ingredients of the source cases by the ones required by the user, and is encoded by a substitution $\sigma = A \rightsquigarrow B$, meaning that "$A$ has to be substituted by B". An adaptation has a cost, denoted by $cost(A \rightsquigarrow B)$. The lesser the cost of adaptation of R_1 is, the higher is the similarity of R_1 with Q. The final result of TAAABLE for Q, ordered by increasing similarity, is R_1 (no adaptation), R_2 with the adaptation $GreenOnion \rightsquigarrow Leek$, R_3 with the adaptation $Endive \rightsquigarrow Leek$, R_2 is retrieved before R_3 because $GreenOnion$ is closer to $Leek$ than $Endive$ (i,e. $cost(GreenOnion \rightsquigarrow Leek) < cost(Endive \rightsquigarrow Leek)$). R_4 is not retrieved because the cost of substitution of $Salmon$ by $Leek$ is too high to adapt R_4 to Q.

TAAABLE uses also adaptation knowledge (AK) of the form $(context, \sigma)$ [24]: *context* is the recipe or the class of recipes on which the substitution σ can be applied. For example, the AK where $Salmon$ could be replaced with $Leek$ in R_4 is represented by $(R_4, Salmon \rightsquigarrow Leek)$. Using this AK, R_4 which was not proposed by the query generalization process is now proposed.

Example of Reasoning with the Meta-knowledge Approach

A meta-knowledge container is added to take into account reliability. It will be used to modify the list of answers returned by TAAABLE. Table 2 shows the reliability of recipes and reliability of adaptations for a specific user: Tom.

Using MKM, unreliable KUs (e.g. which are lower than a given reliability threshold) will be filtered and will not participate to the reasoning process. For example, with a threshold fixed to 0.6, the recipe R_1 and the adaptations A_1 are eliminated, for the user Tom.

The answers can be ranked according to the reliability of the KUs that are involved, for example, when using the average aggregation function, in the following order: R_4 with A_3, R_3 with A_2. Indeed, with the average aggregation function, the ranking score of R_4 with A_3, which is 0.8, is higher than the ranking score of R_3 with A_2, which is 0.75. R_2 with A_1 is not proposed because A_1 cannot be applied anymore. These ranked answers are specific to Tom, because the reliability of a knowledge for Tom is influenced by the trust score he has towards users of the system who produced the knowledge.

7   Conclusion

In this paper, we presented an approach for reasoning on partially reliable e-community knowledge by contrast to consensual and validated knowledge in classical CBR systems. The approach proposes to associate meta-knowledge about reliability of knowledge, based on user evaluations, and a model has been presented.

We have illustrated the interest of the approach with an example in the community cooking domain. Ongoing work consists in implementing the use case presented in section 6 in the framework of the TAAABLE project. A collaborative web space which allows to manage cases, AK, and users, and in which users may evaluate these KUs and other users has already been developed. The functions for computing the different meta-knowledge scores must yet be integrated in this collaborative work space. Experiments with the TAAABLE e-community will be driven at short term in order to evaluate if using MKM and knowledge from an e-community provides similar or better results than those obtained using the classical architecture exploiting consensual knowledge.

References

1. Naudet, Y., Latour, T., Vidou, G., Djaghloul, Y.: Towards a novel approach for high-stake decision support system based on community contributed knowledge base. In: 10th International Conference on Intelligent Systems Design and Applications (ISDA), pp. 730–736 (2010)
2. Skaf-Molli, H., Desmontils, E., Nauer, E., Canals, G., Cordier, A.E., Lefevre, M.: Knowledge Continuous Integration Process (K-CIP). In: 21st World Wide Web Conference - Semantic Web Collaborative Spaces Workshop, pp. 1075–1082 (2012)
3. Richards, D.: A social software/web 2.0 approach to collaborative knowledge engineering. Information Sciences 179(15), 2515–2523 (2009)
4. Hendricks, V.F., Pritchard, D.: New waves in epistemology. New Waves in Philosophy. Palgrave Macmillan (2008)
5. Wang, R.Y.: A product perspective on total data quality management. Commun. ACM 41(2), 58–65 (1998)
6. Even, A., Shankaranarayanan, G.: Utility-driven assessment of data quality. SIG-MIS Database 38(2), 75–93 (2007)

7. Marsh, S.P.: Formalising Trust as a Computational Concept. PhD thesis, University of Stirling (1994)
8. Cook, K.S., Hardin, R., Levi, M.: Cooperation Without Trust? The Russell Sage Foundation Series on Trust. Russell Sage (2007)
9. Grandison, T., Sloman, M.: Trust management tools for internet applications. In: First International Conference on Trust Management, pp. 91–107 (2003)
10. Golbeck, J.A.: Computing and applying trust in web-based social networks. PhD thesis, University of Maryland (2005)
11. Castelfranchi, C., Falcone, R.: Principles of Trust for MAS: Cognitive Anatomy, Social Importance, and Quantification. In: Third International Conference on Multi Agent Systems (ICMAS 1998), p. 72 (1998)
12. Artz, D., Gil, Y.: A survey of trust in computer science and the Semantic Web. Web Semantics: Science, Services and Agents on the WWW 5(2), 58–71 (2007)
13. Jøsang, A., Ismail, R.: The beta reputation system. In: Proceedings of the 15th Bled Electronic Commerce Conference, pp. 324–337 (2002)
14. Abdul-Rahman, A., Hailes, S.: Supporting Trust in Virtual Communities. In: HICSS, p. 9 (2000)
15. Knap, T., Mlýnková, I.: Revealing beliefs influencing trust between members of the czech informatics community. In: Datta, A., Shulman, S., Zheng, B., Lin, S.-D., Sun, A., Lim, E.-P. (eds.) SocInfo 2011. LNCS, vol. 6984, pp. 226–239. Springer, Heidelberg (2011)
16. Quijano-Sánchez, L., Bridge, D., Díaz-Agudo, B., Recio-García, J.A.: A case-based solution to the cold-start problem in group recommenders. In: Agudo, B.D., Watson, I. (eds.) ICCBR 2012. LNCS, vol. 7466, pp. 342–356. Springer, Heidelberg (2012)
17. Huynh, T.D., Jennings, N.R., Shadbolt, N.R.: An integrated trust and reputation model for open multi-agent systems. Autonomous Agents and Multi-Agent Systems 13(2), 119–154 (2006)
18. Neisse, R., Wegdam, M., Van Sinderen, M., Lenzini, G.: Trust management model and architecture for context-aware service platforms. In: On the Move to Meaningful Internet Systems: CoopIS, DOA, ODBASE, GADA, and IS, pp. 1803–1820 (2007)
19. Cordier, A., Fuchs, B., Lana de Carvalho, L., Lieber, J., Mille, A.: Opportunistic acquisition of adaptation knowledge and cases - The IaKa Approach. In: Althoff, K.-D., Bergmann, R., Minor, M., Hanft, A. (eds.) ECCBR 2008. LNCS (LNAI), vol. 5239, pp. 150–164. Springer, Heidelberg (2008)
20. Leake, D.B., Whitehead, M.: Case provenance: The value of remembering case sources. In: Weber, R.O., Richter, M.M. (eds.) ICCBR 2007. LNCS (LNAI), vol. 4626, pp. 194–208. Springer, Heidelberg (2007)
21. Saaya, Z., Smyth, B., Coyle, M., Briggs, P.: Recommending case bases: Applications in social web search. In: Ram, A., Wiratunga, N. (eds.) ICCBR 2011. LNCS, vol. 6880, pp. 274–288. Springer, Heidelberg (2011)
22. Schafer, J.B., Frankowski, D., Herlocker, J., Sen, S.: The adaptive web. In: Brusilovsky, P., Kobsa, A., Nejdl, W. (eds.) Adaptive Web 2007. LNCS, vol. 4321, pp. 291–324. Springer, Heidelberg (2007)
23. Badra, F., Bendaoud, R., Bentebitel, R., Champin, P.-A., Cojan, J., Cordier, A., Després, S., Jean-Daubias, S., Lieber, J., Meilender, T., Mille, A., Nauer, E., Napoli, A., Toussaint, Y.: Taaable: Text Mining, Ontology Engineering, and Hierarchical Classification for Textual Case-Based Cooking. In: ECCBR Workshops, Workshop of the First Computer Cooking Contest, pp. 219–228 (2008)
24. Gaillard, E., Lieber, J., Nauer, E.: Adaptation knowledge discovery for cooking using closed itemset extraction. In: The Eighth International Conference on Concept Lattices and their Applications, CLA 2011 (2011)

On the Plan-Library Maintenance Problem
in a Case-Based Planner

Alfonso Emilio Gerevini[1], Anna Roubíčková[2],
Alessandro Saetti[1], and Ivan Serina[1]

[1] Dept. of Information Engineering, University of Brescia, Brescia, Italy
[2] Faculty of Computer Science, Free University of Bozen-Bolzano, Bolzano, Italy
{gerevini,saetti,serina}@ing.unibs.it, anna.roubickova@stud-inf.unibz.it

Abstract. Case-based planning is an approach to planning where previous planning experience stored in a case base provides guidance to
solving new problems. Such a guidance can be extremely useful when
the new problem is very hard to solve, or the stored previous experience
is highly valuable (because, e.g., it was provided and/or validated by
human experts) and the system should try to reuse it as much as possible. However, as known in general case-based reasoning, the case base
needs to be maintained at a manageable size, in order to avoid that the
computational cost of querying it excessively grows, making the entire
approach ineffective. We formally define the problem of case base maintenance for planning, discuss which criteria should drive a successful policy
to maintain the case base, introduce some policies optimizing different
criteria, and experimentally analyze their behavior by evaluating their
effectiveness and performance.

1   Introduction

It is well known that AI planning is a computationally very hard problem [9]. In
order to address it, over the last two decades several syntactical and structural
restrictions that guarantee better computational properties have been identified
(e.g., [3,4]), and various algorithms and heuristics have been developed (e.g.,
[7,16]). Another complementary approach, that usually gives better computational performance, attempts to build planning systems that can exploit additional knowledge not provided in the classical planning domain model. This
knowledge is encoded as, e.g., domain-dependent heuristics, hierarchical task
networks and temporal logic formulae controlling the search, or it can be automatically derived from the experiences of the planning system in different forms.

Case-based planning (e.g., [8,15,17,21]) follows this second approach and concerns techniques that improve the overall performance of the planning system by
reusing its previous experiences (or "cases"), provided that the system frequently
encounters problems similar to those already solved and that similar problems
have similar solutions. If these assumptions are fulfilled, a well-designed case-
based planner gradually creates a plan library that allows more problems to be
solved (or higher quality solutions to be generated) compared to using a classical domain-independent planner. Such a library is a central component of a

S.J. Delany and S. Ontañón (Eds.): ICCBR 2013, LNAI 7969, pp. 119–133, 2013.

case-based planning system, which needs a policy to maintain the quality of the library as high as possible in order to be efficient. Even though this problem has been studied in the context of case-based reasoning, comparable work in the planning context is still missing.

In this paper, we define the assumptions underlying the case-based methodology in the context of planning, which characterize the typical distribution of the cases in a plan library. Then we formalize the problem of maintaining the plan library, we introduce criteria for evaluating its quality, and we propose different policies for maintaining it. Such policies are experimentally evaluated and compared using a recent case-based planner, OAKplan [17], and considering some benchmark domains from the International Planning Competitions [10].

2   Preliminaries

In this section, we give the essential background and notation of (classical) planning problem [9], as well as some basic concepts in case-based reasoning and planning. A *planning problem* is a tuple $\Pi = \langle \mathcal{F}, \mathcal{I}, \mathcal{G}, \mathcal{A} \rangle$ where: $\mathcal{F}$ is a finite set of ground atomic propositional formulae; $\mathcal{I} \subseteq \mathcal{F}$ is the set of atoms that are true in the initial state; $\mathcal{G} \subseteq \mathcal{F}$ is a set of literals over $\mathcal{F}$ defining the problem goals; $\mathcal{A}$ is a finite set of actions, where each $a \in \mathcal{A}$ is defined by a set $pre(a) \subseteq \mathcal{F}$ forming the preconditions of a, and a set $eff(a) \subseteq \mathcal{F}$ forming the effects of a.

A *plan* π for a planning problem Π is a partially ordered set of actions of Π. A plan π *solves* Π if the application of the actions in π according to their planned order transforms the initial state to a state $\mathcal{S}_g$ where the goals $\mathcal{G}$ of Π are true ($\mathcal{G} \subseteq \mathcal{S}_g$). Classical generative planning is concerned with finding a solution plan for a given planning problem.

Case-based planning (CBP) is a type of case-based reasoning, exploiting the use of different forms of planning experiences concerning problems previously solved and organized in *cases* forming a case base or *plan library*. The search for a solution plan can be guided by the stored information about previously generated plans in the case base, which may be adapted to become a solution for the new problem. When a CBP system solves a new problem, a new case is generated and possibly added to the library for potential reuse in the future. In order to benefit from remembering and reusing past plans, a CBP system needs efficient methods for retrieving analogous cases and for adapting retrieved plans, as well as a case base of sufficient size and coverage to yield useful analogues.

In our work, we focus on planning cases that are planner-independent and consist of a planning problem Π and a solution plan π of Π. A plan library is a set of cases $\{\langle \Pi_i, \pi_i \rangle \mid i \in \{1, \ldots, N\}\}$, constituting the experience of the planner using this library. In our approach, the relevant information of each library case is encoded using a graph-based representation called planning problem graph (for a detailed description about this representation, the interested reader can see [17]). A case-based planner follows a sequence of steps typical in CBR [1]:

– *Retrieve* - querying the library to identify cases suitable for reuse and select the best one(s) of these;

- *Reuse* - adapting the retrieved plans to solve the new problem;
- *Revise* - testing the validity of the adapted plan in the new context by, e.g., a (simulated) execution of the plan repairing it in case of failures;
- *Retain* - possibly storing the new problem and the corresponding solution plan into the library,

where the first three steps composed the *adaptation phase* and the fourth one the *maintenance phase*. The general CBR schema may differ depending on various implementation choices, e.g., the retrieval phase may provide one or more cases to be reused, or, the reuse may discard the proposed cases for insufficient quality and generate a solution from scratch, i.e., behave like classical generative planner.

3   Related Work

The topic of case base maintenance has been of a great interest in the case-based reasoning community for the last two decades. However, the researchers studying case-based planning have not paid much attention to the problem of case base maintenance yet. Therefore, the related work falls mostly in the field of CBR, where most of the proposed systems handle classification problems.

Leake and Wilson [11] defined the problem of case-based maintenance as "an implementation of a policy to revise the case base to facilitate future reasoning for a particular set of performance objectives". Depending on the evaluation criteria, they distinguish two types of CB maintenance techniques — the *quantitative criteria* (e.g., time) lead to performance-driven policies, while the *qualitative criteria* (e.g. coverage) lead to competence-driven policies.

The quantitative criteria are usually easier to compute; among these policies belong the very simple random deletion policy [13] and a policy driven by a case utility metric [14], where the utility of a case is increased by its frequent reuse and decreased by costs associated with its maintenance and matching.

The most used qualitative criterion corresponds to the notion of "competence" introduced by Smyth [18]. Intuitively, the elements are removed from the case base in reverse order w.r.t. their importance, where the importance of a case is determined by the case "coverage" and "reachability". The two notions capture how many problems the case solves and how many cases it is solved by. Note, however, that differently form our approach to CBP, in his work Smyth considers systems without an underlying generic generative solver. With such a solver the case-based system can solve any problem independently of the quality of the case base, and the notion of competence needs to be reconsidered.

The notion of competence was also used to define footprint deletion and footprint-utility deletion policies [19]. Another extension is the RC-CNN algorithm [20], which compresses the case base using the compressed-nearest-neighbor algorithm and ordering derived by relative coverages of the cases. Furthermore, Leake and Wilson [12] suggested replacing the relative coverage by relative performance of a case. Zhu and Yang [23] however claim that the competence-driven policies of Smyth and his collaborators do not ensure competence preservation. They propose a case addition policy which mimics a greedy algorithm for set covering, adding always the case that has the biggest coverage until the whole original case base is covered or the limit size is reached. The policies we propose in this paper differ from the approach of Zhu and Yang [23] mainly in the condition guiding the selection of the cases to keep in the case base — Zhu and Yang evaluate the utility of each case based on the frequency of its reuse in comparison with the frequency with which its neighbors are reused. Moreover, their policy does not consider the quality with which the original case base is covered, whereas the weighted approach proposed here does. In a sense, we generalize the work of Zhu and Yang and adapt it for using it in the context of planning.

Muñoz-Avila studied the case retention problem in order to filter redundant cases [15], which is closely related to the problem studied here. However, the policy proposed in [15] is guided by the case reuse effort, called *the benefit of the retrieval*, required by a specific "derivational" case-based planner to solve the problem. Intuitively, a case c is kept only if there is no other case in the case base that could be easily adapted by the planner to solve the problem represented by c. In our approach, the decision of keeping a case is independent from adaptation cost of the other cases. Moreover, the policy proposed in [15] can decide only about problems solved by the adopted derivational case-based planner; while the policies studied in this paper are independent from the planner used to generate the solutions of the cases.

4   Case Base Maintenance

The core idea of CBP is providing a complementary approach to traditional generative planning under some assumptions. Coming from the field of case-based reasoning, the world needs to be *regular* and problems need to *recur*. The regularity of the world requires that similar problems have similar solution. Such an assumption obviously links together the similarities between problems and between solutions, which (among others) provides a guarantee that a retrieved case containing a problem similar to the new problem to solve will provide a good solution plan for the reuse.[1] The later assumption (the problems recur) is meant to ensure that a case base contains a good reuse candidate to retrieve.

In addition to the assumptions on the problems and their solutions, there are assumptions coming from the design of the case-based methodology — a case base needs to represent as many various experiences the system has made as possible, while remaining of manageable size. The interplay between these two parameters has a significant impact on the observed performance of the case-based planner, because a too large case base requires a vast amount of time to be queried, whereas even well designed retrieval algorithm fails to provide a suitable case to the reuse procedure if such a case is not present in the case base.

These assumptions seem quite reasonable and not overly restrictive, however, they are not very formal. In the following, we propose some definitions that allow us to formalize the assumptions and to define a maintenance policy.

---

[1] During the retrieval, the system has no information about the solution it is looking for and so it needs to decide solely based on the properties of the problem.

4.1   The Maintenance Problem

The case base maintenance is responsible for preserving and improving the quality of the case base. So far the planning community has focused the research on CBP mostly on the problems related to the reuse and retrieval steps. The retention usually settled with one of the extremes — either maintaining everything or using a pre-built case base which is fixed during the lifetime of the system.

To design a procedure for the maintenance, we start by deciding which parameters of the case base define its quality, and so which criteria should guide the maintenance policy in determining which experiences to keep and which to discard. Obviously, an important criterion is the variety of problems the case base can address, which is also referred to as the case base *competence* [18], and its interplay with the size, or cardinality, of the case base.

However, the notion of competence used in CBR cannot be directly adapted to the use in the planning context. Differently from CBR, where a case usually either can or cannot be adapted to solve a problem, the reuse procedure of planning systems can change any unfit part of the stored solution, or it can even disregard the whole stored solution and attempt to find a new solution from scratch. In modern case-based planners, any case can be used to solve any problem; the system will decide how much the new solution deviates from the stored one, how expensive the reuse is, and therefore, *how useful* the stored solution is. Consequently, we define some criteria for guiding the maintenance.

There are two different kinds of maintenance policies — an *additive policy*, which considers inserting a new case into the case base when a new solution is found/provided, and a *removal policy*, which identifies cases that can be removed without decreasing the quality of the case base too much. Hence, we formalize the general maintenance problem as a two-decision problem.

Definition 1 (Case Base Maintenance Problem)
- *Given a case base $\mathcal{L} = \{c_i \mid c_i = \langle \Pi_i, \pi_i \rangle, i \in \{1, \ldots, n\}\}$, decide for each $i \in \{1, \ldots, n\}$ whether the case c_i should be removed from $\mathcal{L}$.*
- *Given a new case $c = \langle \Pi, \pi \rangle$, $c \notin \mathcal{L}$, decide whether c should be added to $\mathcal{L}$.*

In our work, we focus on the removal maintenance, as we see it as the most critical one — in the absence of a policy to decide which elements to add, we can simply add every new case until the case base reaches a critical size, and then employ the removal maintenance policy to obtain a small case base of good coverage. The alternative approach, which adds "useful" cases until the case base reaches a critical size, may perform not that well, as, differently from the first approach, it needs to estimate the distribution of the future problems, whereas the first approach operates on known past data.

We start by considering when a case can address another problem, or rather *how well* it can do so. Intuitively, planning can be interpreted as a search problem in the space of plans (e.g., [9]), where classical planners start from an empty plan, while case-based planners start from a plan retrieved from the plan library (see, e.g., LPG-Adapt [6]). We can define the *distance* between the stored solution and the solution of the new problem as the minimum number of actions that need to be added/removed in order to convert the stored plan to the new one. Let $\wp$ denote the space of plans for a given planning domain. Then, a distance function $d_a : (\wp \times \wp) \to [0, 1]$ measures the distance of any two plans $\pi_i, \pi_j \in \wp$, where the greater distance indicates greater effort needed during the *adaptation phase*. However, computing such a function can be very hard as, in the worst case, it can be reduced to searching for the solution plan from an empty plan (i.e., to the classical planning problem), which is known to be PSPACE-hard [4]. Therefore we define $d_a(\pi_i, \pi_j)$ as the number of actions that are in π_i and not in π_j plus the number of actions that are in π_j and not in π_i, normalized over the total number of actions in π_i and π_j [22], that is, $d_a(\pi_i, \pi_j) = \frac{|\pi_i - \pi_j| + |\pi_j - \pi_i|}{|\pi_i| + |\pi_j|}$, unless both plans are empty, in which case their distance is 0.

Clearly, if the case-based system needs to revert to an empty plan and search from there, the provided case is not considered useful. Hence, we can say that a case can be useful to solve a problem if the distance between the corresponding plans is not bigger than the distance from an empty plan. The distance from the empty plan estimates the effort the case-based system needs to spend in order to generate a solution from scratch, which is equivalent to the estimated work a generative planner requires in order to find a solution as the reuse procedures often perform the same kind of search as generative planners do. Consequently, it is not worth trying to reuse more distant plans than the empty one.

Definition 2. *Given a finite value $\delta \in \mathbb{R}$, we say that a case $c_i = \langle \Pi_i, \pi_i \rangle$ can be useful to solve problem Π, that is $\mathbf{addresses}_\delta(c_i, \Pi)$, if there exists a solution plan π for Π such that $d_a(\pi_i, \pi) < \delta$.*

Note that the definition of $addresses_\delta(c_i, \Pi)$ heavily relies on the distance between the solutions, and completely disregards the relation of the relative problems. However, also the structural properties of the problems play a considerable role, as the case retrieval step is based on the planning problem descriptions. Therefore, we also use a distance function d_r that is intended to reflect the similarity of the problems. Let $\mathcal{P}$ denote the space of problems in a given planning domain, $\Pi \in \mathcal{P}$ be a new problem, and $\Pi' \in \mathcal{P}$ be a problem previously solved. Assuming that the matching between the objects of Π and Π' has been already performed, and that $\mathcal{I}' \cup \mathcal{G} \neq \emptyset$, a problem distance function $d_r : (\mathcal{P} \times \mathcal{P}) \to [0, 1]$ is defined as follows [17]: $d_r(\Pi', \Pi) = 1 - \frac{|\mathcal{I}' \cap \mathcal{I}| + |\mathcal{G}' \cap \mathcal{G}|}{|\mathcal{I}| + |\mathcal{G}|}$, where $\mathcal{I}$ and $\mathcal{I}'$ ($\mathcal{G}$ and $\mathcal{G}'$) are the initial states (sets of goals) of Π and Π', respectively. If $\mathcal{I}' \cup \mathcal{G} = \emptyset$, $d_r(\Pi', \Pi) = 0$.

The smaller distance d_r between two problems is, the more similar they are; consequently, by the regular world assumption, they are more likely to have similar solutions, and so it is useful to retrieve from the case base the case for a problem that is mostly similar to the problem to solve. We can say that d_r *guides the retrieval phase* while d_a *estimates the plan adaptation effort*. The maintenance policy should consider both distances, in order not to remove important cases, but also to support the retrieval process. Therefore, we combine the two functions, obtaining distance function $d : ((\mathcal{P} \times \wp) \times (\mathcal{P} \times \wp)) \to [0, 1]$ measuring distance between cases. The combination of d_r and d_a allows us to assign different importance to the similarity of problems and their solutions, depending on the application requirements.[2]

The assumption of regular world presented at the beginning of this section was using a notion of similarity between problems and solutions, neither providing details on how the similarity should be interpreted, nor specifying *which* solutions are considered. This is an important concern when a problem may have several significantly different solutions. We formalize this assumption keeping the notion of similarity undetailed to preserve the generality of the definition, but establishing the quantification over the solutions as follows:

Definition 3 (Regular World Assumption). *If Π and $\Pi' \in \mathcal{P}$ be two similar planning problems of a planning domain with plan space $\wp$ such that $\Pi \neq \Pi'$. If the world is **regular**, then $\forall \pi \in \wp$ that is a solution for Π $\exists \pi' \in \wp$ that is a solution for Π' such that π and π' are similar, and $\forall \pi' \in \wp$ that is a solution for Π' $\exists \pi \in \wp$ that is a solution for Π such that π' and π are similar.*

We interpret problem and plan similarity using the distance functions d_r and d_a. Specifically, we consider two problems Π, Π' similar iff $d_r(\Pi, \Pi') < \delta$, and two solutions π, π' similar iff $d_a(\pi, \pi') < \delta'$, where δ and δ' are reals whose specific values depend on the particular distance function and on the planning domain considered. Next, we formalize the notion of problem recurrence:

Definition 4 (Recurring Problems Assumption). *For every new problem a case-based planner encounters, it is likely that a similar problem has been already encountered and solved.*

Case-based planning relies on the two assumptions of Def. 3-4 to be fulfilled. If that indeed is the case, since in our approach every encountered problem is simply added to the case base, when a new problem Π is encountered, the case base likely contains a case $c_i = \langle \Pi_i, \pi_i \rangle$ such that $addresses_\delta(c_i, \Pi)$ holds for some (small) value of δ, and, by Def. 3, the case-based planning system can produce solutions similar to the previous ones for Π by reusing those. Moreover, by Def. 4, it can be expected that the cases in the case base create groups of elements, that we call case *clusters*, similar to each other and that could be reduced to smaller groups without significant loss of information.

Definition 4 does not define *how much* likely similar problems are assumed to be encountered. This means that there can be different degrees of problem recurrence. In the strongest case, all new encountered problems are similar to a problem in the case base. Different degrees of problem recurrence lead to differently structured plan libraries in terms of case clusters, which can affect the performance of a plan library maintenance policy exploiting them.

In our work, case similarity is interpreted by means of the distance function d, i.e., c is similar to c' iff $d(c, c') < \delta''$, where a specific value of $\delta'' \in \mathbb{R}$ depends on the specific implementation choices as well as on the domain.

---

[2] We use a linear combination $d = \alpha \cdot d_r + (1 - \alpha) \cdot d_a$, where $\alpha = 0.5$ for domains that exhibit very regular behavior (e.g. Logistics), while we used $\alpha = 1$, i.e., $d = d_r$, for domains where the solutions generated for the case base using planner TLplan [2] differed even for similar problems (e.g. ZenoTravel), being quite irregular w.r.t. d_a.

4.2   Maintenance Policies

Instead of maximizing competence as an absolute property of a case base, the maintenance is guided by minimizing the amount of knowledge that is lost in the maintenance process, where removing a case from the library implies losing the corresponding knowledge, unless the same information is contained in some other case. The following notions of case covering and case base coverage are defined to capture this concept:

Definition 5. *Given a case base $\mathcal{L}$ and a case distance threshold $\delta \in \mathbb{R}$, we say that a case $c_i \in \mathcal{L}$ covers a case $c_j \in \mathcal{L}$, that is, **covers**(c_i, c_j), if $d(c_i, c_j) \leq \delta$.*

Definition 6. *Let $\mathcal{L}, \mathcal{L}'$ denote two case bases and let C denote the set of all cases in $\mathcal{L}$ that are covered by the cases in $\mathcal{L}'$, i.e., $C = \{c_i \in \mathcal{L} \mid \exists c_i' \in \mathcal{L}', \text{covers}(c_i', c_i)\}$. The coverage of $\mathcal{L}'$ over $\mathcal{L}$, denoted **coverage**$(\mathcal{L}', \mathcal{L})$, is defined as $\frac{|C|}{|\mathcal{L}|}$.*

We can now formally define the outcome of an algorithm addressing the plan library maintenance problem — it should be a case base $\mathcal{L}'$ that is smaller than the original case base $\mathcal{L}$, but that contains very similar experiences. Under such conditions, we say that $\mathcal{L}'$ reduces $\mathcal{L}$:

Definition 7. *Case base $\mathcal{L}'$ reduces case base $\mathcal{L}$, denoted as **reduces**$(\mathcal{L}', \mathcal{L})$, if and only if $\mathcal{L}' \subseteq \mathcal{L}$ and $coverage(\mathcal{L}', \mathcal{L}) = 1$.*

In the previous definition, we may set additional requirements on $\mathcal{L}'$ to find a solution that is optimal in some ways. For example, we may want to minimize the size of $\mathcal{L}'$, or we may try to maximize the quality of the coverage. The structure of the policy remains the same — it constructs $\mathcal{L}'$ by selecting the cases that satisfy a certain *condition optimizing* $\mathcal{L}'$. Such a condition corresponds to a specific criterion the maintenance policy attempts to optimize.

Random Policy [18]. This policy reduces the case base by randomly removing cases [13], which is easy to implement and fast to compute. However, the coverage of the reduced case base $\mathcal{L}'$ over the original case base $\mathcal{L}$ cannot be guaranteed.

Distance-Guided Policy. Due to the assumption of recurring problems, we expect that the problems in the library can be grouped into sets of problems that are similar (close in the sense of d_r) to each other. Consequently, by the assumption of regular world, for a problem Π' there exists a solution π' that is similar to the solution π of a stored case $c = \langle \Pi, \pi \rangle$ where Π is similar to Π'. Case $c' = \langle \Pi', \pi' \rangle$ is similar to c (close in the sense of d) and its inclusion in the case base introduces some redundancy because of its similarity with c.

We propose a distance-guided policy that attempts to remove the cases that are mostly redundant. Intuitively, these cases are those having their distance from other cases too small. In particular, the distance-guided policy identifies the cases to remove by exploiting the notion of *average minimum distance* δ_μ in the case base. Given a case $c_i \in \mathcal{L}$, the minimum distance case c_i^* of c_i is a case in $\mathcal{L}$ such that $d(c_i, c_i^*) < d(c_i, c_j), \forall c_j \in \mathcal{L} \setminus c_i^*$. The distance guided policy keeps a case c_i in the case base if and only if $d(c_i, c_i^*) \geq \delta_\mu$, where δ_μ is defined as follows: $\delta_\mu = \Sigma_{c_i \in \mathcal{L}} \frac{d(c_i, c_i^*)}{|\mathcal{L}|}$.[3]

The distance-guided policy is clearly better informed than the random policy, and it can recognize cases of high importance for the coverage of the case base (e.g., isolated elements that are dissimilar to any other case). The better information is however reflected by increased computational complexity – the distance-guided policy needs to consider the distance between all pairs of cases in order to find the closest one; therefore it requires quadratic number of distance evaluations, resulting in run-time of $\mathcal{O}(|\mathcal{L}|^2 \cdot t_d)$, where t_d denotes the time needed to compute the distance between two cases.

Coverage-Guided Policy. The distance-guided policy can preserve the knowledge in the case base better than the random policy does. However, it is not optimal, as some information is missed when only pairs of cases are considered. We generalize the approach by considering *all* the cases that may contain redundant information at once. For that we define the notion of neighborhood of a case c with respect to a certain similarity distance value δ, denoted $n_\delta(c)$.

The idea of the case neighborhood is to group elements which contain redundant information and hence that can be reduced to a single case. The case neighborhood uses a value of δ in accordance with Def. 5. Note that such a value, together with the structure and distribution of the cases in the case base, influences the cardinality of the case neighborhoods and therefore determines the amount by which $\mathcal{L}$ can be reduced.

Definition 8 (Case neighbourhood). *Given a case base $\mathcal{L}$, a case $c \in \mathcal{L}$ and a similarity distance threshold $\delta \in \mathbb{R}$, the neighborhood of c is $n_\delta(c) = \{c_i \in \mathcal{L} \mid d(c, c_i) < \delta\}$.*

The Coverage-Guided policy is concerned with *finding a set $\mathcal{L}'$ of cases such that the union of all their neighborhoods covers all the elements of the given case base $\mathcal{L}$*, or, using the terminology of Def. 7, finding a case base $\mathcal{L}'$ such that $reduces(\mathcal{L}', \mathcal{L})$ holds.

There are many possible ways to reduce a case base in accordance with this policy, out of which some are more suitable than others. We introduce two criteria for reducing the case base that we observed can significantly influence the performance of a case-based system adopting the coverage-guided policy: *minimizing the size* of the reduced case base, which has a significant impact on the efficiency of the retrieval phase, and *maximizing the quality* of the coverage of the reduced case base, which influences the adaptation costs. Considering the first criterion, the optimal result of the coverage-guided policy takes account of the number of elements in the reduced set:

Definition 9 (Cardinality Coverage-Guided Policy). *Given a similarity threshold value $\delta \in \mathbb{R}$ and a case base $\mathcal{L}$, find a reduction $\mathcal{L}'$ of $\mathcal{L}$ with minimal cardinality.*

---

[3] The isolated cases are excluded in the computation of δ_μ. A case c_i is considered isolated if distance $d(c_i, c_i^*) = 0.5$.

Algorithm: CoverageBasedPolicy$(\mathcal{L}, \delta)$

Input: a case base $\mathcal{L} = \{c_i \mid i \in 1 \leq i \leq n\}$, a threshold $\delta \in \mathbb{R}$.[5]
Output: a case base $\mathcal{L}'$ reducing $\mathcal{L}$

1. $\mathcal{L}' \leftarrow \emptyset$;
2. $Uncovered \leftarrow \mathcal{L}$;
3. **repeat**
4.   select $c_i \in Uncovered$ that satisfies $condition(c_i)$;
5. $Uncovered \leftarrow Uncovered \setminus n_\delta(c_i)$;
6. $\mathcal{L}' \leftarrow \mathcal{L}' \cup \{c_i\}$;
7. **until** $Uncovered = \emptyset$;
8. **return** $\mathcal{L}'$;

Fig. 1. A greedy algorithm computing a Coverage-Based Policy approximation

Concerning the second criterion, consider three cases c, c_1, c_2 so that $d(c, c_1) < d(c, c_2) < \delta$. By Def. 5, c covers both c_1 and c_2, however, the expected adaptation cost of c_1 is lower than the cost of c_2, and therefore c_1 is better covered. The quality of the case base coverage can intuitively be defined as the average distance from the removed cases to the closest kept case (*average coverage distance*). Regarding the coverage quality, the optimal result of the coverage-guided policy is a case base $\mathcal{L}'$ reducing $\mathcal{L}$ with minimal average coverage distance. Note, however, that if only the coverage distance was considered, then $\mathcal{L} = \mathcal{L}'$ would be a special case of optimal reduced case base. Therefore, the quality measure to optimize needs to be more complex in order to take account of the size of the reduced case base. In particular, given a reduction $\mathcal{L}'$ of $\mathcal{L}$, we define the *uncovered neighborhood* $U_\delta(c)$ of an element $c \in \mathcal{L}$ as its neighbors in $\mathcal{L} \setminus \mathcal{L}'$, i.e., $[U_\delta(c) = \{c_j \in \mathcal{L} \mid c_j \in \{n_\delta(c) \cap \mathcal{L} \setminus \mathcal{L}'\} \cup \{c\}\}$. Then, we define the cost of a case c as a real function $v_\delta(c) = \left(\frac{\Sigma_{c_j \in U_\delta(c)} d(c, c_j)}{|U_\delta(c)|} + p \right)$. The first term within the brackets indicates the average coverage distance of the uncovered neighbors; the second term, $p \in \mathbb{R}$, is a penalization value that is added in order to favorite reduced case bases with fewer elements and to assign a value different from 0 also to isolated cases in the case base.[4] The sum of these costs for all the elements of a reduced set $\mathcal{L}'$ defines the cost $\mathcal{M}_\delta(\mathcal{L}')$ of $\mathcal{L}'$, i.e., $\mathcal{M}_\delta(\mathcal{L}') = \Sigma_{c \in \mathcal{L}'} v_\delta(c)$. The policy optimizing the quality of the case base coverage can then be defined as follows:

Definition 10 (Weighted Coverage-Guided Policy). *Given a similarity threshold value $\delta \in \mathbb{R}$ and a case base $\mathcal{L}$, find a reduction $\mathcal{L}'$ of $\mathcal{L}$ that minimizes $\mathcal{M}_\delta(\mathcal{L}')$.*

Unfortunately, computing the reduction of Def. 10 can be computationally very expensive. Therefore, we propose to compute an approximation of the reduced case base of this policy using the greedy algorithm described by Fig. 1. This algorithm has two variants that depend on how line 4 is implemented and corresponds to the two proposed versions of the coverage-guided policy. For the

---

[4] In our experiments, we use $p = max_{c_i \in \mathcal{L}} d(c_i, c_i^*)$.
[5] If the δ value is not provided, the algorithm uses the average minimum distance (δ_μ).

Cardinality Coverage-Guided Policy, the *condition* test at line 4 of the algorithm is used to select the uncovered element c_i with greatest $|U_\delta(c_i)|$ in order to maximize the number of uncovered elements in $n_\delta(c_i)$ that can be covered by inserting c_i into $\mathcal{L}'$. While, for the Weighted Coverage-Guided Policy, in order to optimize the quality of the reduced case base $\mathcal{L}'$, the condition of line 4 is used to select the uncovered element c_i with the minimum $\frac{v_\delta(c_i)}{|U_\delta(c_i)|}$ value, where the $v_\delta(c_i)$ value is scaled down by $|U_\delta(c_i)|$ to favor the cases that cover higher number of still-uncovered elements.

5   Experimental Results

The policies presented in the previous sections have been implemented in a new version of the CBP system OAKPlan [17]. In our experiments, the plan retrieved by OAKPlan is adapted using planner LPG-Adapt [7]. The benchmark domains considered in the experimental analysis are the available domains DriverLog, Logistics, Rovers, and ZenoTravel from the 2nd, 3rd and 5th International Planning Competitions. [10]

For each considered domain we generated a plan library with $\sim$5000 cases. Specifically, each plan library contains a number of case clusters ranging from 34 (for Rovers) to 107 (for ZenoTravel), each cluster c is formed by using either a large-size competition problem or a randomly generated problem $\overline{\Pi}_c$ (with a problem structure similar to the large-size competition problems) plus a random number of cases ranging from 0 to 99 that are obtained by changing $\overline{\Pi}_c$. Problem $\overline{\Pi}_c$ was modified either by randomly changing at most the 10% of the literals in its initial state and set of goals, or adding/deleting an object to/from the problem. The solution plans of the planning cases were computed by planner TLPlan [2]. TLPlan exploits domain-specific control knowledge to speedup the search significantly, so that large plan libraries can be constructed by using a relatively small amount of CPU time. In our libraries, plans have a number of actions ranging from 68 to 664. For each considered domain, we generated 25 test problems, each of which derived by (randomly) changing problem of a cluster randomly selected among those in the case base. Note that the cases in the library are grouped into clusters and test problems were generated from the library problems because the aim of our experimental analysis is studying the effectiveness of the proposed techniques for domains with recurring problems. We experimentally compared nine specific maintenance policies:

- three random policies, R_{50}, R_{75}, and R_{90}, that remove a case from the full plan library with probability 0.50, 0.75, and 0.90, respectively;
- three distance-guided policies, D_1, D_2, and D_3, that remove the mostly redundant cases from (a) the full plan library, (b) the library obtained from D_1, and (c) the library obtained from D_2, respectively;
- three coverage-guided policies, C_1, C_2, and C_3, that compute a reduced case base by using the greedy algorithm in Fig. 1 with (a) the full plan library, (b) the library obtained from D_1, and (c) the library obtained from D_2, respectively.

Table 1. Evaluation of nine reduced plan libraries. Gray boxes indicate the best results.

Domain	Random policy			Distance-Guided policy			Coverage-Guided policy		
	R_{50}	R_{75}	R_{90}	D_1	D_2	D_3	C_1	C_2	C_3
DriverLog									
Case-base size	2617	1368	566	3152	2253	1727	2318	1222	684
Coverage	0.776	0.628	0.501	0.972	0.857	0.733	0.956	0.681	0.502
#Uncovered	1177	1957	2623	146	754	1404	231	1679	2617
Avg. uncov. dist.	0.067	0.112	0.171	0.017	0.035	0.058	0.022	0.069	0.130
Logistics									
Case-base size	2615	1274	462	2826	1443	659	2767	1283	460
Coverage	0.888	0.767	0.572	1	0.996	0.874	1	0.999	0.862
#Uncovered	583	1213	2226	0	21	658	0	5	720
Avg. uncov. dist.	0.036	0.064	0.103	0.018	0.043	0.057	0.018	0.042	0.059
Rovers									
Case-base size	2107	1012	518	2130	1358	1165	1758	1018	720
Coverage	0.599	0.358	0.227	0.672	0.476	0.387	0.586	0.375	0.294
#Uncovered	1730	2770	3336	1416	2261	2646	1786	2696	3044
Avg. uncov. dist.	0.131	0.218	0.295	0.065	0.120	0.156	0.080	0.150	0.210
ZenoTravel									
Case-base size	2493	1242	479	2718	1729	1240	2588	1205	538
Coverage	0.989	0.959	0.873	1	1	0.993	1	0.999	0.999
#Uncovered	56	202	632	0	0	36	0	1	1
Avg. uncov. dist.	0.027	0.046	0.067	0.014	0.027	0.036	0.015	0.031	0.042

Table 1 compares the reduced plan libraries by using the nine considered maintenance policies in terms of size, coverage (using $\delta = 0.1$) w.r.t. the full plan library, number of elements of the full plan library that are not covered by the reduced plan libraries, and average distance from any uncovered case to the closest case in the reduced plan library. Obviously, the closer the coverage is to 1, or, equally, the lower the number of uncovered cases is, the better the maintenance policy is. Moreover, since a high-quality policy should remove only redundant cases, lower values of the average minimum distance from the uncovered cases indicates better plan libraries.

While the size of the case bases obtained by C_1 and D_1 is often comparable with the case base obtained by using R_{50}, C_1 and D_1 are always better (and usually much better) than R_{50} in terms of coverage, number of uncovered elements and average minimum distance from the uncovered cases. Similarly, while the sizes of the case bases are often comparable, C_2 and D_2 are better than R_{75}, and C_3 and D_3 are better than R_{90}. The results in Table 1 also confirm the fact that the random policy may remove important cases, since, for instance, the number of uncovered elements is often high, while the other policies can compute reduced case bases of comparable size but with fewer uncovered elements. Moreover, it is interesting to note that the case bases with the best coverage are computed by D_1, although those obtained by C_1 have a similar coverage while contain fewer cases. For ZenoTravel, even if the case bases obtained through the distance-guided policies contain many more cases, the coverage of the case bases obtained by the coverage-guided policies is similar to or better than by the distance-guided policies.

Table 2. Performance of OAKPlan. Gray boxes indicate the best results.

Domain	Full library	Random policy			Distance-Guided policy			Coverage-Guided policy		
		R_{50}	R_{75}	R_{90}	D_1	D_2	D_3	C_1	C_2	C_3
DriverLog										
Avg. CPU seconds	19.8	11.1	6.7	5.7	12.9	9.4	7.5	9.6	5.6	3.3
Speed score	3.76	6.83	11.93	20.39	5.90	8.08	10.01	7.86	14.30	22.70
Avg. plan stability	0.801	0.817	0.787	0.751	0.818	0.823	0.811	0.812	0.829	0.839
Logistics										
Avg. CPU seconds	34.4	22.2	22.2	15.0	21.5	13.9	10.2	21.3	12.7	8.9
Speed score	6.13	9.83	13.16	17.62	10.42	15.84	22.44	10.55	17.31	24.88
Avg. plan stability	0.953	0.945	0.920	0.912	0.952	0.950	0.950	0.952	0.952	0.952
Rovers										
Avg. CPU seconds	160.3	82.8	44.5	78.0	67.3	27.0	19.7	53.3	18.6	17.6
Speed score	3.54	6.27	10.36	15.02	7.58	15.73	20.25	8.40	21.76	22.30
Avg. plan stability	0.975	0.956	0.953	0.870	0.970	0.969	0.969	0.969	0.970	0.969
ZenoTravel										
Avg. CPU seconds	28.2	29.8	21.9	21.7	22.7	21.3	20.6	22.7	19.3	17.1
Speed score	13.19	12.62	16.66	16.92	16.19	17.10	17.94	16.13	20.16	24.14
Avg. plan stability	0.892	0.853	0.866	0.831	0.892	0.882	0.877	0.897	0.897	0.897

Table 2 shows the performance of planner OAKPlan using the full plan library and the reduced libraries derived by the nine considered maintenance policies for the considered domains DriverLog, Logistics, Rovers and ZenoTravel in terms of average CPU seconds, average plan stability and IPC speed score (defined below). Given a library plan π' and a new plan π computed for solving a test problem, the *plan stability* of π with respect to π' can be defined as $1 - d_a(\pi, \pi')$, where d_a is the plan distance function defined in Section 4.1. Having a high value of plan stability can obviously be very important in plan adaptation, because, e.g., high stability reduces the cognitive load on human observers of a planned activity by ensuring coherence and consistency of behaviors [6]. A good reduced case base should allow the planner to produce stable solutions. Given two compared policies and a problem set, the average CPU time and plan stability for each policy is computed over the test problems in the set that are solved by both the compared policies.

The *speed score function* was first introduced and used by the organizers of the 6th International Planning Competitions [5] for evaluating the relative performance of the competing planners, and since then it has been a standard method for comparing planning systems performances. The speed score for a maintenance policy m is defined as the sum of the speed scores assigned to m over all the considered test problems. The speed score for m with a planning problem Π is defined as: 0 if Π is unsolved using policy m and $T_{\Pi}^*/T(m)_{\Pi}$ otherwise, where T_{Π}^* is the lowest measured CPU time to solve problem Π and $T(m)_{\Pi}$ denotes the CPU time required to solve problem Π using the case base reduced through policy m. Higher values of the speed score indicate better performance.

The results in Table 2 indicate that OAKPlan using the libraries reduced through the compared distance-guided and coverage-guided policies is always faster than using the full library. Moreover, even the use of the simple random policies makes OAKPlan almost always faster than using the full library.

Concerning plan stability, the plans computed using the libraries reduced by the distance-guided and the coverage-guided policies are always on average as much stable stable as the plans computed using the full library. Surprisingly, for DriverLog and ZenoTravel, OAKPlan with policy C_3 computes plans that are even more stable than with the full library. The rationale of this is related to the use of LPG-Adapt in OAKPlan: since LPG-Adapt is based on a stochastic local search algorithm, it may happen that LPG-Adapt computes plans that are far from the library plans even if there exist solution plans similar to some of them.

OAKPlan using C_1 and D_1 is always on average faster than using R_{50}, while the size of the case bases is often comparable, except for D_1 and domain DriverLog. For DriverLog, OAKPlan using D_1 is slower than using R_{50}, because the library reduced by D_1 is much bigger than the one reduced by R_{50}. Moreover, OAKPlan using C_1 and D_1 computes plans that are almost always on average more stable than using R_{50}. The performance gaps of C_2 and D_2 w.r.t. R_{75}, and of C_3 and D_3 w.r.t. R_{90} are similar. Finally, in terms of average CPU time, speed score, and average plan distance, the coverage-guided policies perform almost always better than, or similarly to, the distance-guided policies.

6   Conclusion

In this work, we have addressed the problem of maintaining a plan library for case-based planning by proposing and experimentally evaluating some maintenance policies of the case base. The investigated policies optimize different quality criteria of the reduced case base.

The random policy, that is also used in general case-based reasoning, does not optimize any criterion but is very fast to compute. We have introduced two better informed policies, the distance-guided and the coverage-guided policies, which attempt to generate reduced case bases of good quality. Since computing such policies can be computationally hard, we have proposed a greedy algorithm for effectively computing an approximation of them. An experimental analysis shows that these approximated policies can be much more effective compared to the random policy, in terms of quality of the reduced case base and performance of a case-base planner using them.

There are several research directions to extend the work presented here. We intend to study in detail additional distance functions to assess the similarity between problems and solutions, to develop and compare additional policies, to investigate alternative methods for efficiently computing good policy approximations, and to extend the experimental analysis with a larger set of benchmarks. Moreover, current work includes determining the computational complexity of the two proposed (exact) coverage-guided policies, that we conjecture are both NP-hard.

References

1. Aamodt, A., Plaza, E.: Case-based reasoning: foundational issues, methodological variations, and system approaches. AI Communications 7(1), 39–59 (1994)
2. Bacchus, F., Kabanza, F.: Using temporal logic to express search control knowledge for planning. Artificial Intelligence 116(1-2), 123–191 (2000)
3. Bäckström, C., Chen, Y., Jonsson, P., Ordyniak, S., Szeider, S.: The complexity of planning revisited – a parameterized analysis. In: 26th AAAI Conf. on AI (2012)
4. Bäckström, C., Nebel, B.: Complexity results for SAS+ planning. Computational Intelligence 11, 625–655 (1996)
5. Fern, A., Khardon, R., Tadepalli, P.: The first learning track of the int. planning competition. Machine Learning 84(1-2), 81–107 (2011)
6. Fox, M., Gerevini, A., Long, D., Serina, I.: Plan stability: Replanning versus plan repair. In: 16th Int. Conf. on AI Planning and Scheduling (2006)
7. Gerevini, A., Saetti, A., Serina, I.: Planning through stochastic local search and temporal action graphs. JAIR 20, 239–290 (2003)
8. Gerevini, A., Saetti, A., Serina, I.: Case-based planning for problems with real-valued fluents: Kernel functions for effective plan retrieval. In: 20th European Conf. on AI (2012)
9. Ghallab, M., Nau, D.S., Traverso, P.: Automated planning - theory and practice. Elsevier (2004)
10. Koenig, S.: Int. planning competition (2013), http://ipc.icaps-conference.org/
11. Leake, D.B., Wilson, D.C.: Categorizing case-base maintenance: Dimensions and directions. In: Smyth, B., Cunningham, P. (eds.) EWCBR 1998. LNCS (LNAI), vol. 1488, pp. 196–207. Springer, Heidelberg (1998)
12. Leake, D.B., Wilson, D.C.: Remembering why to remember: Performance-guided case-base maintenance. In: Blanzieri, E., Portinale, L. (eds.) EWCBR 2000. LNCS (LNAI), vol. 1898, pp. 161–172. Springer, Heidelberg (2000)
13. Markovitch, S., Scott, P.D., Porter, B.: Information filtering: Selection mechanisms in learning systems. In: 10th Int. Conf. on Machine Learning, pp. 113–151 (1993)
14. Minton, S.: Quantitative results concerning the utility of explanation-based learning. Artificial Intelligence 42(2-3), 363–391 (1990)
15. Muñoz-Avila, H.: Case-base maintenance by integrating case-index revision and case-retention policies in a derivational replay framework. Computational Intelligence 17(2), 280–294 (2001)
16. Richter, S., Westphal, M.: The lama planner: Guiding cost-based anytime planning with landmarks. JAIR 39, 127–177 (2010)
17. Serina, I.: Kernel functions for case-based planning. Artificial Intelligence 174(16-17), 1369–1406 (2010)
18. Smyth, B.: Case-base maintenance. In: Mira, J., Moonis, A., de Pobil, A.P. (eds.) IEA/AIE 1998. LNCS, vol. 1416, pp. 507–516. Springer, Heidelberg (1998)
19. Smyth, B., Keane, M.T.: Adaptation-guided retrieval: Questioning the similarity assumption in reasoning. Artificial Intelligence 102(2), 249–293 (1998)
20. Smyth, B., McKenna, E.: Footprint-based retrieval. In: Althoff, K.-D., Bergmann, R., Branting, L.K. (eds.) ICCBR 1999. LNCS (LNAI), vol. 1650, p. 343. Springer, Heidelberg (1999)
21. Spalazzi, L.: A survey on case-based planning. AI Review 16(1), 3–36 (2001)
22. Srivastava, B., Nguyen, T.A., Gerevini, A., Kambhampati, S., Do, M.B., Serina, I.: Domain independent approaches for finding diverse plans. In: 20th Int. Joint Conf. on AI (2007)
23. Zhu, J., Yang, Q.: Remembering to add: Competence-preserving case-addition policies for case-base maintenance. In: 16th Int. Joint Conf. on AI (1998)

Learning Feature Weights from Positive Cases

Sidath Gunawardena, Rosina O. Weber, and Julia Stoyanovich

The iSchool, Drexel University, Philadelphia, Pennsylvania, USA
{sidath.gunawardena,rosina,stoyanovich}@drexel.edu

Abstract. The availability of new data sources presents both opportunities and challenges for the use of Case-based Reasoning to solve novel problems. In this paper, we describe the research challenges we faced when trying to reuse experiences of successful academic collaborations available online in descriptions of funded grant proposals. The goal is to recommend the characteristics of two collaborators to complement an academic seeking a multidisciplinary team; the three form a collaboration that resembles a configuration that has been successful in securing funding. While seeking a suitable measure for computing similarity between cases, we were confronted with two challenges: a problem context with insufficient domain knowledge and data that consists exclusively of successful collaborations, that is, it contains only positive instances. We present our strategy to overcome these challenges, which is a clustering-based approach to learn feature weights. Our approach identifies poorly aligned cases, i.e., ones that violate the assumption that similar problems have similar solutions. We use the poorly aligned cases as negatives in a feedback algorithm to learn feature weights. The result of this work is an integration of methods that makes CBR useful to yet another context and in conditions it has not been used before.

Keywords: Case Alignment, Case Cohesion, Density Clustering, Multidisciplinary Collaboration, Recommender Systems, Single Class Learning, Subspace Clustering.

1 Introduction

Case-based Reasoning (CBR) enables the reuse of experiences to perform a variety of reasoning tasks based on learning from a collection of those experiences. Advances in information technology are increasing the types and quantities of experiences that are available, providing new avenues for CBR applications [23]. Sometimes new data is made available that poses novel challenges to using the CBR methodology.

One crucial step in adopting CBR is to design a similarity measure that will support the most accurate solutions possible. As widely discussed in the literature (e.g., [2]), the quality of the CBR solution depends on accurately representing the relative relevance of the features used to represent cases.

The two main approaches for assigning feature weights are domain knowledge and feedback algorithms. The algorithms use feedback to adjust feature weights such that cases from the same class are made more similar and cases from different classes are made less similar [2]. As is typical of learning algorithms, a dense dataset containing

S.J. Delany and S. Ontañón (Eds.): ICCBR 2013, LNAI 7969, pp. 134–148, 2013.

both positive and negative instances is required. In this problem context the data has only positive instances posing the challenge of how to learn feature weights.

The goal of this paper is to enable the use of CBR in the absence of ideal conditions to adopt it, i.e., when domain knowledge is insufficient and only positive instances are available to learn feature weights. We explore the feasibility of using feedback algorithms with only positive instances. Our approach considers cases based on how they are distributed in the problem and solution spaces.

Our strategy is based on the premise that a problem context suitable for CBR is one where similar problems have similar solutions. This has been explored through measures such as alignment [21] and cohesion [16]. Likewise, our approach seeks to identify cases that are well aligned versus cases that are poorly aligned. In our approach we use the well and poorly aligned cases to play the respective roles of positive and negative instances to learn feature weights with feedback algorithms.

To identify these cases, we employ clustering methods based on the intuition that the difference between well and poorly aligned cases is revealed when outliers are identified in the problem and solution spaces. In [12] we show that our clustering-based approach consistently identifies poorly aligned cases that are low in alignment [21] and cohesion [16].

We organize this paper as follows. We start by explaining the problem that motivates this research in Section 2. The following sections describe the steps of our investigation. As these sections vary greatly in their content, we include in each section related and background work instead of having one single section. Then, in Section 3, we describe our approach to *learning feature weights from only positive cases*. In Section 4, we present measures of case alignment and compare to our approach for finding well and poorly aligned cases. In Section 5, we discuss the clustering methods for determining aligned cases. In Section 6, we present studies with different case bases in support of the quality and generalization of our approach. In Section 7, we implement our approach on case bases that have negatives to evaluate how the negatives learned from our approach performs in comparison. In Section 8, we investigate Single Class Learning and how it performs in comparison with and in complement to our approach. Section 9 summarizes, concludes, and presents future work.

2   Motivating Problem

This section describes the motivation for this work in detail: the dataset, the problem context, case representation, and evaluation.

2.1   Recommending Characteristics of Academic Collaborators

The problem context is a user (i.e., collaboration seeker) who is an academic seeking to engage in multidisciplinary research. The solution is a configuration (i.e., set of features) of two collaborators that, together with the seeker, will form a collaboration. This recommendation of multiple members differentiates this problem from similar problems in group recommendation [24].

For a new collaboration seeker, the task is to find the most similar member of an existing collaboration, and then replace that member with the seeker to create a recommendation. These three form a collaboration that resembles a configuration that has been successful in securing funding. The experiences of successful multidisciplinary collaborations we use are grants that were awarded funding. A detailed description of how the grant case base was assembled can be found in [10].

A particularity of such data is that it does not include grants that were not funded. Thus, the experiences are all positive instances of collaboration. It should be noted that there is no evidence that a collaboration would not be successful simply because it is not in the data.

Furthermore, these experiences are not in traditional problem-solution form of many recommender systems [3, 4]. An n-member collaboration can take the form of a case by identifying each member in turn as the problem part of the case (the collaboration seeker). When the seeker is the problem, the remaining n-1 members comprise the solution. As each collaboration can be perceived from the perspective of each member, an n-member collaboration can be reconfigured as m cases, where n = m. Each collaboration produces as many cases as the number of its members. For the purpose of this work, we consider collaborations of three members only. Thus, for a 3-member collaboration the number of cases created is three. Our dataset has 66 three-member collaborations, and thus is transformed into 198 cases in the case base (henceforth referred to as the collaboration case base).

Table 1. A Collaboration Case

Features	Collaboration members	Feature Name	Description	Example values
Problem Features	Collaboration Seeker	Title	Academic Title	{Full Professor, Associate Professor, Assistant Professor}
		Research Interest	Areas of research	Genetics, Mechanical Engineering, Economics, …
		Institution Type	Highest degree granted by inst.	{Doctoral, Masters, Bachelors}
Solution Features	Recommended Characteristics Collaborator #1	Title, Research Interest, Inst. Type	Same as problem features	
	Recommended Characteristics Collaborator #2	Title, Research Interest, Inst. Type	Same as problem features	

The features used in the experiments are title, research interest and institution type as shown in (**Table** 1). The recommendation consists of the characteristics of two collaborators who, when combined with the seeker, create a combination of these features that are consistent with successful collaborations. The selection of these three features to describe this problem is discussed in [13]. We next describe how we evaluate the quality of the recommended solutions.

2.2   Quality of Solutions

Given that this work is of investigative nature, we evaluate the quality of the solutions via Leave-One-Out Cross-Validation (LOOCV). For the collaboration case base, accuracy is measured based on the number of edits required to transform the solution suggested by the algorithm into the solution of the left out case. Each feature that needs to be changed is one edit. Related features count as one half edit (e.g., if Assistant Professor is changed to Associate Professor). The solutions have a total of six features, thus the range of this measure is 0 (a perfect match) to 6 (completely inaccurate). For clarity this is expressed as a percentage. For example, a distance of two edits implies that 4 out of the 6 features are a match, giving an accuracy of 66.7%.

3   Learning Feature Weights from Positive Instances

When both positive and negative instances are present, feedback algorithms use the correct and incorrect solutions to learn weights. Lacking negative instances due to the availability of the data, we seek an alternative to play the role of negative instances.

Our strategy is based on the intuition that if a standard premise for CBR holds i.e., that similar problems have similar solutions [17] then cases that do not follow this premise can be used as negatives in order to learn weights. Based on previous related works (e.g., [21]) we refer to cases where similar problems have similar solutions as *well aligned*. Cases that do not meet this premise are not well aligned, i.e., they are *poorly aligned* cases. Poorly aligned cases have no, or very few, close neighbors either in the problem space or the solution space (or both). Our approach is premised on determining which cases are poorly aligned and using them in the role of negative instances in a feedback algorithm to learn weights.

The rationale behind this clustering-based approach is that the *poorly aligned cases perform the same function as negative instances in the context of the feedback algorithm*. They allow us to learn weights that when presented with a new case will be able to find a similar case in the subset of cases that are well aligned.

The poorly aligned cases are still legitimate cases, using them in the role of negative instances is exclusively for the purposes of learning feature weights via a feedback algorithm. We note that these poorly aligned cases can bring valuable diversity to the case base. Even though we have established that they are poorly aligned, we including these cases in the evaluation to provide a consistent overall evaluation and avoids the tradeoff between the number of removed cases and accuracy.

Cases that are poorly aligned occur in areas of low density in either the problem or solution space (or both). We seek to identify therefore areas of the problem and solution space that are more versus less dense. Density clustering is the recommended method for this task [25]. Density clustering creates clusters of high density areas, with those points in low density areas being classified as outliers. The outliers in the context of the case base are poorly aligned as they do not have sufficient neighbors to form, or to be included in, a cluster. We discuss the particulars of the clustering methods in more detail in Section 5, and for now speak in general terms.

We apply a clustering algorithm first in the problem space and then in the solution space. In each space the clustering algorithm will identify cases that have neighbors and flag those that do not have neighbors as outliers. Our assumption is that if a case is an outlier in either of these spaces it is a case that is poorly aligned. Our clustering-based approach provides proxy for negatives to allow the use of feedback algorithms to learn weights in datasets with only positive cases and is described in **Fig. 1**.

```
Given a case base CB
Cluster cases in the problem space
Label outliers as negative
Cluster cases in the solution space
Label outliers as negative
Label remaining unlabeled cases as positive
Apply feedback algorithm to learn weights using labeled positives and
negatives
Evaluate average accuracy of resulting solutions via LOOCV for all
cases in CB
```

Fig. 1. Approach to Use Feedback Algorithm with Positive Cases

We provide evaluation of this approach in later sections. Prior to the evaluation, we next demonstrate that this clustering-based approach is consistent with related methods in the CBR literature for computing measures of alignment and cohesion.

4   Identifying Aligned Cases

We premise our approach on determining which cases in a case base are well aligned and which are poorly aligned. Note that when we, in this paper, refer to aligned cases we are not referring to a specific measure, but to the general concept of problems that are neighbor having solutions that are also neighbors.

Poorly aligned cases may occur due to diversity of cases, or even corruption from errors within the case [20]. However, even when every case is a legitimate experience, some cases may have a greater or a lesser or even a negative impact on the performance of the case base [26]. Cases that have solutions dissimilar to their nearest neighbors can create noise in the case-base [23] and particularly for classification tasks such cases can lead to misclassification [7]. Given the impact of poorly aligned cases, several measures have been used to determine which cases poorly aligned.

Case cohesion [16] quantifies how similarly a case behaves to its nearest neighbors in terms of both its problem and its solution. This measure requires two similarity thresholds, one for the problem space and one for the solution space to determine which cases are similar enough to be considered neighbors. Case alignment [21] considers the similarity of both the problem and the solution spaces for a pre-specified number of neighboring cases.

These two previous measures locally compare an individual case to its neighbors. The method presented by [28] determines aligned cases by ranking, for each case, its nearest neighbors in the problem and solution space by decreasing similarity. Rank correlation provides a measure of the level of alignment by determining how similar the two rankings are. The Global Alignment MEasure (GAME) [6] provides a single measure of how well the entire case base is aligned by measuring, for the case base as a whole, the extent to which problems and solutions overlap.

We verify our method by demonstrating that cases identified by density clustering as poorly aligned have low cohesion [16] and alignment scores [21], using the collaboration case base described in Section 2. We calculate the cohesion scores for each case in the case base and then select the bottom 5% of cases, i.e. the cases with the lowest cohesion scores. We then determine what proportion of that set of cases is identified as poorly aligned by the clustering-based approach. This process is the repeated for the bottom 10% of cases. These studies are detailed in [12] and we present a summarized result in **Fig. 2**. In [12] we experimented with different parameters for cohesion and density clustering and best results are presented here in **Fig. 2a**. We repeat the process using the alignment scores, and show the best results in **Fig. 2b**.

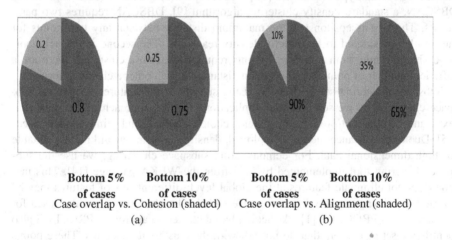

Bottom 5%	Bottom 10%	Bottom 5%	Bottom 10%
of cases	of cases	of cases	of cases

Case overlap vs. Cohesion (shaded)   Case overlap vs. Alignment (shaded)

(a)   (b)

Fig. 2. Overlap of Poorly Aligned Cases with Low Cohesion and Low Alignment Cases

These results indicate that a major proportion of the cases identified by density clustering as poorly aligned also have low alignment and cohesion scores. This verifies that density clustering is consistent with analogous measures in the literature.

5   The Selection of Clustering Methods

In our approach (see pseudocode in **Fig. 1**), we identify poorly aligned cases by clustering on the problem and solutions spaces. Here, we implement our approach using two clustering methods: density clustering and subspace clustering. We explore which method is more suitable based on the dimensionality of the data.

5.1   Density Clustering and Subspace Clustering

It is not our goal to test the specific merits of different implementations of density or subspace clustering algorithms; we use two representative implementations: DBSCAN and PROCLUS. We highlight the differences between these in **Table 2**.

Table 2. Chracteristics of Clustering Algorithms Used

	Required Parameters	Cluster Shape	Feature Usage	Solution Format
DBSCAN	Neighborhood size, Epsilon	Less Regular	All	Multi-valued
PROCLUS	Number of clusters, average number of dimensions	More Regular	Varying Subsets	Single-valued

Density clustering creates clusters of high density areas, with those points in low density areas being classified as outliers. Density clustering was chosen as it is recommended for this type of problem [25]. For demonstrating density clustering, we use DBSCAN, a standard density clustering algorithm [9]. DBSCAN requires two parameters. The first is epsilon (ε), the maximum distance between any two points for them to be considered to be directly density reachable. The second is the neighborhood size, the minimum number of points required to form a cluster. If there are a sufficient number of points within epsilon distance of each other a cluster is formed.

Subspace clustering selects clusters using subsets of the features. We select subspace clustering as we see potential applications for this approach in big data contexts, and subspace clustering is a natural choice for such high dimensional spaces [15]. Due to the distance measure employed, density clustering would be less suitable for high dimensional data. For demonstrating subspace clustering, we use the subspace clustering tool implemented by [22] from the WEKA package [14]. This method does not eliminate features at the global level; different sets of features may be selected as relevant for the different clusters [15]. The clustering algorithm used for the subspace is PROCLUS [1] a k-mediod based clustering approach. PROCLUS also identifies a set of outliers that do not fall with the clusters it generates. These points do not lie close to the mediods identified.

5.2   Comparing Clustering Methods

We use two case bases the first is low dimensional, and the second is high dimensional. The two case bases have both positive and negative cases. However, to test our approach we ignore the solution classes when we learn feature weights through our approach. We only use the solution for evaluation purposes. The first is a case base of football plays (**Table 3**).

Table 3. Football Case Base

	Feature Name	Description	Example Values
	Time	Time remaining (mins)	{60-33, 32-30, 29-3, 2-0}
	Down	Period of play	{1,2,3, 4}
Problem	Distance	Distance to get a first down	{Short, Medium, Long}
Features	Field Position	Position on field (yards)	{1-15, 16-60, 61-99 }
	Score	Current score differential	{within 7 pts, ahead by more than 7 pts, behind by more than 7 pts}
Solution	Play	The football play executed	Pass Deep, Run Left, Punt, …

The five problem features of this case base reflect a game state of an American Football game. The solution is the play that was executed in the given game state. The case base consists of 106 cases. The density clustering uses a distance matrix that is the unweighted sum of the similarity between features. As the features are ordinal, if the features are one step apart (e.g. Short Distance vs. Medium Distance) the distance is 0.5, otherwise the distance is 0 or 1 if the features match or do not. To implement the subspace clustering algorithm, the ordinal features are converted to numeric (e.g., short, medium, and long to 1, 2 and 3).

The second case base consists of business project management cases (**Table 4**), for brevity only a seven of the 23 features is shown. The problem features describes the characteristics of a project. The solution is a binary feature signifying whether the project was a success or a failure. There are 88 cases, with 67 successful projects and 21 failed ones. The density clustering uses a distance matrix that is the unweighted sum of the similarity between features. The similarity for all features is Boolean.

Table 4. Project Management Case Base

	Feature Name	Example Values
Problem	Project manager was given full authority	{Yes, No}
Features	The project began with a committed champion	{Yes, No}
	The sponsor was involved with decisions	{Yes, No}
	Developers were involved in the estimates	{Yes, No}
	There was well defined scope	{Yes, No}
	The requirements were complete and accurate	{Yes, No}
	Customers had realistic expectations	{Yes, No}
Solution	Project outcome	{Success, Failure}

The implementation of the approach follows the same form that is described in **Fig. 1**. First the cases are clustered in the problem space using the chosen clustering method and any outliers are labeled as negatives. Then this process is repeated for the solution space. The unlabeled cases are then labeled as positives, and a feedback algorithm is used to learn the feature weights. In this and subsequent experiments we use a genetic algorithm to learn weights, but this approach can be used with any feedback algorithms.

The quality of the solutions selected by each set of feature weights is evaluated through LOOCV. For both case bases, accuracy is measured by the percentage of correctly classified cases using the solutions as class labels. For evaluation all the cases are used, including the poorly aligned cases. We investigate the following hypotheses:

H1: the average accuracy from using density clustering will be greater than using subspace clustering when the dimensionality of the data is low.

H2: the average accuracy from using subspace clustering will be greater than using density clustering when the dimensionality of the data is high.

Table 5. Average Accuracy, % of Correct Classifications

Case Base	Density Clustering	Subspace Clustering
Football	70%	67%
Project Management	84%	91%

From **Table 5** we see that H1: density clustering performs better in the context of low dimensional data, and H2: the subspace clustering performs better with higher dimensional data. These observations are preliminary, and we use them only as a guide for our further experiments.

Table 6. Percentage of Cases Identified as Poorly Aligned

Case Base	Density Clustering	Subspace Clustering
Football	20%	25%
Project Management	6%	5%

The two clustering method also behave similarly in the proportion of cases identified as poorly aligned (**Table 6**). In the next section we explore the quality of the solutions arising from the use of the feature weights learned via our approach

6   Quality and Generalization of the Approach

In this section, we first present the quality of our proposed approach by showing how it improves the average accuracy of the recommendations for the collaboration case base (Section 2). Second, we present how the approach generalizes by showing how it improves average accuracy over no feature weights for different case bases. Based on the results of our investigations in Section 5, we use density clustering when the dimensionality is low, and subspace clustering when dimensionality is high.

We investigate the hypothesis H3: the average accuracy using the clustering-based approach is statistically significantly better than using no feature weights. We show, for the collaboration case base, the average accuracy when we implement the approach described in **Fig. 1** versus no feature weights (**Table 7**).

The average accuracy is measured as described in section 2.2, and is expressed as a percentage that denotes how well the recommendation matches the case left out by the LOOCV. As this is one of several comparisons we will do using these results, a one-way ANOVA test is used to determine if there is a significant difference between the means of the various methods ($\alpha = 0.05$). The post hoc analysis is via the Bonferroni Correction using $\alpha = 0.0083$ to maintain the significance level of 0.05.

Table 7. Average Accuracy, (* $\alpha = 0.05$)

Case Base	Clustering Method	Average Accuracy	No Feature Weights
Collaboration	Density	67.2%*	63.5%

Next we show results of the density-based and subspace implementations of our approach for the different case bases. Again we compare the average accuracy from using the approach versus having no feature weights. In **Table 8** we show the results for the two classification case bases already described in the previous section. The average accuracy is now computed as the the percentage of correct classifications. We test the difference in error rates at $\alpha = 0.05$ for statistical significance [8]. This is a separate experiment than the previous so a Bonferroni Correction of $\alpha = 0.025$ is used.

Table 8. Average Accuracy, (* $\alpha = 0.05$)

Case Base	Clustering Method	Average Accuracy	No Feature Weights
Football	Density	*70%	48%
Project	Subspace	*91%	76%

We show that H3: using the clustering-based approach produces an average accuracy that is statistically significantly higher than when using no feature weights across all three case bases. Based on our previous experiments subspace clustering is used in high dimensional spaces. The asterisks indicate statistical significance ($\alpha = 0.05$).

7   Quality of the Approach with Negative Instances

The main reason for the proposed clustering-based approach is the need to learn feature weights when only positive cases are available. In this section, we compare the quality of the resulting solutions by using the clustering-based approach in case bases where negative instances are available. We learn feature weights via a feedback algorithm using the actual labeled negatives. We compare the resulting accuracy to the accuracy obtained from learning weights using the poorly aligned cases as negatives using the approach as presented in **Fig. 1**.

In order to demonstrate the quality of approach, we propose H4: average accuracy using the clustering-based approach is NOT statistically significantly different from the average accuracy using weights learned with labeled negatives in case bases that have negatives. We test the difference in error rates for statistical significance to determine if there is any diffence between the average accuracies ($\alpha = 0.05$).

Table 9. Percentage of Correctly Classified Cases (* $\alpha = 0.05$)

Case Base	Clustering Method	Clustering-based	Labeled Negatives
Football	Density	70%	68%
Project	Subspace	91%	85%

In **Table 9** we show the average accuracy for the case bases computed as the percentage of correct classifications for the clustering-based approach compared to using actual negatives. These results confirm H4 showing that the clustering-based approach produces feature weights of quality not significantly different to when negatives are available. We next compare our approach to an alternative method for reasoning with positive instances.

8   Comparison to Alternative Method for Reasoning with Positive Instances

In previous sections, we explored studies to substantiate the approach we proposed in Section 3. In this section we consider an alternative machine learning method for learning from positive instances.

8.1   Single Class Learning

Single Class Learning (SCL) [18, 19] is used for some classification tasks where the data has only positive examples. SCL is used in a context of datasets with a small number of positive and a large set of unlabeled instances. In contexts such as web page classification [27], or gene classification [5], obtaining complete training data to be used in conjunction with supervised methods may be problematic. SCL uses the characteristics of the positive instances to build rules to identify likely negative instances in the set of unlabeled instances. SCL methods have been implemented with support vector machines [19] and expectation maximization (EM) [18].

To apply this method in a problem context where there are only positive instances, we start from all the possible problem and solution combinations in the respective spaces of problem and solution features. Then we consider the problems and solutions not represented in the case base as unlabeled data. Next we assume the cases with combinations of features that are very different from the positive instances to be instances that are likely negative. These can then be translated into rules to identify negative instances based on their feature values. A problem-solution whose feature combination matches one of those rules is interpreted as a negative instance.

We only learned rules for the features rank and institution type. The large number of allowable values and the taxonomic nature of the feature research interest prevented its inclusion. An example of a learned rule might identify as a negative instance a collaboration of two assistant professors and one associate professor who are faculty at non-research institutions. Thus, SCL operates differently compared to our approach. SCL determines whether an entire recommendation is a negative instance or not. It considers the combination of the features in both problem and solution spaces.

It is technically possible to represent the combinations of characteristics of negatives learned from the SCL in the form of cases. However, because we only included the rank and institution features, it would not be possible to use this representation to learn the weights for all three features.

8.2   Recommending Characteristics of Collaborators with SCL

A preliminary method using the ideas behind SCL in our motivating case base is given in [12]; we expand the treatment here, by implementing an SCL-based approach using the EM algorithm based on [18].

We implement it by randomly selecting 10% of the original positive cases to act as 'spies' on the unlabeled data. Thus we have two class labels, the original positives P and the spies S. The EM algorithm is used to classify unlabeled data as either S or P. The intuition is that the instances that are likely negatives will be classified as S, but with the lowest probability as being in class S. We take those instances and classify them as possibly negative instances N, we then return the spies S to their original classification P. We again have two classes: S, the original positive instances, and N the possible negatives. We rerun the EM algorithm using these new classes. The unlabeled instances that now show a high probabilistic classification to be in class N are chosen as likely negative instances.

We use the resulting combinations of problem and solution deemed as likely negatives as rules combined with CBR using no feature weights to obtain recommendations. This way we can compare the average accuracy of the recommendations produced by the clustering-based approach against the average accuracy of recommendations produced with rules from the SCL-based approach.

Furthermore, as our goal is to find the most accurate solution to our motivating problem context, we also examine the knowledge learned from SCL and use it in combination with our approach. The combination of the two methods is straight forward. From the recommendation produced by the clustering-based approach (again we refer to **Fig. 1**), we eliminate the solutions that violate the rules produced by SCL. To this end we investigate the following hypotheses:

H5: average accuracy using the clustering-based approach is NOT statistically significantly different from the average accuracy learned using the SCL-based approach.

H6: A combination of the two approaches will result in average accuracy that is statistically better than the average accuracy of each individual approach.

In **Table 10** we present the average accuracy resulted from the SCL-based approach, from the clustering-based approach, and the combination of the two methods. A one-way ANOVA test is used to determine if there is a significant difference between the means of the different methods ($\alpha = 0.05$), post hoc analyses is done using a Bonferroni Correction of $\alpha = 0.0083$.

Table 10. Average Accuracy (* $\alpha = 0.05$)

Clustering-based Approach	SCL-based Approach	Combined
67.2%	67.0%	70.8%*

The results in **Table 10** confirm H5, there is no statistically significant difference in accuracy between the clustering-based and the SCL-based approaches, (p = 0.059). H6 is also confirmed; the combined clustering-based and SCL-based approach outperforms the individual approaches to a statistically significant extent (p = 0.000).

9   Conclusions and Future Work

We have shown through our experiments (Section 6) that our clustering-based approach makes it feasible to learn feature weights in problem contexts with only positive instances and insufficient domain knowledge. We are motivated by a context with only positive instances: recommending characteristics of collaborators. Note that the goal is to recommend the characteristics of the collaborators, not specific individuals. Preliminary studies on additional case bases suggest generality of the approach.

Our approach relies on using poorly aligned cases in the role of negatives to learn feature weights with a feedback algorithm. We show our approach is consistent with other alignment methods from the literature (Section 4).

The proposed approach requires the use of a clustering method. We showed that the choice of the clustering method varies depending on the dimensionality of the targeted case base. Our preliminary tests confirm the intuition that density clustering should be superior for low dimensional case bases, while subspace clustering should be more suitable for high dimensional ones (Section 5).

In order to further determine the quality of our proposed approach, we also apply it in case bases that do have negative instances to compare the results. We found that our clustering-based approach leads to comparable accuracy to when the actual negative instances with state of the art methods are used (Section 7).

We compared our approach to an alternative to deal with lack of negatives, SCL. SCL assigns as a positive or negative instance a problem and solution combination. We do not learn weights using the knowledge from the SCL-based approach as we did not learn rules about all features. Thus, if we converted the learned knowledge into class labels for the cases, we would not be able to learn weights for all features. Therefore, we chose to use the rules directly to determine whether a recommendation is consistent or violates any rules. We compared our approach to SCL combined with CBR using no feature weights and find the performances to be comparable.

The similarity function recommends the solution of the candidate most similar to the target collaboration seeker. Due to the existence of poorly aligned cases, it is possible that the recommendation is a configuration that does not exist in the case base. Thus, the negative instances learned from the SCL-based approach can be used as rules to enhance the similarity function to prevent the recommendation of configurations that are not present in the case base that may be negatives. This is seen by the superior performance of the combined approach over other approaches.

We seek to investigate our approach further by developing it for collaborations larger than three members and by examining additional clustering methods such as bi-clustering. We also wish to further investigate a representation of cases that would allow us to use the knowledge learned from the SCL-based approach to learn feature weights. To further develop our contribution in the collaboration sphere, the knowledge learned is being investigated via a survey with reviewers with experience in grant funding. We see this approach to determining poorly aligned cases assisting in automating the determination of the suitability of CBR in the context of big data.

Acknowledgements. This work is supported in part by U.S. EPA STAR Program and the U.S. Dept. of Homeland Security; Grant # R83236201. The authors wish to thank Joe Healy, Alinafe Matenda, & Joe Smith for the use of the football dataset in our experiments. We also thank our reviewers for their comments that helped improve the paper.

References

1. Aggarwal, C.C., Wolf, J.L., Yu, P.S., Procopiuc, C., Park, J.S.: Fast Algorithms for Projected Clustering. ACM SIGMOD Record 28(2), 61–72 (1999)
2. Aha, D.W.: Feature weighting for lazy learning algorithms. In: Liu, H., Motoda, H. (eds.) Feature Extraction, Construction and Selection: A Data Mining Perspective, pp. 13–32. Kluwer, Norwell (1998)
3. Adomavicius, G., Tuzhilin, A.: Toward the Next Generation of Recommender Systems: A Survey of the State-Of-The-Art and Possible Extensions. IEEE Transactions on Knowledge and Data Engineering 17(6), 734–749 (2005)
4. Burke, R.: Hybrid Recommender Systems: Survey and Experiments. User Modeling and User-adapted Interaction 12(4), 331–370 (2002)
5. Calvo, B., López-Bigas, N., Furney, S.J., Larrañaga, P., Lozano, J.A.: A Partially Supervised Classification Approach to Dominant and Recessive Human Disease Gene Prediction. Computer Methods and Programs in Biomedicine 85(3), 229–237 (2007)
6. Chakraborti, S., Cerviño Beresi, U., Wiratunga, N., Massie, S., Lothian, R., Watt, S.: Visualizing and Evaluating Complexity of Textual Case Bases. Advances in Case-Based Reasoning, 104–119 (2008)
7. Delany, S.J.: The Good, the Bad and the Incorrectly Classified: Profiling Cases for Case-Base Editing. In: McGinty, L., Wilson, D.C. (eds.) ICCBR 2009. LNCS, vol. 5650, pp. 135–149. Springer, Heidelberg (2009)
8. Dietterich, T.G.: Approximate Statistical Tests for Comparing Supervised Classification Learning Algorithms. Neural Computation 10(7), 1895–1923 (1998)
9. Ester, M., Kriegel, H.-P., Sander, J., Xu, X.: A Density-Based Algorithm for Discovering Clusters In Large Spatial Databases with Noise. In: Simoudis, E., Han, J., Fayyad, U.M. (eds.) Proceedings of the Second International Conference on Knowledge Discovery and Data Mining, pp. 226–231. AAAI Press, Menlo Alto (1996)
10. Gunawardena, S., Weber, R.O.: Blueprints for Success Guidelines for Building Multidisciplinary Collaboration Teams. In: Filipe, J., Fred, A.L.N. (eds.) ICAART 2012 Proceedings of the 4th Intl. Conference on Agents and Artificial Intelligence, pp. 387–399. SciTePress (2012)

11. Gunawardena, S., Weber, R.O.: Reasoning with Organizational Case Bases in the Absence Negative Exemplars. In: ICCBR 2012: 2nd Workshop on Process-Oriented Case-Based Reasoning, pp. 35–44 (2012)
12. Gunawardena, S., Weber, R.O.: Applying CBR principles to Reason without Negative Exemplars. In: FLAIRS 2013 (in press, 2013)
13. Gunawardena, S., Weber, R.O., Agosto, D.E.: Finding that Special Someone: Interdisciplinary Collaboration in an Academic Context. Journal of Education for Library and Information Science 51(4), 210–221 (2010)
14. Hall, M., Frank, E., Holmes, G., Pfahringer, B., Reutemann, P., Witten, I.H.: The WEKA Data Mining Software: An Update. ACM SIGKDD Explorations Newsletter 11(1), 10–18 (2009)
15. Kriegel, H.P., Kröger, P., Zimek, A.: Clustering High-Dimensional Data: A Survey on Subspace Clustering, Pattern-Based Clustering, and Correlation Clustering. ACM Transactions on Knowledge Discovery from Data (TKDD) 3(1), 1–58 (2009)
16. Lamontagne, L.: Textual CBR Authoring Using Case Cohesion. In: Proceedings of the 2006 Workshop on Textual CBR, pp. 33–43 (2006)
17. Leake, D.B. (ed.): Case-Based Reasoning: Experiences, Lessons, and Future Directions. AAAI Press/MIT Press, Menlo Park, CA (1996)
18. Liu, B., Lee, W.S., Yu, P., Li, X.: Partially Supervised Classification of Text Documents. In: Proceedings of the Nineteenth International Conference on Machine Learning (2002)
19. Liu, B., Dai, Y., Li, X., Lee, W.S., Yu, P.S.: Building text classifiers using positive and unlabeled examples. In: Third IEEE International Conference on Data Mining, pp. 179–186. IEEE (2003)
20. Massie, S., Craw, S., Wiratunga, N.: When Similar Problems Don't Have Similar Solutions. In: Weber, R.O., Richter, M.M. (eds.) ICCBR 2007. LNCS (LNAI), vol. 4626, pp. 92–106. Springer, Heidelberg (2007)
21. Massie, S., Wiratunga, N., Craw, S., Donati, A., Vicari, E.: From anomaly reports to cases. In: Weber, R.O., Richter, M.M. (eds.) ICCBR 2007. LNCS (LNAI), vol. 4626, pp. 359–373. Springer, Heidelberg (2007)
22. Müller, E., Günnemann, S., Assent, I., Seidl, T.: Evaluating clustering in subspace projections of high dimensional data. Proceedings of the VLDB Endowment 2(1), 1270–1281
23. Plaza, E.: Semantics and experience in the future web. Advances in Case-Based Reasoning, 44–58 (2008)
24. Quijano-Sánchez, L., Bridge, D., Díaz-Agudo, B., Recio-García, J.A.: A Case-Based Solution to the Cold-Start Problem in Group Recommenders. In: Agudo, B.D., Watson, I. (eds.) ICCBR 2012. LNCS, vol. 7466, pp. 342–356. Springer, Heidelberg (2012)
25. Richter, M.M., Weber, R.O.: Case-based reasoning: a textbook. Springer, Berlin (in press, 2013)
26. Smyth, B., McKenna, E.: Footprint-based retrieval. In: Althoff, K.-D., Bergmann, R., Branting, L.K. (eds.) ICCBR 1999. LNCS (LNAI), vol. 1650, pp. 343–357. Springer, Heidelberg (1999)
27. Yu, H., Han, J., Chang, K.C.-C.: PEBL: Web Page Classification Without Negative Examples. IEEE Trans. Knowledge and Data Engineering 16(1), 70–81 (2004)
28. Zhou, X.F., Shi, Z.L., Zhao, H.C.: Reexamination of CBR hypothesis. In: Bichindaritz, I., Montani, S. (eds.) ICCBR 2010. LNCS, vol. 6176, pp. 332–345. Springer, Heidelberg (2010)

User Perceptions of Relevance and Its Effect on Retrieval in a Smart Textile Archive

Ben Horsburgh[1,2], Susan Craw[1,2], Dorothy Williams[3],
Simon Burnett[3], Katie Morrison[3], and Suzanne Martin[4]

[1] School of Computing Science & Digital Media
[2] IDEAS Research Institute
[3] IMaGeS Research Institute
Robert Gordon University, Aberdeen, Scotland, UK
{b.horsburgh,s.craw,d.williams,k.morrison}@rgu.ac.uk
[4] School of Textiles & Design
Heriot-Watt University, Galashiels, Scotland, UK
s.e.martin@hw.ac.uk

Abstract. The digitisation of physical textiles archives is an important process for the Scottish textiles industry. This transformation creates an easy access point to a wide breadth of knowledge, which can be used to understand historical context and inspire future creativity. The creation of such archives however presents interesting new challenges, such as how to organise this wealth of information, and make it accessible in meaningful ways. We present a Case Based Reasoning approach to creating a digital archive and adapting the representation of items in this archive. In doing so we are able to learn the important facets describing an item, and therefore improve the quality of recommendations made to users of the system. We evaluate this approach by constructing a user study, which was completed by industry experts and students. We also compare how users interact with both an offline physical case base, and the online digital case base. Evaluation of our representation adaptation, and our comparison of physical and digital archives, highlights key findings that can inform and strengthen the process for creating new case bases.

Keywords: Learning Refined Representation, Digitisation of Physical Archives, User Evaluation.

1 Introduction

The textiles industry is an important part of the local economy, history, and future of Scotland. Many prominent companies, manufacturing textiles for well-known designers, have existed for over 100 years. This heritage and experience is important, providing companies with a competitive edge over international rivals. However, while design processes and manufacturing technology have been kept up to date, the archiving of knowledge has suffered.

Typical archives throughout the textiles community are kept physically in storage rooms, similar to that in Figure 1. These physical archives can be difficult

S.J. Delany and S. Ontañón (Eds.): ICCBR 2013, LNAI 7969, pp. 149–163, 2013.

Fig. 1. Typical Archive Room

to take full advantage of, unless someone knows exactly what they are looking for, and where it is. Sharing the knowledge held in separate archives can also be problematic, with many unaware of what may even be in an archive. There is wide interest across the community in how this knowledge can be made more accessible, and used as a source of inspiration for new textile designs.

In collaboration with Johnstons of Elgin, we have investigated how a physical archive can be transformed into a digital one. The first stage of this transformation is to understand the nature of the collection they have, and identify the important information available. Having obtained this information it is possible to create digital versions of physical assets, using photographs and descriptions. However, while this process makes information available, it does not organise it in a meaningful way. An index that highlights the important features of each digital asset, and facilitates the searching and browsing of assets, is required.

Case based reasoning (CBR) provides a structured way of modelling and learning from the past experience of users. Through capturing user behaviour and interactions, there is an opportunity to capture implicit knowledge about the content of the archive, that may improve the retrievals and recommendations. This implicit knowledge may be used to modify the asset index, and thus highlight the important features of each asset.

In this paper we discuss related work on the digitisation of physical archives, and how CBR has been used in retrieval and recommender systems. Our approach to understanding and selecting content for digitisation is then discussed. This approach includes two workshops with Johnstons of Elgin, a major textiles designer. We discuss our creation of the digital archive, and how cases can be defined to describe the selected assets. We then develop an implicit learning method that is able to refine the case representations based on user interactions. Next, we describe our user experiments, which evaluate our implicit learning method, and draw comparisons between the stakeholder and end-user understanding of and interaction with the assets. We present the results of these user experiments, and finally draw some conclusions from this study.

2   Related Work

Although companies are only just realising the potential of digitally archiving their assets, several studies have investigated how this may be achieved. Evans [1] discusses the underlying requirements for the 'perfect' fashion archive, moving from physical to digital. A key finding of this work is that such archives do not necessarily need to contain all assets, but the organisation and interaction of key assets is important. Paterson [2] examines the success of digitising the House of Fraser fashion and textile archive. In this project, assets were added to a digital library, but catalog-book style indexes were relied upon to navigate the collection. The author concludes that for such an archive to be successful, more sophisticated indexing techniques and interactions are required. This view is further supported by Brown [3], who notes that sophisticated searching functionality is required for the successful creation of a digital archive.

CBR has previously been used successfully to provide searching functionality for e-commerce systems [4,5]. Such systems are somewhat similar to a textiles archive, however user motivation may differ. Case based reasoning has also been used to construct recommender systems, helping users to navigate a digital collection of items [6]. One main advantage of introducing CBR into a recommender system is that each item is no longer static in the collection. As users interact with the system, creating new problem queries and new solution recommendations, the system adapts to take advantage of this knowledge [7]. Such dynamic behaviour is critical to the successful creation of our digital archive.

Milne et al. [8] introduce this dynamic behaviour in an image retrieval system, by modifying the case representation weights after each successful query. Case weights that are consistently not aligned with associated queries, are diminished, thus removing noise from case representations and improving the overall system. Ontañón and Plaza [9] define a similarity measure based on the anti-unification of two structured cases being compared. This approach further highlights how shared information, in this example between two cases, may be used to further inform a CBR system.

3   Selection of Content for Archive

At the onset of this project no case base, or even digital versions of assets were available. The initial phase therefore was to engage with the owners of the data, explore, and formally define what each of the assets meant to them. To achieve this two formal workshops were conducted with the stakeholders: the first workshop to understand what an asset is, and how it may be described; the second workshop to understand how the stakeholders define relationships between assets. The results of these two workshops allowed us to create our initial digital case base.

3.1   Stakeholder Workshop 1: Asset Descriptions

The objective of the first user workshop was to identify what exactly assets are, and how they can be described. Staff from the company creating the archive were asked to contribute assets that they considered to be relevant to a digital archive. These contributions were initially in the form of one or more photographs, and a total of 150 items were contributed. From these 150 assets, justifications for each to be included in the archive were presented, and the 30 most interesting assets were selected to focus on by popular vote.

Example assets that were selected for inclusion in the archive range from textiles related, such as looms, tartan, and bale hook, to company related, such as history / heritage, and original skills and crafts. The images associated with each asset were clipped into plant pots, as illustrated in Figure 2, allowing everyone at the workshop to easily interact with and move around each of the items.

Fig. 2. Items Contributed by Stakeholders

The next step in this workshop was to gain an understanding of how each item may be described. With the assets clearly laid out to encourage discussion, a dialog with the stakeholders began to identify some key themes and commonalities amongst the assets. These themes were then translated into a set of labels, which the stakeholders used to categorise the collection. The 4 labels that emerged were future, luxury, heritage, and sentimental. The participants were then all given plant labels and post-it notes. The plant labels were used to annotate the items with key terms that small groups in the workshop discussed and defined, and the post-it notes allowed people to provide more detailed free-text descriptions. The outcome of this workshop formed the initial structure of our case representation for an item, consisting of the image, labels and tags, and free text.

3.2   Stakeholder Workshop 2: Asset Relationships

Having selected meaningful assets and obtained suitable descriptions, the second workshop aimed at understanding the structure of the collection as a whole. One important goal towards creating a useful archive is that it is intuitive and easy to navigate in a meaningful way. In CBR systems, this is most commonly achieved by introducing a meaningful semantic similarity measure.

To understand how the assets relate to each other, and gain insight into what similarity means, the items were again placed on the floor in their plant pots. The participants were split into their 4 respective departments; Retail, HR & Finance, Production, and Design. Each department was then asked to use coloured tape to form a map of how each item relates to another, illustrated in Figure 3. Each colour indicates how the stakeholders define relationships within three separate interest groups: future, heritage, and tourist. These groups were considered because they were believed to be the three main interest areas that end users could be classified into.

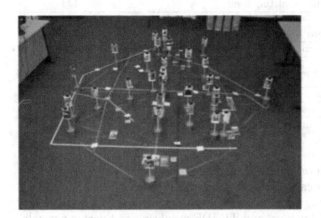 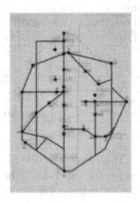

Fig. 3. Item map created by retail team (left) and transcription (right)

The construction of the maps was initiated by selecting start points from the 30 most popular assets. The edges were then added based on how the stakeholders believed each of the interest groups would navigate the archive, moving through related nodes. All of the assets were available for inclusion, but typically only between 30 and 40 were mapped. There was a general consensus amongst the 4 departments as to what assets were included. However, the way in which they were included differed between each department. For example, the production department produced a map which resembled a production line, while the retail team designed a map which was much more exploratory, similar to how someone may browse a shop. Although there were differing views across departments regarding which assets were relevant to each other, these maps provide interesting data which may be compared with usage patterns of the end-users.

4   Creating an Initial Case Base

The outcome of the workshop exercises was a definitive set of assets that are used to construct the initial case base. We represent this archive of assets as a case base of cases. The final set of features that were used for each case are illustrated in Table 1. There are three types of features in our case structure; free text, tags, and images. Free-text features were processed by tokenising, stemming, and removing stop words from the data. These processed tokens were then used to construct a single term-frequency vector for each separate feature. The vectors for each feature were weighted using TF-IDF, and finally normalised.

Table 1. Case Representation

Type	Feature	Description	Vocabulary Size
Free Text	Title	The name of the item	144
Free Text	Description	Post-it note description	393
Free Text	Justification	Reason the item was chosen	344
Free Text	Other	Any other relevant descriptive information	72
Free Text	Facts	Interesting facts about the item	160
Tag	Aspects	Part of company the item relates to	14
Tag	Labels	Future, luxury, heritage, or sentimental	4
Tag	Terms	Key descriptive tags	59
Image	Main Image	SIFT Image features for main image	100

Tags were not tokenised or stemmed, and no stop words were removed. The reason for this is that the vocabulary sizes are much smaller, and each instance of the feature was typically only between one and three words long. Term-frequency vectors were again created for tags, and TF-IDF weighting and vector normalisation applied.

Images were indexed using the Scale Invariant Feature Transform (SIFT) algorithm [10], which detects and describes local features within an image. These local features were then clustered using the k-means algorithm, and a cluster-frequency vector created. The SIFT algorithm was chosen to describe images because it is well proven across many domains, and after clustering provides a representation structure which is similar to our textual representations.

The selection of assets from workshops, with free text, tags, and labels, together with the image provides the cases for the digital archive case base.

4.1 Content Similarity

We develop two methods for users to interact with the archive system: querying the items directly, and through a recommender system. Users provide a search query by typing search terms into a query box. These search terms must then be structured so that they may be used to access our case index. To achieve this, we construct a new temporary case from the query. In this case, each feature dimension matching a tokenised search term is incremented by 1, and all other dimensions are set to 0. For example, if the ith dimension of the label feature describes the frequency of term 'tartan rug', and the user queries for 'red tartan rug', then the term frequency is 2.

The recommender system uses a query-by-example approach, and thus the case describing the asset a user is currently viewing is used as the search query. To query our case base, using either a temporary or example case, we average the cosine similarities between each pair of individual feature vectors, calculated as:

$$\text{similarity}(Q, R) = \frac{\sum_{f=1}^{F} \frac{Q_f \cdot R_f}{|Q_f| \cdot |R_f|}}{F} \tag{1}$$

where Q is the query case, R is the potential retrieval, F is the number of features in a case, and Q_f and R_f are the fth feature-vector of the query and retrieval cases.

5 Implicit Learning Method

With an archive case base constructed, features extracted, and similarity measure defined, the next stage is to refine methods used to navigate and interact with the archive. Although the workshop map could possibly be used as an initial refinement for similarity, the map covers only 20% of the assets used. To overcome this problem with missing edges, we propose a learning method which takes advantage of the implicit feedback created as each user interacts with the system.

5.1 Learning Feature Dimension Weights

The similarity measure proposed provides a good starting point to allow users to interact with the system. However, the system is simply using all of the information provided, whether it is relevant or not. The term frequencies of each feature dimension are based on the views of a relatively small number of participants at the workshops, and may be biased towards their opinions as employees. To overcome this problem we develop a new refinement method, which is able to learn implicitly which parts of a feature-vector are most important.

As each user interacts with the system, each query-retrieval pair they follow is stored. As each new user clicks on a retrieval, the feature-vectors of the retrieved case are refined. In the case of a recommendation, where the query is an existing case, the query case is also refined. This refinement is based on the information which is common to both query and recommendation, and how often the pair appears within the stored user interactions.

Let p denote the number of times a query-retrieval pair has previously been successful, and t denote the number of times a query has been successful. The strength s of refinement is calculated as

$$s = 1 + \log\left(\frac{p}{t} + 1\right) \tag{2}$$

which is in the range 1 to 1.69. Refinement is then applied to information which is common to both the query and retrieval as

$$Q_{fi} = \begin{cases} s \cdot Q_{fi} & \text{if } Q_{fi} > 0 \text{ and } R_{fi} > 0 \\ Q_{fi} & \text{otherwise} \end{cases} \qquad (3)$$

and

$$R_{fi} = \begin{cases} s \cdot R_{fi} & \text{if } Q_{fi} > 0 \text{ and } R_{fi} > 0 \\ R_{fi} & \text{otherwise} \end{cases} \qquad (4)$$

where f is the feature-vector being refined, and i is the ith dimension of feature vector f. The conditions '$Q_{fi} > 0$ and $R_{fi} > 0$', and '$R_{fi} > 0$ and $Q_{fi} > 0$' assure that only feature dimensions which are shared are refined. After the weights are modified, the vectors are re-normalised. This refinement process is illustrated for a single feature-vector in Figure 4.

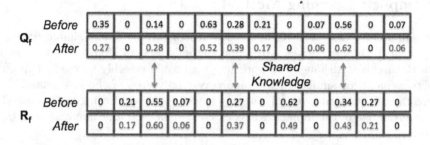

Fig. 4. Implicit Learning of Feature Dimension Weights

Normalising each feature-vector after increasing the weights on shared knowledge, means that the weights of unshared knowledge will be decreased. The amount of information in the feature-vector remains constant, but is moved from one dimension to another. The effect of increasing the weights of shared dimensions in a feature vector will mean that the query-retrieval pair will become more similar to each other. However, this modification may also have further effects throughout the entire search space, pushing both query and retrieval to be more similar to some unknown items, and less similar to other items. This behaviour is desirable, since it is a consequence of refining an item's representation to more accurately reflect how it relates to the collection.

6   User Experiments

To evaluate the methods discussed, an archive website was developed and made available to selected users online. The goal of this website was to measure and evaluate several objectives. Firstly, we wish to evaluate the implicit learning method that has been developed. However, we are also interested in the user engagement with the archive system. In this section the website is described, and the evaluation method for our user trial is presented.

6.1   User Engagement with the Archive Website

The system that was developed allowed users to navigate the archive in several different ways, shown in Figure 5. A search box was provided in the top right corner to allow free-text searching of the archive. As a user views an item in the archive they are also provided with a set of recommended items, below the item they are currently viewing. In Figure 5, a user is viewing the 'Dye Pot' item, and this item is used as a query for the recommendation list. These recommendations were generated using the similarity measure defined in Equation (1), and as users followed recommendations the implicit learning defined in Equation (4) was applied.

A recent trend in many catalog-based websites is to allow users to bookmark, or 'favourite' items that they want to return to. To gain a fuller insight into how users decide to interact with our archive system therefore, we also implement this feature. As users browse the collection, they can add any item to a favourites bar which is accessible from every page.

Fig. 5. Smart Textiles Archive System

6.2   Recommendation Quality

To evaluate the effect of our implicit learning method we measure how effective the recommender system is. Ideally, we would hope that as the items become more refined, the most relevant recommendations will appear closer to the top of the ranked recommendation list. To measure this we therefore calculate the average recommendation rank of a single query item as

$$\text{average recommendation rank} = \frac{\sum_{j=1}^{J} \text{rank}(j)}{J} \tag{5}$$

where j is an instance of the query item being used, J is the number of times the query has ever been used, and $\text{rank}(j)$ is the position that the recommendation a user clicked on was presented. We report the mean average recommendation rank across all queries, obtained at varying levels of refinement.

7 Results: Effects of Implicit Learning

The archive was made available online for 1 month, and invitations were sent to both industry experts and students from the textiles field. Over this trial period, 8 industry experts from 5 separate companies, and 11 students from 2 universities participated in the study. Each user was asked to complete several investigative tasks, for example, finding out about a certain type of material using the archive.

Figure 6 shows the effect that our implicit learning method has on recommendation rank. The vertical axis shows the average position of the recommendation that was clicked on by a user. The horizontal axis shows the number of times that the query item has been modified by the implicit learning method.

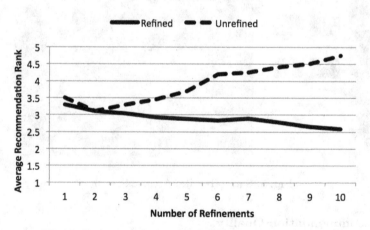

Fig. 6. Recommendation Rank After Implicit Learning

The solid black line in Figure 6 shows the average recommendation rank achieved after each refinement using our implicit learning method. After a single refinement to the query, the average recommendation rank is 3.3, meaning that on average, a user will click on either the 3rd or 4th recommendation in our ordered list. As the system is used more, and more refinements are made, implicit learning has the effect of lowering this average rank. After 10 refinements the average recommendation position the user clicks on is 2.6.

The dashed line in Figure 6 shows the ranking that recommendations would have if the refined representation is not used. This line is not flat as expected, because all clicks were logged for recommendations made using our refined representation. The difference between the solid and dashed line does however illustrate the power of our implicit learning method. After 10 refinements the average rank of a good recommendation is reduced by 1 position in the ranked list. This makes the browsing experience for the user easier, and helps them to find the items they are interested in more quickly.

Reducing the average rank of a recommendation is not only because the query has been refined; we apply our implicit learning method to both the query and recommendation. This is an important contributing factor to the results observed in Figure 6. When a user is viewing an item, and follows a recommendation, the recommended item is refined. This item then becomes the query for the recommendations provided on the page which is loaded, and the more meaningful refined representation can be used. This illustrates the power of refining a representation, compared to simply re-ordering results based on previous cases.

8 Results: User Engagement with Archive

Further to evaluating our implicit learning method, we are also interested in how users engage with the digital archive. In our workshops to establish the initial case base, the stakeholders categorised each item as primary, secondary, and supplementary. The stakeholders also constructed a map of items, based on the relationships that they considered to be relevant to someone browsing the archive. These workshops provide a wealth of information that can be compared to how the industry expert and students engaged with the digital archive.

8.1 Physical and Digital Relationship Maps

Using the user behaviour logged by our digital archive, we are able to construct the recommendation map illustrated in Figure 7. Each node in the map represents an item, and each edge represents a followed recommendation. The direction of the arrows represents the direction query to recommendation, where the arrow points to the recommended item. The position of each node in the map is determined using multi-dimensional scaling, where each item constitutes a new dimension, and the distance between items is inversely proportional to the number of times the query-recommendation pair occurred.

In the recommendation map three distinct clusters can be observed, each annotated in Figure 7. Perhaps unsurprisingly these clusters emerge as some of the major themes discussed throughout the workshop phase. The company is a textiles manufacturer, and the importance of history and heritage was agreed upon by all stakeholders. The design / process cluster occurs as a result of the distinct nature of the production and design teams' contributions.

The brightness of each node in the map indicates how frequently the item was viewed, where lightest is most frequent. This helps to identify hubs within the items, which are used frequently as both a query and a recommendation. This shading also helps to identify the items that may be considered primary, secondary, and supplementary, in a similar manner to the offline workshops.

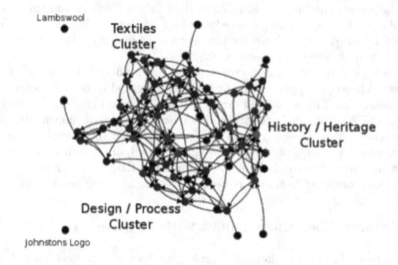

Fig. 7. Map of Followed Recommendations

Figure 8 illustrates the map created by the stakeholders during the workshop. The position of each node in the map is frozen to match the positions defined in Figure 7, facilitating comparison of the offline and online item relationships. One thing that stands out instantly between the two maps is that the stakeholder map contains many fewer edges. The reason for this was primarily due to time constraints, and the stakeholders focussed on the primary assets.

The length of edges in the stakeholder map are in general very long. This indicates that the relationships of items as perceived by the stakeholder does not match the relationship engaged with by the online users. The history / heritage cluster is well understood by the stakeholder, illustrated by the shorter edges, but there is confusion between the design and textiles clusters. Within the company, the realisation that product and design process should be considered as separate areas of interest was not made. From a business perspective these two areas are very closely connected, but to online users they are not. Finally, it can be observed that some nodes have no edges in the online map. This illustrates a further misunderstanding from the stakeholder perspective of how an end user would want to traverse the digital archive.

These observations of differences between the stakeholder workshop and online usage further illustrate the importance of our representation refinement method. At the point of putting the system online, each item begins with an unrefined representation, that has been constructed by the item owners. While this representation correctly describes an item, it does not necessarily reflect a description

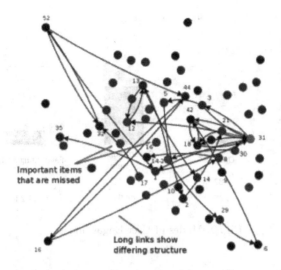

Fig. 8. Map Created by Stakeholders

that is meaningful to end users. The creation of knowledge can be influenced by company processes and perceptions, which we have shown may not match those of an end user. Implicit learning is an essential part of refining the initial representations to produce high quality recommendations that are meaningful to the end users.

8.2 Modes of Engagement

In our description of the system developed we mentioned that there were 3 differing modes of engagement with the archive: querying, recommendations, and a favouriting system. The reason for these three modes was to learn more about how users of archive systems choose to engage with such systems. At an industry event, we interviewed some attendees from the textiles community about how they envisage archive systems. The large majority of responses were that users would have a specific thing they are looking for, and would therefore primarily engage with the query system. When discussing how effective they believed a recommender system would be, some of the participants were strongly dismissive of such systems, and believed they would never be appropriate for such archives.

As each of the industry expert and student users interacted with the archive, their mode of interaction was logged. Figure 9 shows the level of each mode of engagement, as a percentage of total engagement with the system. Contrary to what the interviewed users initially predicted, the primary mode of engagement with the archive is through the recommender system, accounting for 73% of all engagement with the archive. In comparison, only 10% of engagement was through the query system, and 17% through the favouriting system.

These results highlight the importance that a recommender system plays in many online applications, where users may not know exactly what they are

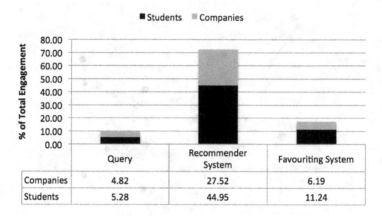

Fig. 9. Modes of User Engagement

looking for. Feedback collected from users indicated that they were using the archive as a form of inspiration: starting with an item they recognise found through the query system, and then exploring the similar items suggested by the recommender system. This places the recommender system as a key component of the archive, and as such implicit learning is essential to take advantage of the end user interests to improve the system.

9   Conclusions

We have conducted a study of how a physical archive can be transformed into a digital archive. This study includes how stakeholders of the archive understand their knowledge, and how this knowledge may be transformed into a meaningful digital system. Through the workshop phase of the project, important insights and knowledge about the physical collection was learned, which facilitated the construction of the archive.

However, results show that knowledge learned at the workshop phase is biased towards the stakeholder perception of the items and their relationships. We have developed a new implicit learning method that can refine the knowledge learned from the workshop phase, and adapt the representations of an item to reflect end user interests. Our method strengthens the shared knowledge between queries and recommendations, thus strengthening the knowledge that is important to their relationship. Over time this process refines the entire collection so that the representations are weighted appropriately, and a more meaningful similarity measure can be calculated. The result of this refinement process is that relevant items appear closer to the top of ranked recommendation lists, thus enabling the end user to find interesting relevant items more quickly.

Investigations of user engagement with our archive highlights the importance of a recommender system to exploratory and inspirational systems. Quite often the user will have a vague idea of what they are looking for, but do not know how to describe it as a well defined query. This conflicted with the engagement patterns predicted by stakeholders, who knew their data well enough that perhaps a query system may have worked well internally. However, our study further highlights the need for a retrieval system to learn what is important to the end-user base, managed successfully by our implicit learning method.

Acknowledgements. The authors wish to acknowledge AHRC for its support through its Digital Transformations Research Development award for this project (AH/J013218/1) and the industrial collaborator Johnstons of Elgin for their contribution of data and stakeholder workshops.

References

1. Evans, C.K.: Developing the perfect fashion archive. In: 1st Global Conference: Fashion Exploring Critical Issues (2009)
2. Paterson, C.: Selling fashion: realizing the research potential of the house of fraser archive, university of glasgow archive services. Textile History 40(2), 170–184 (2009)
3. Brown, C.: Digitisation projects at the university of dundee archive services. Program: Electronic Library and Information Systems 40(2), 168–177 (2006)
4. Nisanbayev, Y., Ko, I., Abdullaev, S., Na, H., Lim, D.: E-commerce applications of the hybrid reasoning method. In: International Conference on New Trends in Information and Service Science, NISS 2009, pp. 797–801. IEEE (2009)
5. Vollrath, I., Wilke, W., Bergmann, R.: Case-based reasoning support for online catalog sales. IEEE Internet Computing 2(4), 47–54 (1998)
6. Bridge, D., Göker, M., McGinty, L., Smyth, B.: Case-based recommender systems. The Knowledge Engineering Review 20(03), 315–320 (2005)
7. Lorenzi, F., Ricci, F.: Case-based recommender systems: A unifying view. In: Mobasher, B., Anand, S.S. (eds.) ITWP 2003. LNCS (LNAI), vol. 3169, pp. 89–113. Springer, Heidelberg (2005)
8. Milne, P., Wiratunga, N., Lothian, R., Song, D.: Reuse of search experience for resource transformation. In: Workshop Proceedings of the 8th International Conference on Case-Based Reasoning, pp. 45–54 (2009)
9. Ontañón, S., Plaza, E.: On similarity measures based on a refinement lattice. In: McGinty, L., Wilson, D.C. (eds.) ICCBR 2009. LNCS, vol. 5650, pp. 240–255. Springer, Heidelberg (2009)
10. Lowe, D.G.: Object recognition from local scale-invariant features. In: The Proceedings of the Seventh IEEE International Conference on Computer Vision, vol. 2, pp. 1150–1157. IEEE (1999)

Case-Based Goal-Driven Coordination
of Multiple Learning Agents

Ulit Jaidee[1], Héctor Muñoz-Avila[1], and David W. Aha[2]

[1] Department of Computer Science & Engineering, Lehigh University, Bethlehem, PA 18015
[2] Navy Center for Applied Research in AI, Naval Research Laboratory (Code 5514),
Washington, DC 20375
{ulj208,munoz}@lehigh.edu, david.aha@nrl.navy.mil

Abstract. Although several recent studies have been published on goal reasoning (i.e., the study of agents that can self-select their goals), none have focused on the task of learning and acting on large state and action spaces. We introduce GDA-C, a case-based goal reasoning algorithm that divides the state and action space among cooperating learning agents. Cooperation between agents emerges because (1) they share a common reward function and (2) GDA-C formulates the goal that each agent needs to achieve. We claim that its case-based approach for goal formulation is critical to the agents' performance. To test this claim we conducted an empirical study using the Wargus RTS environment, where we found that GDA-C outperforms its non-GDA ablation.

Keywords: Goal-driven autonomy, case-based reasoning, multi-agent systems.

1   Introduction

Goal reasoning is the study of introspective agents that can reason about what goals they should dynamically pursue (Klenk *et al.*, in press). Goal-driven autonomy (GDA) (Muñoz-Avila *et al.*, 2010; Molineaux *et al.*, 2010) is a model of goal reasoning in which an agent revises its goals by reasoning about discrepancies it encounters during plan execution monitoring (i.e., when its expectations are not met) and their explanation.

GDA agents have not been designed to learn and act with large state and action spaces. This can be a problem when applying them to real-time strategy (RTS) games, which are characterized by large state and action spaces. In these games, agents control multiple kinds of units and structures, each with the ability to perform certain actions in certain states, while competing versus an opponent who is controlling his own units and structures. To date, GDA agents that learn to play RTS games can be applied to only limited scenarios (e.g., Jaidee *et al.*, 2011) or control only a small set of decision-making tasks within a larger hard-coded system that plays the full game (e.g., (Weber *et al.*, 2012)).

To address this limitation, we introduce GDA-C, a partial GDA agent (i.e., it implements only two of GDA's four steps) that divides the state and action space among multiple reinforcement learning (RL) agents, each of which acts and learns in

S.J. Delany and S. Ontañón (Eds.): ICCBR 2013, LNAI 7969, pp. 164–178, 2013.
© Springer-Verlag Berlin Heidelberg 2013

the environment. Each RL agent performs decision making for all the units with a common set of actions. For example, in an RTS game, it will assign one RL agent to control all footmen, which is a melee combat unit, and another RL agent to control the barracks, which is a building that produces units (e.g., footmen).

That is, each RL agent α_k is responsible for learning and reasoning on a space of size $|S_k|\,|\mathcal{A}_k|$, where S_k is agent α_k's set of states and $\mathcal{A}_k$ is its set of actions. Thus, GDA-C's overall memory requirement, assuming n RL agents, is $|S_1\|\mathcal{A}_1| +...+ |S_n\|\mathcal{A}_n|$. This is a substantial reduction in memory requirements compared to a system that must reason with a space of size $|S\|\mathcal{A}|$, where $S = \cup_{1\leq i<n} S_i$ and $\mathcal{A} = \cup_{1\leq i<n}\mathcal{A}_i$ (i.e., all combinations of states and actions).

Cooperation among GDA-C's agents emerges as a result of combining two factors: (1) all its agents share a common reward function and (2) it uses case-based reasoning (CBR) techniques to acquire/retain and reuse/apply its goal formulation knowledge.

We claim that agents which share the same reward function, augmented with coordination provided by GDA-C, outperform agents that coordinate by sharing only the reward function. To test this claim we conducted an empirical evaluation using the Wargus RTS environment in which we compared the performance of GDA-C versus CLASS$_{QL}$ (Jaidee *et al.*, 2012), an ablation of GDA-C where the RL agents coordinate by sharing only the same reward function. We first compared GDA-C and CLASS$_{QL}$ indirectly by testing both against the built-in AI in Wargus, a proficient AI that comes with the game and is designed to be competitive versus a mid-range player. We also compared their performance in direct competitions. Our main findings are:

- Versus the Wargus built-in AI, GDA-C outperformed CLASS$_{QL}$
- GDA-C also outperformed CLASS$_{QL}$ in most direct comparisons

Our paper continues as follows. In Section 2 we describe related work, and then present a formalization of the problem we are studying in Section 3. Section 4 discusses the RL agents and Section 5 presents the GDA-C algorithm. Section 6 discusses the states and actions defined in Wargus, while Section 7 presents the empirical evaluation. Finally, Section 8 concludes with future work suggestions.

2   Related Work

Weber et al. (2012) report on EISBot, a system that can play a complete RTS game. EISBot plays complete games by using six managers (e.g., for building an economy, combat), only one of which uses GDA (i.e., it selects which units to produce). The GDA system GRL (Jaidee *et al.*, 2012) plays RTS game scenarios were each side starts with a fixed number of units. No buildings are allowed and hence no new units can be produced, which drastically reduces the GRL's state and action space. In contrast to these and other GDA systems that play RTS games (e.g., Weber *et al.*, 2010), GDA-C controls most aspects of an RTS game by assigning units and buildings of the same type to a specialized agent.

Many GDA systems manage expectations that are predicted outcomes from the agent's actions. Most work on GDA assumes deterministic expectations (i.e., the same outcome occurs when actions are taken in the same state). These expectations are computed in a number of ways. Cox (2007) generates instances of expectations by using a given model of abstract explanation patterns. Molineaux et al. (2011) use planning operators to define expectations. Borrowing ideas from Weber et al. (2012), GDA-C uses vectors of numerical features to represent the states and expects that actions will increase their values (e.g., sample features include total gold generated or number of units, both of which a player would like to increase). When this does not happen (i.e., when this constraint is violated), a *discrepancy* occurs.

When most GDA algorithms detect a discrepancy between an observed and an expected state, they formulate new goals in response. Some systems use rule-based reasoning to select a new goal (Cox, 2007), while others rank goals in a priority list and use truth–maintenance techniques to connect discrepancies with new goals to pursue (Molineaux *et al.*, 2010). Interactive techniques have also been used to elicit new goals from a user (Powell *et al.*, 2011). GDA-C instead learns to rank goals by using RL techniques based on the performance of the individual agents.

GDA-C has some characteristics in common with GRL (Jaidee *et al.*, 2012), which also uses RL for goal formulation. However, GRL is a single agent system and, unlike GDA-C, cannot scale to play complete RTS games.[1]

3 Multi-agent Setting

The task we focus on is to control a set Γ of agents $\alpha_1,...,\alpha_n$, where each belongs to one class c_k in $C = \{c_1, c_2, ..., c_n\}$. Each class c_k has its own set of class-specific states S_k. The collection of all states is denoted by S (i.e., $S = \bigcup_{1 \leq k \leq n} S_k$). Each agent α_k can execute actions in $\mathcal{A}_k$ for every class specific state.

A stochastic policy is a mapping $\pi_k: S_k \to \{(a, p) | a \in \mathcal{A}_k, p \in [0,1]\}$. That is, for every state $s \in S_k$, $\pi_k(s)$ defines a distribution $\{(a_1, p_1), ..., (a_n, p_n)\}$, where a_i is an action in $\mathcal{A}_k$ and p_i is the expected return from taking action a_i in state s and following policy π_k thereafter. The return is a function of the rewards obtained. For example, the return can be defined as the summation of the future rewards. Our goal is to find an optimal policy $\pi_k^*: S_k \to \{(a, p) | a \in \mathcal{A}_k, p \in [0,1]\}$ such that π_k^* maximizes the expected return.

It is easy to prove that, given a collection of n independent policies $\pi_1,...,\pi_n$ where each π_k maximizes the returns for class k, then $\pi = (\pi_1,...,\pi_n)$ is an optimal policy. As we will see in Section 4, GDA-C uses this fact by running n RL agents, one for each class c_k. If each converges to an optimal policy, their n-tuple policies will be an optimal policy for the overall problem. This results in a substantial reduction of the memory requirement compared to a conventional RL agent that is attempting to learn a combined optimal policy $\pi^* = (\pi_1,...,\pi_n)$ where each π_i must reason on all states and actions. This conventional RL agent will require $|S| \times |\mathcal{A}|$ space, where $S = \bigcup_{1 \leq i < n} S_i$ and $\mathcal{A} = \bigcup_{1 \leq i < n} \mathcal{A}_i$ (i.e., counting all combinations of state n-tuples times all combinations of n-tuple actions).

---

[1] This means that the player starts with limited resources, units, and structures but can (1) harvest additional resources, (2) build any structure, (3) train any unit, (4) research any technology, and (5) control the units to defeat an opponent.

In contrast the n agents $\alpha_1, ..., \alpha_n$ each attempt to learn an optimal policy π^*_k, which requires $|S_1 \times \mathcal{A}_1| + ... + |S_{n-1} \times \mathcal{A}_n|$ space (i.e., adding the memory requirements of each individual agent α_k). The following inequality holds:

$$|S \times \mathcal{A}| \geq |S_1 \times \mathcal{A}_1| + ... + |S_{n-1} \times \mathcal{A}_n|,$$

assuming that $\forall_{i,j} (i \neq j) (\mathcal{A}_i \cap \mathcal{A}_j = \{\} \wedge S_i \cap S_j = \{\})$. This is common in RTS games where the actions that a unit of a certain type can take are disjoint from the actions of units of a different type. Under these assumptions, and for $n \geq 2$, the expression on the right is substantially lower than the expression on the left. For example, assuming $\forall_k$ $|S_k| = t$ and $|\mathcal{A}_k| = m$, then the LHS is equal to $(n \times t \times n \times m)$ whereas the RHS is equal to $(n \times t \times m)$. That is, the space saved is $(1 - \frac{1}{n}) \times 100\%$. The following table summarizes some of the savings for these assumptions:

Table 1. Space saved by GDA-C compared to a conventional RL agent

n	% of saved space	n	% of saved space	n	% of saved space
1	0	4	75	10	90
2	50	5	80	20	95

In our work, we use Q-learning (Sutton and Barto, 1998) to control each of the α_k agents. Thus, our baseline system consists of n Q-learning agents that are guaranteed, after a number of iterations, to converge to an optimal policy. We refer to this baseline system as $CLASS_{QL}$ because each Q-learning (QL) agent controls a class of units in Wargus.

4 Case Bases and Information Flow in the GDA-C Agent

We now discuss how case-based reasoning techniques are used in GDA-C to manage goals on top of $CLASS_{QL}$. Figure 1 depicts a high-level view of the information flow in GDA-C, which embeds the standard RL model (Sutton and Barto, 1998). GDA-C has two threads that execute in parallel. First, the GDA thread selects a goal, which in turn determines the policy that each RL agent will use and refine. Second, the $CLASS_{QL}$ thread performs Q-learning to control each of the α_k agents.

The two case bases, *Policies* and *GFCB*, are learned from previous instances (e.g., previously played Wargus games). Given a policy π, a *trajectory* is a sequence of states $< s_0, ..., s_m >$ visited when following π from the starting state s_0. Any such state in this trajectory is a goal that can be achieved by executing π. The policy is assigned the last state in a trajectory as its goal. The case base *Policies* is a collection of pairs (g, π_g), where π_g is a policy that should be used when pursuing goal g. GDA-C stores such pairs as it encounters them.

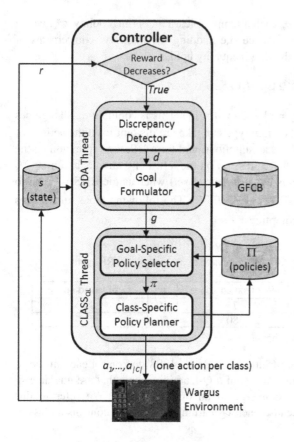

Fig. 1. Information flow in GDA-C

The other case base assists with goal formulation. When a discrepancy d occurs between the expected state X and the actual state observed by the **Discrepancy Detector**, this discrepancy is passed to the **Goal Formulator**, which uses GFCB to formulate a new goal. **GFCB** maintains, for each (current) goal discrepancy pair, (g,d), a collection $\{(g_1,v_1),..,(g_m,v_m)\}$, where g_i is a goal to pursue next and v_i is the expected return of pursuing it. It outputs the next goal g to achieve.

The **Goal-Specific Policy Selector** selects a policy π based on the current goal g. The **Class-Specific Policy Learner** learns policies for new goals and refines the policies of existing goals. It uses Q-learning to update the Q-table entry $Q(s,a)$, given current state s and action taken a, as well as next state s' and next reward r (Sutton and Barto, 1998).

In many environments, there is no optimal policy for all situations. For example, in an adversarial game, a policy might be effective against one opponent's strategy but not versus others. By changing the goal when the system is underperforming, GDA-C changes the policy that is being executed, thereby making it more likely to adjust to different strategies.

We now provide formal definitions for the GDA process. Here we assume a state is represented as a vector $s = (v_1, \dots, v_n)$ of numeric features, where v_i is a value of a feature f_i. Borrowing ideas from Weber et al. (2012), the agent uses **optimistic expectations**. An expectation is **optimistic** iff $v'_i \nless v_i$, where *expectation* $e = (v'_1, \dots, v'_n)$ and *previous state* $s = (v_1, \dots, v_n)$. We use optimistic expectation *implicitly* in our algorithm. That is, if the previous state is $s = (v_1, \dots, v_n)$ and, after executing an action, we reach a current state $s' = (v'_1, \dots, v'_n)$ such that, for some k, $v'_k < v_k$ holds, then a **discrepancy** occurs. We represent a discrepancy as a vector of Boolean values $d = (b_1,\dots,b_n)$, where b_k is *true* iff $v'_k < v_k$ holds. Basically, the agent expects that actions will not decrease the features' values. As we will see in Section 6, our state model consists of numeric features (e.g., the numbers of our own units) whose values the agent expects will remain the same or increase, but not decrease.

5   The GDA-C Algorithm

GDA-C coordinates the execution of a set of RL agents and how they learn. GDA-C uses an online learning process to update the Policies and GFCB case bases. Each GDA-C agent has its own individual Q-table. All q-values in Q-tables are initialized to zero. In each iteration of the algorithm, only some **units** (i.e., class instances such as peasants and archers) will be ready to execute a new action because others may be busy. Every unit records the state when it starts executing its current action. This is necessary for updating values in Q-tables. Below we present the pseudo-code of GDA-C, followed by its description.

GDA-C (Δ, Π, GFCB, $\mathcal{C}$, $\mathcal{A}$, ε, g_0) =
1: $s' \leftarrow$ GETSTATE(); $d' \leftarrow$ CALCULATEDISCREPANCY(s', s'); $\pi \leftarrow \Pi(g_0)$; $g \leftarrow g_0$
2: //-------- GDA thread --------
3:      **while** episode continues
4:          $s \leftarrow$ GETSTATE()
5:          WAIT(Δ)
6:          $r \leftarrow U(s) - U(s')$ // s' is the prior state
7:          **if** $r < 0$ **then**
8:              $d \leftarrow$ CALCULATEDISCREPANCY(s', s)
9:              GFCB $\leftarrow$ Q-LEARNINGUPDATE(GFCB, d', g, d, r)
10:             $g \leftarrow$ GET(GFCB, d, ε)   // ε-greedy selection
11:             $\pi \leftarrow \Pi(g)$
12:             $s' \leftarrow s$; $d' \leftarrow d$
13: //-------- CLASS$_{QL}$ thread --------
14:     **while** episode continues
15:         $s \leftarrow$ GETSTATE()
16:         **parallel for** each class $c \in \mathcal{C}$ // this loop controls agent α_c
17:             $s_c \leftarrow$ GETCLASSSTATE(c, s)
18:             $\mathcal{A}_c \leftarrow$ GETCLASSACTIONS($\mathcal{A}, c$); $A \leftarrow$ GETVALIDACTIONS($\mathcal{A}_c, s_c$)
19:             $\pi_c \leftarrow \pi(c)$
20:             **for** each instance $u \in c$ // this loop controls each unit or instance of class c
21:                 **if** u is a new instance **then**
22:                     $s'_u \leftarrow s_c$; $a'_u \leftarrow$ do-nothing
23:                 **if** instance u finished its action **then**
24:                     $r_u \leftarrow$ U(s_c) – U(s'_u)   // U(s) is the utility of state s
25:                     $\pi_c \leftarrow$ Q-LEARNINGUPDATE($\pi_c, s'_u, a'_u, s_c, r_u$)
26:                     $a \leftarrow$ GETACTION($\pi_c, \varepsilon, s_c, A$)
27:                     EXECUTEACTION(a)
28:                     $s'_u \leftarrow s_c$; $a'_u \leftarrow a$
29: **return** Π, GFCB

GDA-C has two threads that execute in parallel and begin simultaneously when a game episode starts. The GDA thread (lines 3-12) selects a goal, which in turn determines the policy $\pi = (\pi_1,...,\pi_n)$ that each RL agent will use and refine. The CLASS$_{QL}$ thread (Lines 14-28) performs Q-learning control on each of the α_k agents. When the GDA thread is deactivated (which is how our baseline system CLASS$_{QL}$ works), the CLASS$_{QL}$ thread refines the *same* policy from the beginning of the episode to the end. When the GDA thread is activated, the policy that CLASS$_{QL}$ refines is the *most recent* one selected by the GDA thread.

GDA-C receives as input a constant number Δ (a delay before selecting the next goal), a policy case base Π, a goal formulation case base (GFCB), a set of classes $\mathcal{C}$, a set of actions $\mathcal{A}$, a constant value ε (for *ε-greedy* selection in Q-learning, whereby the action with the highest value is chosen with a probability $1-\varepsilon$ and a random action is chosen with a probability ε), and the initial goal g_0.

The GDA Thread: The variable s' is initialized by observing the current state, d' is initialized with a null discrepancy (e.g., CalculateDiscrepancy(s', s')), and a policy π is retrieved from Π for the initial goal g_0 (all in Line 1). While the episode continues (Line 3), the current state s is observed (Line 4). After waiting for Δ time (Line 5), the reward r is obtained by comparing the utilities of current state s and previous state s' (Line 6). Our utility function calculates, for a given state, the total "hit-points" of the controlled team's units and subtracts those of the opponent team. When a unit is "hit" by other units, its hit-points will be decreased. A unit "dies" when its hit-points decrease to zero. If the reward is negative (Line 7), a new goal (and hence a new policy) will be selected as follows. First, the discrepancy d between s' and s is computed (Line 8). GFCB is then updated via Q-learning, taking into account previous discrepancy d', current goal g, discrepancy d, and reward r (Line 9). Then ε-greedy selection is used to select a new goal g from GFCB with discrepancy d (Line 10). Next, a new policy π is retrieved from Π for goal g (Line 11). Policy π will be updated in the CLASS$_{QL}$ thread. Finally, previous state s' and discrepancy d' are updated (Line 12).

The CLASS$_{QL}$ Thread: While the episode continues (Line 14), the current state s is updated (Line 15). For each class c in the set of classes $\mathcal{C}$ (Line 16), the class-specific state s_c is acquired from s (Line 17). Agents from different classes have different sets of actions that they can perform. Therefore, a set of valid actions A must be obtained for each class s_c (Line 18). π_c is initialized with the policy for class c, which depends on the overall policy π updated in the GDA thread (Line 19). For each instance (or *unit*) u of class c (Line 20), if u is a new instance, initialize its state and action (Line 21-22). If u finished its action then calculate the reward r_u and update the policy π_c via Q-learning (Line 23-25). A new action is selected based on policy π_c using ε-greedy action selection (Line 26). Finally, the action is executed and the previous state s'_u and previous action a'_u are updated (Lines 27-28).

When the episode ends, GDA-C will return the policy case base Π and the goal formulation case base GFCB (Line 29).

Although at any point each agent α_k is following and updating a policy π_k, this does *not* mean that all units controlled by α_k will execute the same action. This is due to a combination of three factors. First, even when two units u and u' start executing the same action at the same time, there is no guarantee that they will finish at the same time. For example, if the action is to move u and u' to a specific location L, one of them might be hindered (e.g., engaged in combat with an enemy unit). Hence, u and u' might reach L at different times and therefore the subsequent actions they execute might differ because the state may have changed between the times that they arrive at L. Second, actions are stochastic (chosen with the ε-greedy method). Third, the policies are changing over time as a result of Q-learning or even altogether as a result of the GDA thread. Therefore, at different times, even if in the same state, units might perform different actions.

6   States and Actions in Wargus

In this paper, we use Wargus in our experiments. Wargus is a widely used testbed for adversarial environments (e.g., (Aha *et al.*, 2005; Judah *et al.*, 2010; Mehta *et al.*, 2009; Ontañón and Ram, 2011)). In Wargus decision making must be conducted in real time. Wargus follows a rock-paper-scissors model for unit-versus-unit combat. For example, archers are strong versus footmen but weak versus knights. For these reasons, Mehta et al. (2009) argue that Wargus is a good research testbed for studying agent-based control methods. Each type of unit defines a unique class c so that every unit in that class can execute a set of actions $\mathcal{A}_c$. For example, an Archer can shoot an enemy from a distance while Gryphon Rider can fly across any barriers. Analogously, we also model each type of building (e.g., a Blacksmith, which can improve a unit's defense and damage, and a Barracks, which produces units such as Archers and Footmen for a specified amount of resources) as a class. In total, we modeled the following 12 classes:

1. Town Hall
2. Blacksmith
3. Lumber Mill
4. Church
5. Barrack
6. Knight
7. Footman
8. Archer
9. Ballista
10. Gryphon Rider
11. Gryphon Aviary
12. Peasant Builder

Each unit type has a different state representation. To reduce the number of states, we discretized features (*italicized* below) with many values (e.g., we used 18 bins for gold, where bin 1 means 0 gold and bin 18 corresponds to more than 4000). We also measure the distances from an enemy's units to the controlled player's camp using Manhattan distance. The features of the state representations per class are:

- Town Hall: *food, peasants*
- Blacksmith, Lumber Mill and Church: *gold, wood*
- Barrack: *gold, food, footmen, archer, ballista, knight*
- Knight, Footman, Archer, Ballista and Gryphon Rider: *our footmen, enemy footmen, number of enemy town halls, enemy peasants, enemy attackable units*

that are stronger than our footmen, enemy attackable units that are weaker than our footmen

- Gryphon Aviary: *gold, food, gryphon rider*
- Peasant Builder: *gold, wood, food, number of barracks, lumber mill built?,*[2] *blacksmith built?, church built?, gryphon built?, path to a gold mine?, town hall built?*

CLASS$_{QL}$ (and, hence, GDA-C) reasons with composite actions such as "knight attack enemy camp", which are composed of several primitive actions such as selecting a building in the enemy camp, navigating to that building, and attacking it. Below is the list of all possible actions per class (by default every class can perform the action *do-nothing*):

- Town Hall: train peasant, upgrade to keep/castle
- Blacksmith: upgrade sword level 1, same but 2, upgrade human shield level 1, same but 2, upgrade ballista level 1, same but 2
- Lumber Mill: upgrade arrow level 1, same but 2, elven ranger training, ranger scouting, research longbow, ranger marksmanship
- Church: upgrade knights, research healing, research exorcism
- Barrack: train a footman, train an elven archer/ranger, train a knight/paladin, train a ballista
- Knight, Footman, Archer, Ballista, Gryphon Rider: wait for attack, attack the enemy's town hall/great hall, attack all enemy's peasants, attack all enemy's units that are near to our camp, attack all enemy's units that have their range of attacking equal to one, same but more than one, attack all enemy's land units, attack all enemy's air units, attack all enemy's units that are weaker (the enemy's units that have hit-points less than those of us), and attack all enemy's units (no matter what kind)
- Gryphon Aviary: train a gryphon rider
- Peasant Builder: build farm, build barracks, build town hall, build lumber mill, build black smith, build a stable, build a church, and build a gryphon aviary.

Our reward function is:

$$\text{total-hit-points(controlled team)} - \text{total-hit-points(enemy team)}$$

Each unit and building is assigned a number of hit points based on their type (e.g., Paladins have more than Peasants). Games are typically played until either the controlled team or the enemy is reduced to 0 points, at which time it loses the game.

7   Empirical Study

We measured the performance of GDA-C versus its ablation CLASS$_{QL}$ in experiments on small, medium, and large Wargus maps whose sizes are 32×32, 64×64, and 128×128 cells, respectively. In each map, we have two opponent teams (human and orc). Each starts with only one Peasant/Peon (i.e., a unit used to harvest resources and construct new buildings), one Town Hall/Great Hall, and a nearby gold mine. Each competitor

---

[2] The question mark signals that this is a *binary* feature.

also starts on one side of a forest that divides the map into two parts. We added this forest and walls to provide opponents with sufficient time to build their armies. Otherwise, our algorithms will learn an efficient early attack (called a "rush"), which will end the game when the opponents have produced only a few units or buildings.

7.1   Experimental Setup

We conducted two experiments. In the first, we compared the performance of each algorithm (i.e., GDA-C or CLASS$_{QL}$) against Wargus's built-in AI. The built-in AI in Wargus is quite good; it provides a challenging game to an average human player. In the second, we instead compared their performance in a direct competition. We use five adversaries (defined below) and the Wargus' built-in AI to train and test each algorithm. These adversaries can construct any type of unit unless otherwise stated:

- *Land Attack*: This tries to balance offensive/defensive actions with research. It builds only land units.
- *Soldier's Rush*: This attempts to overwhelm the opponent with cheap military units early in the game.
- *Knight's Rush*: This attempts to quickly research advanced technologies, and launch large attacks with the strongest units in the game (knights for humans and ogres for orcs).
- *Student Scripts*: We included the top two competitors that were created by students for a classroom tournament.

To ensure there is no bias because of the landscape, we swapped the sides of each team in each round. Also, to prevent race inequities, in each round each team plays once with each race (i.e., human or orc).

In Experiment 1, we trained GDA-C and CLASS$_{QL}$ by playing one game versus each of the five adversaries. We then tested GDA-C and CLASS$_{QL}$ by playing one game against the Wargus's built-in AI. The performance metric is:

$$(wins(\text{GDA-C}) - wins(\text{built-in AI})) - (wins(\text{CLASS}_{QL}) - wins(\text{built-in AI})),$$

where *wins(A)* is the number of wins for team *A*. For Experiment 2, we trained GDA-C and CLASS$_{QL}$ with all five adversaries and then tested them in combat against each other. We report results for the average of ten runs, where the performance metric is:

$$wins(\text{GDA-C}) - wins(\text{CLASS}_{QL})$$

In Experiment 1, the matches pitting the two algorithms versus the built-in AI took place after training GDA-C and CLASS$_{QL}$ against each of the other five adversaries for n games, where we varied $n = 0,1,2,...,N$. Similarly, in Experiment 2 the matches pitting GDA-C versus CLASS$_{QL}$ took place after training them against each of the

Table 2. The average time of running a game for both experiments

Map size	One game	Experiment 1	Experiment 2
small	31 sec	25 hours	38 hours
medium	3 min 27 sec	115 hours	172 hours
large	11 min 28 sec	191 hours	286 hours

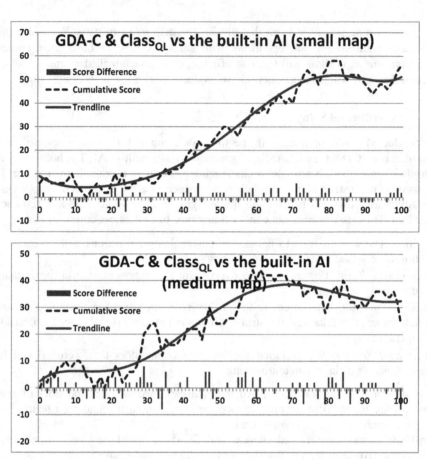

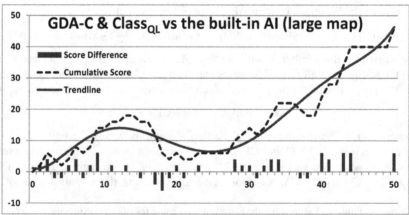

Fig. 2. The results of Experiment 1: The relative performance of GDA-C versus CLASS$_{QL}$ playing against the built-in Wargus AI on the three maps

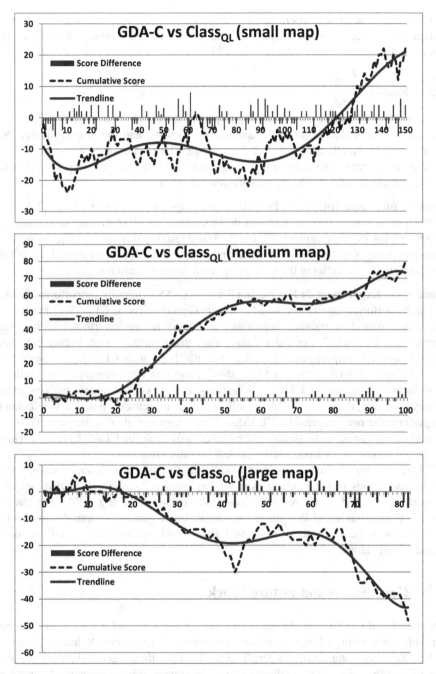

Fig. 3. The results of Experiment 2: GDA-C versus CLASS$_{QL}$ on the small, medium, and large maps

adversaries for n games, where again $n = 0,1,2,...,N$. The total number N of games varied as indicated in the results. Table 2 shows the running times for the experiments.

7.2   Results

Figures 2 and 3 display the results for Experiments 1 and 2, respectively. For both experiments each data point is the average of 10 tests, and the graphs display the results for the small, medium, and large maps. There are two curves: the score difference for each data point and the cumulative score difference up to that data point. The x-axis refers to the training iteration number.

Results for Experiment 1: For all three maps, both GDA-C and CLASS$_{QL}$ outperform the built-in AI (not shown in the graphs) but GDA-C does so at a higher rate than CLASS$_{QL}$, as shown in Figure 2. These results illustrate the effectiveness of changing policies as GDA-C does when underperforming compared to sticking to the current policies and refining them by using reinforcement learning.

Results for Experiment 2: For the small map CLASS$_{QL}$ initially outperforms GDA-C but its performance improves and it eventually outperforms CLASS$_{QL}$. From $x = 110$ (i.e., after 110 training iterations), it begins to outperform CLASS$_{QL}$ and surpasses it by $x = 117$. For the medium map, the algorithms start evenly but then GDA-C quickly outperforms CLASS$_{QL}$. For the large map CLASS$_{QL}$ outperforms GDA-C. We ran further iterations (not shown) and this trend continues. We believe that for the large map, CLASS$_{QL}$ is learning a very good strategy, perhaps even optimal for the map, and GDA-C will continue to retrieve policies that cannot outperform the one executed by CLASS$_{QL}$. This suggests that, at some point, GDA-C should deactivate its GDA thread and continue only with the CLASS$_{QL}$ thread. How we would identify such a point is a topic left for future research.

There is a lot of fluctuation in individual data points. For example, despite the cumulative trends in the medium map for Experiment 2, which show that GDA-C outperforms CLASS$_{QL}$, the reverse occasionally occurs (e.g., at $x = 70$). The reason for this fluctuation is that Wargus is a stochastic environment that introduces a lot of randomness in the outcomes of individual actions and, hence, in the overall outcome of individual games.

8   Conclusions and Future Work

We introduced GDA-C, an algorithm that divides the state and action spaces among multiple, cooperating RL agents, where each agent uses Q-learning to learn a different policy for controlling units of a single class. Because these agents share a common reward function, they can coordinate. GDA-C augments this coordination by using a partial goal-driven autonomy (GDA) agent to retrieve previously stored policies for the RL agents to apply and further revise. Our experiments demonstrate that GDA-C outperforms its ablation, CLASS$_{QL}$, in most situations.

For future work we want to explore two directions. First, we plan to make the state representation more general so it does not depend on the expectation that the feature's values must increase. To do this, we will borrow ideas from our previous GDA research (e.g., (Jaidee *et al.*, 2011; 2012)), in which we used more general state representations. Second, we will examine alternative GDA agents. GDA-C does not include two steps that are common to the GDA model, namely *discrepancy explanation* and *goal management*. We will assess the utility of generating explanations of discrepancies for GDA-C. That is, recent research on GDA (Molineaux *et al.*, 2012) has demonstrated the value of using discrepancy explanations to determine which goals to select, and this may also be true for our studies. Alternative methods for goal management also exist. GDA-C simply replaces one goal with another, without considering, for example, whether the initial goal should simply be delayed. We will study more comprehensive strategies for goal management in our future research.

Acknowledgements. This work was supported in part by NSF grant 1217888.

References

Aha, D.W., Molineaux, M., Ponsen, M.: Learning to win: Case-based plan selection in a real-time strategy game. In: Muñoz-Ávila, H., Ricci, F. (eds.) ICCBR 2005. LNCS (LNAI), vol. 3620, pp. 5–20. Springer, Heidelberg (2005)

Cox, M.T.: Perpetual self-aware cognitive agents. AI Magazine 28(1), 23–45 (2007)

Jaidee, U., Muñoz-Avila, H., Aha, D.W.: Integrated learning for goal-driven autonomy. In: Proceedings of the Twenty-Second International Joint Conference on Artificial Intelligence. AAAI Press, Barcelona (2011)

Jaidee, U., Muñoz-Avila, H., Aha, D.W.: Learning and reusing goal-specific policies for goal-driven autonomy. In: Agudo, B.D., Watson, I. (eds.) ICCBR 2012. LNCS, vol. 7466, pp. 182–195. Springer, Heidelberg (2012)

Judah, K., Roy, S., Fern, A., Dietterich, T.G.: Reinforcement learning via practice and critique advice. In: Proceedings of the Twenty-Fourth AAAI Conference on Artificial Intelligence. AAAI Press, Atlanta (2010)

Klenk, M., Molineaux, M., Aha, D.W.: Goal-driven autonomy for responding to unexpected events in strategy simulations. To Appear in Computational Intelligence (in press)

Mehta, M., Ontañón, S., Ram, A.: Using meta-reasoning to improve the performance of case-based planning. In: McGinty, L., Wilson, D.C. (eds.) ICCBR 2009. LNCS, vol. 5650, pp. 210–224. Springer, Heidelberg (2009)

Molineaux, M., Klenk, M., Aha, D.W.: Goal-driven autonomy in a Navy strategy simulation. In: Proceedings of the Twenty-Fourth AAAI Conference on Artificial Intelligence. AAAI Press, Atlanta (2010)

Molineaux, M., Kuter, U., Klenk, M.: What just happened? Explaining the past in planning and execution. In: Roth-Berghofer, T., Tintarev, N., Leake, D.B. (eds.) Explanation-Aware Computing: Papers from the IJCAI Workshop, Barcelona, Spain (2011)

Molineaux, M., Kuter, U., Klenk, M.: DiscoverHistory: Understanding the past in planning and execution. In: Proceedings of the Eleventh International Conference on Autonomous Agents and Multiagent Systems (pp, pp. 989–996. ACM Press, Valencia (2012)

Muñoz-Avila, H., Aha, D.W., Jaidee, U., Klenk, M., Molineaux, M.: Applying goal directed autonomy to a team shooter game. In: Proceedings of the Twenty-Third Florida Artificial Intelligence Research Society Conference, pp. 465–470. AAAI Press, Daytona Beach (2010)

Ontañón, S., Ram, A.: Case-based reasoning and user-generated AI for real-time strategy games. In: González-Calero, P.A., Gómez-Martín, M.A. (eds.) Artificial Intelligence for Computer Games, Springer, Berlin (2011)

Powell, J., Molineaux, M., Aha, D.W.: Active and interactive discovery of goal selection knowledge. In: Proceedings of the Twenty-Fourth Conference of the Florida AI Research Society, AAAI Press, West Palm Beach (2011)

Sutton, R.S., Barto, A.G.: Reinforcement learning: An introduction. MIT Press, Cambridge (1998)

Weber, B., Mateas, M., Jhala, A.: Applying goal-driven autonomy to StarCraft. In: Proceedings of the Sixth Conference on Artificial Intelligence and Interactive Digital Entertainment. AAAI Press, Stanford (2010)

Weber, B., Mateas, M., Jhala, A.: Learning from demonstration for goal-driven autonomy. In: Proceedings of the Twenty-Sixth AAAI Conference on Artificial Intelligence. AAAI Press, Toronto (2012)

On Deriving Adaptation Rule Confidence
from the Rule Generation Process

Vahid Jalali and David Leake

School of Informatics and Computing, Indiana University
Bloomington IN 47408, USA
{vjalalib,leake}@cs.indiana.edu

Abstract. Previous case-based reasoning research makes a compelling case for the importance of CBR systems determining the system's confidence in its conclusions, and has developed useful analyses of how characteristics of individual cases and the case base as a whole influence confidence. This paper argues that in systems which perform case adaptation, an important additional indicator for solution confidence is confidence in the adaptations performed. Assessing confidence of adaptation rules may be particularly important when knowledge-light methods are applied to generate adaptations automatically from the case base, giving the opportunity to improve performance by astute rule selection. The paper proposes a new method for calculating rule confidence for automatically-generated adaptation rules for regression tasks, when the rules are generated by the common "difference heuristic" method of comparing pairs of cases in a case base, and a method for confidence-influenced selection of cases to adapt. The method is evaluated in four domains, showing performance gains over baseline methods and case based regression without using confidence knowledge.

1 Introduction

Previous research on CBR confidence, has focused largely on how case and case-base characteristics can be used to estimate confidence (e.g., [1]). An interesting question is how confidence can apply to other CBR knowledge containers to improve confidence estimates for results or even to improve solution quality. For example, for any given level of case confidence, selecting high confidence adaptation rules may improve accuracy. This short paper explores assessing confidence of newly-generated rules, based on the confidence of the data used to generate the rules, and exploiting rule and case confidence information to improve performance.

The paper presents a case study for generating and selecting adaptation rules and selecting cases to adapt for case-based regression tasks, i.e., tasks for which the goal is to generate a numerical value. In the basic form of case-based regression, solutions are generated by k-NN, with values computed by simple averaging approaches. To improve performance, the CBR community has developed a number of knowledge-light methods for generating domain-specific adaptations automatically from the case base. For example Hanney and Keane [2] propose an approach based on applying a difference heuristic to pairs of cases, to generate rules which map similar problem differences to similar solution differences. This paper considers whether it is possible to estimate the

S.J. Delany and S. Ontañón (Eds.): ICCBR 2013, LNAI 7969, pp. 179–187, 2013.

confidence of such rules as they are generated, and how such estimates can affect the performance of case-based regression. It reports on an ablation study which assesses the performance of learning and application with confidence considerations, compared to rule learning and application without, and compared to a baseline of k-NN. It also explores how confidence characteristics of domain cases affect performance of the approach. Experimental results show that using case confidences for selecting base cases and using them in ranking adaptation rules can decrease estimation errors, and that the amount of improvement in each domain varies based on the distribution of the confidence level of the cases.

The paper is organized as follows. Section 2 reviews previous research on solution and adaptation rule confidence estimation. Section 3 introduces our method for assessing the adaptation rule confidence based on the rule generation process. Section 4 shows results of evaluations comparing accuracy of case based regression with and without using confidence information. Section 5 presents conclusions and future work.

2   Previous Research

Solution Confidence Estimation: A number of previous projects have proposed methods for estimating confidence in a CBR system's conclusions by considering characteristics of cases and of the problem space as a whole. For example, Cheetham and Price [3,4] explore the problem of assigning confidence to solutions in a CBR system by considering similarity scores of the retrieved cases, the deviation of retrieved solutions, etc. [3]. Delany et al. [5] propose estimating classification confidence based on the similarity between the target case and its k nearest neighbors. Reilly et al. [6] propose a feature-based confidence model for assessing confidence of the proposed values for a feature by recommender systems. Mulayim and Arcos [7] propose a method for identifying areas of the problem space for which cases give uncertain solutions, identifying regions of the case base in which those problems are located, to guide maintenance. Hullermeier's [8] Credible Case-based Inference (CCBI), for regression tasks, estimates solutions based on "credible sets" of cases, i.e., sets of high confidence cases. Craw et al. [9] propose using an auxiliary case based reasoning system to predict solution correctness and confidence. Their confidence estimation method works by retrieving a set of adaptation cases with their associated correctness predictions and combining the predictions.

Considering Adaptation Confidence: Distance Weighting: Distance-weighted k-NN can be seen as using a simple proxy for adaptation confidence when solutions are calculated: If confidence in the contributions from different cases depends on their proximity to a query, distance weighted k-NN takes that adaptation confidence into account by weighting nearby cases more heavily.

Determining Confidence using Rule Frequency: Previous research on adaptation rule generation has considered the space of resulting rules, noting that frequency in the pool of generated adaptation rules may give an indication of the reliability of generated rules. Hanney and Keane's [2] seminal work proposes estimating confidence of rules by their frequency. Wilke et al.'s [10] adaptation learning system takes a similar rule-generation approach, estimating rule certainty based on the degree of generalization applied during rule generation.

Provenance-based Confidence Estimation: Leake and Dial [11] propose a provenance-based method for assessing the quality/confidence of adaptation rules by using feedback propagation. Their method assigns blames to applied adaptation rules based on the reported flaws in a solution via feedback. Minor et al. [12] assess the confidence of adaptation results in workflow domains by using introspection for the modified parts of the adapted solutions. For tracking the adaptation process they use provenance information of each used workflow element.

3   Deriving Rule Confidence from the Rule Generation Process

The method introduced in this paper is an extension to the previous work of the authors on an approach called Ensemble of Adaptations for Regression (EAR) [13]. EAR generates adaptation rules by comparing pairs of cases in a local neighborhood around the input problem. Adaptation rules are built by comparing the problem and solution parts of pair of cases and identifying their differences to generate rules that map the observed differences in problems to the observed differences in solutions.

EAR is a lazy approach to adaptation generation. Given an input problem it selects a set of base cases to adapt and also generates a set of adaptation rules as explained above. For adapting the value of each base case it combines the values of the top r rules that most resemble the differences between the base case and the input problem. The final estimation is generated by combining the adapted values of all selected base cases.

EAR selects cases to adapt based on their distance to the input problem and ranks adaptation rules according to the similarity of their problem parts to the corresponding differences between the base cases and the input problem. However, we hypothesize that using confidence knowledge in these two steps can improve the accuracy of the estimations in domains with uncertainty in the values of the cases.

Our approach, which we call confidence-based case-based reasoning (ConfCBR) estimates adaptation rule confidence based on the quality of inputs to the difference heuristic—The confidence of the cases compared to generate the rule. In addition, it uses confidence knowledge in selecting base cases both to adapt cases and for building adaptation rules. Algorithm 1 summarizes the overall process of ConfCBR. In the algorithm, $NeighborhoodSelection\,(Q, n, CB)$ and $RankRules\,(NewRules, C, Q)$ rank base cases and adaptation rules by using (1) and (2) respectively. The subprocesses are described in more detail in the following discussion.

Selecting Cases from which to Generate New Solutions: We hypothesize that the best solutions will be generated by balancing a tradeoff between case similarity and confidence. (This tradeoff is mediated by the quality of case adaptation. If adaptation were always perfect, we would expect the best results always to be obtained by adapting the most confident case.)

Let P be the set of all possible problems, and CB the cases in the case base. Let $distance : P \times P \to R^+$ measure distance between problem descriptions (for convenience, we will sometimes use the case itself to designate its problem part). Let $conf : CB \to [0, 1]$ compute case confidence. Then for a case C, ConfCBR calculates the ranking value of that case for base case selection by:

Algorithm 1. ConfCBR's method of estimating target values

Input:
Q: input query
n: number of base cases to be used
CB: case base
Output: Estimated solution value for Q

> $CasesToAdapt \leftarrow$ NeighborhoodSelection(Q,n,CB)
> $NewRules \leftarrow$ RuleGenerationStrategy(Q, CB)
> for C in $CasesToAdapt$ do
> $\quad RankedRules \leftarrow RankRules\,(NewRules, C, Q)$
> $\quad ValueEstimate_C \leftarrow CombineAdaptations\,(RankedRules, C)$
> end for
> return CombineVals($\cup_{C \in CasesToAdapt} ValEstimate_C$)

$$rank\,(C, Q) \equiv \frac{conf\,(C)^{\alpha}}{distance\,(C, Q)} \tag{1}$$

where α is a positive real number whose values tune the base case rankings for different domains. If α is equal to zero, cases will be ranked merely based on their distance to the input query. Increasing the value of α has the effect of assigning higher rankings to more confident cases, with large values of α asymptotically approaching assigning equal ranking values (i.e. zero) to cases that are not 100% confident.

Adaptation Rule Ranking: To rank candidate adaptation rules generated from the case base by the difference heuristic, ConfCBR computes a ranking score based on two factors: (1) confidence of the cases from which the rules were generated, and (2) how close the cases from which the rule was generated are to the case to adapt. More specifically, let $R_{i,j}$ be the adaptation rule built from C_i and C_j, and $\Delta\,(C, Q)$ be the difference vector of the features of the query Q. The ranking value of $R_{i,j}$ for adapting the solution of C is:

$$rank\,(R_{i,j}, C, Q) \equiv \frac{(conf\,(C_i) \times conf\,(C_j))^{\beta}}{distance\,(Q_{i,j}, \Delta\,(C, Q))} \tag{2}$$

where β is a positive real number whose values tune the ranking of adaptation rules in different domains.

4 Evaluation

Our experiments address the following questions about ConfCBR:

1. How does ConfCBR's accuracy compare to its accuracy in the ablated conditions (1) rule confidence considered, case confidence ignored, (2) case confidence considered, rule confidence ignored, (3) both confidence factors ignored.

2. How does ConfCBR's accuracy (using confidence, and without) compare to the baseline of distance-weighted k-NN?
3. How do varying case confidence distributions in the case base affect ConfCBR's accuracy?

4.1 Experimental Design

Experiments applied ConfCBR for case-based regression in four sample domains from the UCI repository [14]: MPG, Auto, Hardware and Housing. A data cleaning process removed cases with unknown values and discarded the symbolic features. For each feature, values were standardized by subtracting that feature value's mean from each individual feature value and dividing the result by the standard deviation of that feature. Leave-one-out testing is used for all domains and estimation errors are calculated in terms of Root Mean Squared Error (RMSE). Hill climbing was used to determine the values of α and β for each domain.

Adaptation rules are generated by comparing the top 5% nearest neighbors of the input query, using Euclidean distance in (1) and (2). The top 5% cases are ranked and selected by using (1). These cases are used both as base cases to adapt and the source cases for generating adaptation rules. The case-based regression system then ranks adaptation rules by using (2) and applies a set of rules for adapting the solution of each base case. The number of adaptations applied per base case is also determined by a hill climbing process in all domains. The final estimates are generated by combining adapted solutions of the selected base cases.

The goal of our current study is not to generate case confidence values, but rather, to assess how confidence information can be exploited, once it has been generated. To evaluate our approach under controlled conditions for which the quality of confidence estimations is known, we generated test data whose correctness was characterized by varying known confidence values, as follows. First, we randomly assigned confidence levels to the cases by a Gaussian distribution, with 0.8 and 0.2 used as the mean and standard deviation of the confidence level distributions in all domains except explicitly stated otherwise. The stored values of the cases were then adjusted randomly, according to the assigned confidence values. For example, if 0.9 is assigned to a case as its confidence value, its stored value is increased or decreased by 10% of its value. The original value of the case is used as the "correct" value for assessing performance.

4.2 Performance Comparison

The Effect of Using Confidence Knowledge. Fig. 1 depicts the RMSE for CBR regression (without using confidence knowledge), using confidence knowledge in selecting the base cases only (ConfCBRC), using confidence knowledge only for ranking adaptations only (ConfCBRR) and using confidence knowledge for both selecting base cases and ranking adaptation rules (ConfCBR), for each test domain. As expected, in all domains CBR without confidence considerations shows the worst performance. In three of the four test domains (all except MPG) using confidence knowledge in ranking adaptation rules is more successful in decreasing estimation error compared to using confidence knowledge for selecting the base cases to adapt. In almost all cases, using case confidence knowledge both for selecting base cases and ranking adaptation rules (ConfCBR) provides the most accurate results. Exceptions occur only for one configuration of the Auto domain (when solutions are generated from 5 base cases) and when 5 or more base cases were used in the Hardware domain. In both those cases, confidence-based rule ranking only outperformed the combination, but the combination outperformed rule ranking only for smaller numbers of base cases.

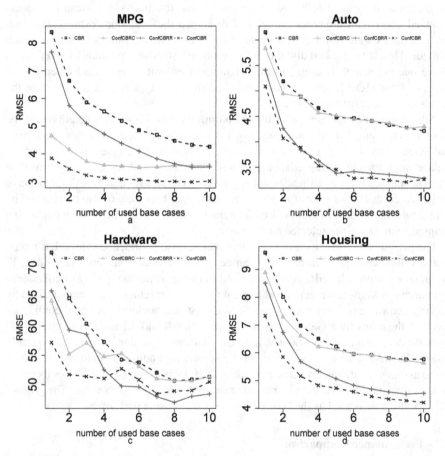

Fig. 1. RMSE comparison for no use of confidence (CBR) only using confidence for ranking base cases (ConfCBRC), only using confidence for ranking adaptation rules (ConfCBRR) and using confidence for both (ConfCBR)

ConfCBR vs. k-NN. To compare the accuracy of ConfCBR with a baseline, we conducted experiments in the test domains using conventional distance weighted k-NN, and distance-weighted k-NN enhanced with case confidence knowledge (ConfkNN). The confidence knowledge in ConfkNN is used for selecting the cases from which the solution will be generated by using (1).

Fig. 2 shows the RMSE of k-NN, Confidence based k-NN (ConfkNN), CBR and ConfCBR in the test domains. For three out of four domains (all except Hardware), the worst performance belongs to the basic CBR approach, which reflects its inability to adjust to varying confidence levels (either of base cases or the cases from which adaptations are built). The largest performance gap between k-NN and ConfCBR is observed for the Hardware domain (ConfCBR performs 33% better than ConfkNN) while this gap is minimized in the MPG domain (ConfCBR only performs 3% better than ConfkNN). In all domains ConfCBR performs better than the baseline methods, showing that CBR enhanced with confidence knowledge is able to generate more accurate estimations compared to the other tested alternative methods.

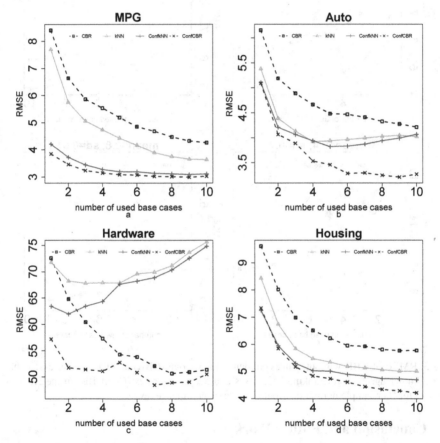

Fig. 2. RMSE of CBR, ConfCBR, k-NN, and k-NN enhanced with confidence knowledge (ConfkNN)

The Effect of Case Confidence Level Distribution on ConfCBR. To assess the effect of case confidence level distribution on ConfCBR we conducted experiments in the Housing domain with different case confidence level distributions. Fig. 3 shows the RMSE of CBR and ConfCBR in the housing domain for four different confidence distributions. As a reference, Part a of Fig. 3 repeats the results of Part a of Fig. 1. However, parts b, c and d show results for three new distributions. Based on comparison of parts a, b and c of Fig. 3 we hypothesize that the difference between the relative performance of CBR and ConfCBR depends more on the standard deviation of the case confidence levels than on their mean value. Part d suggests that for relatively small standard deviations, performance of CBR and ConfCBR is almost identical.

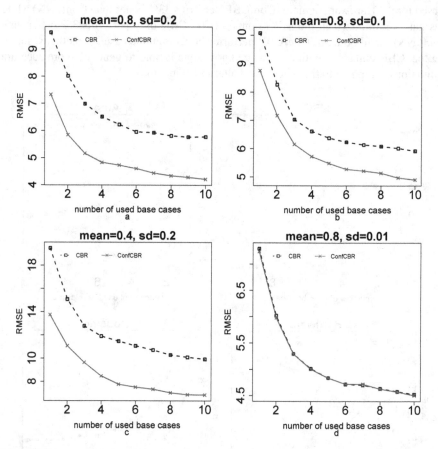

Fig. 3. RMSE of CBR without using confidence knowledge (CBR) and CBR using case confidence knowledge both for ranking base cases and adaptation rules (ConfCBR) in the Housing domain for four sample normal distributions of case confidence levels

5   Conclusion and Future Work

This paper has explored how considerations of case confidence can help to assess the confidence of adaptation rules generated by the difference heuristic, and how adaptation rule and case confidence can be brought to bear not only to assess confidence of solutions, but to generate better solutions. It has introduced a new method, confidence-based case based regression (ConfCBR), which uses confidence knowledge both for selecting base cases and ranking adaptations. Experimental results showed that ConfCBR outperforms a corresponding case-based approach to regression without confidence knowledge and k-NN baseline methods in four sample domains, often by substantial margins. Results also showed that the benefit depends significantly on the distribution of case confidence levels in a case base.

Our current investigation was based on sample data for which confidence levels were artificially created. We are developing methods for estimating the case confidences of new case bases by statistical methods such as outlier detection, and intend to examine ConfCBR performance when case confidence levels are estimated automatically. Another future direction is studying other methods for estimating rule confidence.

References

1. Cheetham, W., Price, J.: Measures of solution accuracy in case-based reasoning systems. In: Funk, P., González Calero, P.A. (eds.) ECCBR 2004. LNCS (LNAI), vol. 3155, pp. 106–118. Springer, Heidelberg (2004)
2. Hanney, K., Keane, M.: The adaptation knowledge bottleneck: How to ease it by learning from cases. In: Leake, D.B., Plaza, E. (eds.) ICCBR 1997. LNCS, vol. 1266, pp. 359–370. Springer, Heidelberg (1997)
3. Cheetham, W.: Case-based reasoning with confidence. In: Blanzieri, E., Portinale, L. (eds.) EWCBR 2000. LNCS (LNAI), vol. 1898, pp. 15–25. Springer, Heidelberg (2000)
4. Cheetham, W., Price, J.: Measures of solution accuracy in case-based reasoning systems. In: Funk, P., González Calero, P.A. (eds.) ECCBR 2004. LNCS (LNAI), vol. 3155, pp. 106–118. Springer, Heidelberg (2004)
5. Delany, S.J., Cunningham, P., Doyle, D., Zamolotskikh, A.: Generating estimates of classification confidence for a case-based spam filter. In: Muñoz-Ávila, H., Ricci, F. (eds.) ICCBR 2005. LNCS (LNAI), vol. 3620, pp. 177–190. Springer, Heidelberg (2005)
6. Reilly, J., Smyth, B., McGinty, L., McCarthy, K.: Critiquing with confidence. In: Muñoz-Ávila, H., Ricci, F. (eds.) ICCBR 2005. LNCS (LNAI), vol. 3620, pp. 436–450. Springer, Heidelberg (2005)
7. Mülâyim, O., Arcos, J.-L.: Understanding dubious future problems. In: Althoff, K.-D., Bergmann, R., Minor, M., Hanft, A. (eds.) ECCBR 2008. LNCS (LNAI), vol. 5239, pp. 385–399. Springer, Heidelberg (2008)
8. Hullermeier, E.: Credible case-based inference using similarity profiles. IEEE Trans. on Knowl. and Data Eng. 19(6), 847–858 (2007)
9. Craw, S., Jarmulak, J., Rowe, R.: Learning and applying case-based adaptation knowledge. In: Aha, D.W., Watson, I. (eds.) ICCBR 2001. LNCS (LNAI), vol. 2080, pp. 131–145. Springer, Heidelberg (2001)
10. Wilke, W., Vollrath, I., Althoff, K.D., Bergmann, R.: A framework for learning adaptation knowledge based on knowledge light approaches. In: Proceedings of the Fifth German Workshop on Case-Based Reasoning, pp. 235–242 (1997)
11. Leake, D.B., Dial, S.A.: Using case provenance to propagate feedback to cases and adaptations. In: Althoff, K.-D., Bergmann, R., Minor, M., Hanft, A. (eds.) ECCBR 2008. LNCS (LNAI), vol. 5239, pp. 255–268. Springer, Heidelberg (2008)
12. Minor, M., Islam, M. S., Schumacher, P.: Confidence in workflow adaptation. In: Agudo, B.D., Watson, I. (eds.) ICCBR 2012. LNCS, vol. 7466, pp. 255–268. Springer, Heidelberg (2012)
13. Jalali, V., Leake, D.: Extending case adaptation with automatically-generated ensembles of adaptation rules. In: Delany, S.J., Ontañón, S. (eds.) ICCBR 2013. LNCS, vol. 7969, pp. 188–202. Springer, Heidelberg (2013)
14. Frank, A., Asuncion, A.: UCI machine learning repository (2010), http://archive.ics.uci.edu/ml

Extending Case Adaptation
with Automatically-Generated Ensembles
of Adaptation Rules

Vahid Jalali and David Leake

School of Informatics and Computing, Indiana University
Bloomington IN 47408, USA
{vjalalib,leake}@cs.indiana.edu

Abstract. Case-based regression often relies on simple case adaptation methods. This paper investigates new approaches to enriching the adaptation capabilities of case-based regression systems, based on the use of ensembles of adaptation rules generated from the case base. The paper explores both local and global methods for generating adaptation rules from the case base, and presents methods for ranking the generated rules and combining the resulting ensemble of adaptation rules to generate new solutions. It tests these methods in five standard domains, evaluating their performance compared to four baseline methods, standard k-NN, linear regression, locally weighted linear regression, and an ensemble of k-NN predictors with different feature subsets. The results demonstrate that the proposed method generally outperforms the baselines and that the accuracy of adaptation based on locally-generated rules is highly competitive with that of global rule-generation methods with much greater computational cost.

1 Introduction

Case-based reasoning (CBR) (e.g., Mantaras *et al.*, 2005) solves new problems by retrieving stored prior cases solving similar problems, and adapting their solutions to fit new circumstances, based on the differences between the new problem and problems addressed by the retrieved case(s). When CBR is applied to synthesis tasks in knowledge-rich domains, an important component of its success is the use of sophisticated case adaptation strategies. However, when CBR approach is applied to regression tasks, reliance on simple case adaptation is common. For example, k-Nearest Neighbor (k-NN) regression approaches often compute target values as a distance-weighted average of the values of the k cases closest to the input problem. Using simple adaptation helps to alleviate the knowledge acquisition problem for case adaptation knowledge for these tasks, and in practice can achieve good performance (e.g., [2]). However, the contrast between extensive focus on case adaptation in other CBR areas and the limited attention to richer adaptation methods for case-based regression raises the question of whether case-based regression performance could be improved by generating richer combination/adaptation rules automatically.

This paper presents new approaches for automatically augmenting adaptation capabilities for case-based regression, using only knowledge contained in the case base. Its

S.J. Delany and S. Ontañón (Eds.): ICCBR 2013, LNAI 7969, pp. 188–202, 2013.

primary contributions are methods for generating adaptation rules from local or global sets of cases, methods for applying ensembles of adaptation rules, and an experimental comparison of alternative strategies for using local and global information in both adaptation rule learning and rule application, which illuminates the relative performance benefits of local and global approaches.

The paper is organized as follows. Section 2 introduces the strategies we consider for generating adaptation rules and selecting the base cases from which the estimations are built. Section 3 introduces our approach, Ensemble of Adaptations for Regression (EAR), a general technique for augmenting k-NN for regression tasks by automatically generating adaptation rules, choosing which of many potentially applicable rules to apply, and using the resulting ensemble of rules for generating new solutions. It also describes the basic parameters of the approach, which adjust its use of local versus global information in selecting cases to adapt and generating adaptation rules from existing cases. Section 4 presents results of an evaluation comparing alternative versions of EAR with k-NN, linear regression, and locally weighted linear regression for estimating solutions in five sample domains. The study shows encouraging results for accuracy and for the ability to rely on local information, compared to more computationally expensive use of extensive global information, which suggests the practicality of lazy learning of adaptation rules based on local information. Section 5 compares related work on using ensemble techniques in CBR and knowledge-light methods for generating and applying adaptations for case-based regression tasks. Section 6 presents conclusions and future work.

2   Learning and Applying Ensembles of Adaptation Rules

Our basic approach to adaptation rule generation builds on the case difference heuristic approach proposed by Hanney and Keane [3] and further explored by others (e.g., [4,5]). The case difference approach builds new adaptation rules from pairs of cases and compares their problem parts (respectively, solution parts), and identifies their differences to generate a candidate rule mapping the observed difference in problems to the observed difference in solutions. For example, for predicting apartment rental prices, if two apartments differ in that one has an additional bedroom, and its price is higher, an adaptation rule could be generated to increase estimated rent when adapting a prior case to predict the price of an apartment with an additional bedroom. Applying the case difference approach depends on addressing questions such as which pairs of cases will be used to generate adaptation rules, how rules will be generated, and how the resulting rule set will be applied to new problems. In the next section, we discuss EAR's strategies for addressing these, and in Section 5 we compare these approaches to previous work.

EAR is a lazy approach to adaptation rule generation. Given an input problem, it generates ensembles of adaptations as needed, based on preselected criteria for (1) selecting a neighborhood of cases in the case base from which to generate solutions by adaptations, and (2) generating rules for adapting each of those cases, ranking the rules for each case and combining the values of the top r rules, and finally, combining the values generated for each of the cases to adapt. This process is summarized in Algorithm 1.

Algorithm 1. EAR's basic algorithm

Input:

Q: input query

n: number of base cases to adapt to solve query

r: number of rules to be applied per base case

CB: case base

Output: Estimated solution value for Q

$CasesToAdapt \leftarrow$ NeighborhoodSelection(Q,n,CB)

$NewRules: \leftarrow$ RuleGenerationStrategy($Q,CasesToAdapt,CB$)

for c in $CasesToAdapt$ **do**

    RankedRules $\leftarrow$ RankRules($NewRules,c,Q$)

    $ValEstimate(c) \leftarrow$ CombineAdaptations($RankedRules, c, r$)

end for

return CombineVals($\cup_{c \in CasesToAdapt} ValEstimate(c)$)

2.1 Selecting Source Cases to Adapt

We consider three general alternatives for selecting the cases to adapt, defined by whether they use highly local, local, or global cases:

1. Nearest: Select only the single nearest neighbor to the query (1-NN)
2. Local: Select the k nearest neighbors to the query (k-NN, for a small value of k greater than 1)
3. Global: Select all cases in the case base

As we discuss in Section 5, adaptation learning methods using *nearest* and *local* case sets have been considered previously in CBR, but the global approach is seldom used. Because the global approach may consider cases quite dissimilar from the input query, its feasibility depends on the quality of the adaptation and combination strategies used.

2.2 Selecting Cases from which to Generate Adaptation Rules

For each case selected to be used as a source case for adaptation, we consider three options for selecting pairs of cases to be used to generate adaptation rules, as listed below. The strategies are described by their names, which have the form StartingCasesEndingCases, where StartingCases describes a set of cases for which rules will be generated, and EndingCases describes the cases to which each of the StartingCases will be compared. Each comparison results in a different rule, so a single starting case may participate in the formation of many rules.

1. Local cases–Local neighbors: Generating adaptation rules by comparing each pair of cases in the local neighborhood of the query.
2. Global cases–Local neighbors: Generating adaptation rules by comparing each case in the case base with its k nearest neighbors
3. Global cases–Global neighbors: Generating adaptation rules by comparing all cases in the case base

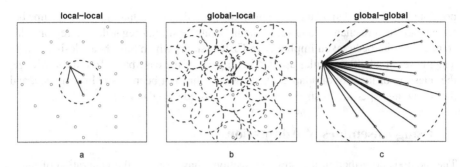

Fig. 1. Illustration of (a) Local cases–Local neighbors, (b) Global cases–Local neighbors, and (c) Global cases–Global neighbors

Figure 1 illustrates the three methods. In each figure, the input problem is at the center. Circles enclosed by dotted lines show neighborhoods of cases from which adaptations will be generated, and a sample point is connected to the cases with which it will be compared to generate adaptation rules.

Potential Tradeoffs. Combining each of the three selection strategies with one of the three adaptation generation strategies gives nine possible approaches. Each approach has potential ramifications for efficiency and accuracy of adaptation.

Ramifications for efficiency: The different methods provide different levels of efficiency. Using Global cases to generate adaptations from Local neighbors requires considering more rules than generating adaptations for local cases only. Using Global cases–Global neighbors, determining each adaptation requires $O(n^2)$ processing for a case base with n cases, which may be infeasible for large case bases.

A related ramification is the potential for lazy learning of adaptation rules as needed. The Global cases–Global neighbors approach requires processing all potential adaptations. If this is applied to the system's initial cases, a rule set can be stored for future use, avoiding re-calculation but potentially requiring considerable storage and—if kept static—not reflecting new cases added to the case base. Local cases–Local neighbors is amenable to a lazy approach with just-in-time generation of adaptation rules, which could enable incremental adaptation rule generation, with adaptation rule generation taking into account any new cases added to the case base in the region of the query.

Ramifications for accuracy: It is more difficult to anticipate the accuracy effects of the strategies. For example, one might hypothesize that generating adaptation rules from local cases would be beneficial because the adaptations are being generated from the same area of the domain space as the input query, making them more likely to properly address the differences between the input query and the base case(s). On the other hand, limiting the scope of adaptations to the context of the input query might sacrifice the benefit of considering distant cases corresponding to relevant adaptations. This raises the interesting question of locality of adaptation knowledge. Even if case characteristics for a particular case base are associated with particular regions of the case base, it is possible that the needed adaptation knowledge is still global: that the relationships between their feature changes and value changes may be similar regardless of region. This stance has long been taken implicitly in CBR systems which have been designed with a single set of adaptation rules applied in all parts of the case base. To our knowledge, the question of locality of adaptation knowledge has not been studied previously, and the following experiments shed some light on this question as well.

3   Using Ensembles of Adaptations

The methods described in the previous sections may result in the generation of many adaptation rules, especially for global–global rule generation. EAR's adaptation rule ensembles are composed of a subset of the selected rules, to increase adaptation efficiency. To select rules, EAR ranks them by the similarity of the current adaptation context and the context in which the rule was generated.

3.1   Defining Adaptation Context

After generating adaptation rules for an input query, EAR attempts to determine which of the generated rules are most relevant. It does this by considering both the similarity of the new query and the case for which the adaptation was generated, and the local adaptation characteristics of the case base, which we call the *adaptation context*. When selecting adaptations to apply to generate a solution for the query, EAR favors adaptations which have been generated for target problems in similar adaptation contexts. When global knowledge is used for generating the adaptations, for example, the cases used to generate an adaptation rule may be quite different from the query, but if the adaptation addressed similar differences, it may still be relevant.

Given a case C in the case base, EAR calculates its adaptation context as a vector based on comparing C to the N cases in a neighborhood containing its nearest neighbor cases. For each case feature, the covariance between the feature and the case solution is calculated over the set of cases in the neighborhood.

Let C_i^j and $Sol(C_i)$ denote the value of the jth feature and the value of the ith case respectively, $CaseMeanVal$ denote the mean of the values of the cases in the neighborhood, and $FeatureMeanVal_j$ represent the mean value of the jth feature of the cases in the neighborhood. Then the jth element of the covariance vector for case C is calculated as follows:

$$Cov_j(C) \equiv \frac{1}{N} \times \sum_{i=1}^{N}(C_i^j - FeatureMeanVal_j)(Sol(C_i) - CaseMeanVal) \quad (1)$$

If f represents the number of features, for any case C, we define $AdaptContext(C)$ to be the vector $(Cov_1, Cov_2, ...Cov_f)$.

3.2   Ranking Adaptation Rules

EAR's adaptation rule ranking considers two factors. The first is the similarity of the pair *query - source case to adapt* and the pair *target case - source case* from which the adaptation rule was generated. The second is the similarity of the adaptation context of the query to the adaptation context of the target case from which the adaptation rule was generated. However, if the adaptations are generated from local cases-local neighbors the second factor is discarded. The first factor favors adaptations generated to adapt similar pairs of cases. For each feature, EAR calculates the per-feature difference, based on a domain similarity metric, and generates a difference vector of those values.

The second factor reflects similarity of the adaptation context (as defined above), and compares the adaptation context vectors of the query and the target case. The rationale is that the same feature difference between two cases may require different adaptations in different parts of the case space, so favoring rules from similar adaptation contexts may improve adaptation results.

EAR's ranking method balances feature differences against adaptation context differences by taking the Hadamard (element-wise) product of the feature difference vector and the adaptation context vector. The ranking score is computed as the Euclidean distance between: (1) the Hadamard product of the adaptation context vector of the case to adapt and the difference vector for the case to adapt and the input query, and (2) the Hadamard product of the context vector of the adaptation rule and the vector representing feature differences of the composing cases of that rule.

More formally, suppose query Q is to be solved by adapting the case C_i. let Δ_i represent the difference vector of the features of the query Q and C_i, and let R_r be the problem part of the rth adaptation rule. Let $\circ$ represent the Hadamard product of two vectors. Then the second component of EAR's rule scoring method is calculated as:

$$d(\Delta_i, R_r) \equiv distance((AdaptContext(C_i) \circ \Delta_i), (AdaptContext(C_r) \circ R_r)) \quad (2)$$

If $D(\Delta_i, R_r)$ is the distance between Δ_i and R_r, then the final ranking of adaptation rules is achieved by using a weighted average of D and d as:

$$RuleScore(R_r) \equiv a \times D(\Delta_i, R_r) + (1-a) \times d(\Delta_i, R_r) \quad (3)$$

where $0 \leq a \leq 1$. The value of a is set to tune the ranking for different domains.

3.3 Estimating the Target Value from a Rule Ensemble

In its simplest form, k-NN estimates the value of a query by averaging the value of its k nearest neighbors. If Q is the query and $Est(Q)$ represents its estimated target value (solution), and $Sol(C_i)$ represents the known solution value of the ith nearest neighbor of Q, then k-NN estimates the value of Q as:

$$Est(Q) \equiv \frac{\sum_{i=1}^{k} Sol(C_i)}{k} \quad (4)$$

For each base case C to be adapted to provide a value for a query, EAR computes a weighted average of the values proposed by each of the n highest-ranked adaptation rules generated for that case by Eq. 3. If $r_i, 1 \leq i \leq n$ are the n top-ranked adaptation rules in order of descending rank score,

$$SuggestedVal(C) = \sum_{i=1,n} \frac{Solution(r_i)}{i} \quad (5)$$

The value for the query is then simply

$$Solution(Q) \equiv \frac{\sum_{i=1}^{k} SuggestedVal(C_i)}{k} \tag{6}$$

4 Experimental Results

We conducted experiments to address the following questions about extending case adaptation with ensembles of automatically-generated adaptation rules:

1. Can using the automatically-generated ensembles of adaptations improve accuracy over using a single adaptation?
2. How does accuracy compare for adaptations generated from local vs. global knowledge?
3. How does EAR's accuracy compare to that of the baseline regression methods locally weighted linear regression and k-NN?
4. How does EAR's accuracy compare to that of case-based regression using standard feature subset ensemble methods?
5. How does EAR's rule process ranking (based on adaptation context and case similarity) affect its performance compared to the baselines of (1) random selection of adaptation rules and (2) considering case similarity only?

4.1 Data Sets and Experimental Design

Our experiments use five data sets from the UCI repository [6]:Automobile (A), Auto MPG (AM), Housing (H), Abalone (AB), Computer Hardware (CH). For all data sets, records with unknown values were removed. To enable comparison with linear regression, only numerical features were used. (Note that if the use of adaptation context in EAR is disabled, it could be used for symbolic features as well; including those potentially would have increased accuracy for EAR when local cases-local neighbors strategy is used for generating the adaptations). For each feature, values were standardized by subtracting that feature's mean value from each individual feature value and dividing the result by the standard deviation of that feature. Table 1 summarizes the characteristics of the test domains.

The experiments estimate the target value for an input query. For the Auto, MPG, Housing, Abalone and Hardware, the respective values to estimate are price, mpg,

Table 1. Characteristics of the test domains

Domain name	# features	# cases	Avg. cases/solution	sol. sd
Auto	13	195	1.1	8.1
MPG	7	392	3.1	7.8
Housing	13	506	2.21	9.2
Abalone	7	1407	176	1.22
Hardware	6	209	1.8	160.83

MEDV (median value of owner-occupied homes in $1000's), rings (for the Abalone data set we only selected cases with rings ranging 1-8), and PRP (published relative performance) respectively. Linear regression and locally weighted linear regression tests used Weka's [7] simple linear regression and locally weighted learning classes respectively. Accuracy is measured in terms of the Mean Absolute Error (MAE) and leave-one-out testing is used for all domains unless explicitly mentioned otherwise.

Hill climbing was used to select the best neighborhood size for each domain based on the training data for calculating adaptation context, for setting the weighting factor a Eqn. 3, and for determining the number of adaptations to consider. The number of adaptations used for different variants of EAR depending on the training data ranges from one for EAR9 to at most 40 for EAR1, EAR2 and EAR3. In all experiments Euclidean distance is used as the distance function in equation 2. Note that the use of contextual information is disabled for versions of EAR that use the local-local strategy to generate adaptations (i.e. EAR1, EAR4 and EAR7).

4.2 Performance Comparison

To address experimental questions 1–3, we conducted tests to compare the results achieved by each of the 9 versions of EAR, k-NN, linear regression (LR) and locally weighted linear regression (LWLR) in the sample domains. Table 2 summarizes the results, which we discuss below. Best values are indicated in bold.

4.3 Discussion of Results

Accuracy from Ensembles vs. Single Adaptations: In the experiments, EAR4 (local, local-local), EAR5 (local, global-local), EAR6 (local, global-global) and EAR9 (global, global-global) usually yield the best results, suggesting the benefit of generating adaptations based on multiple cases and selecting adaptations from their results to combine. For most methods, the tuning process on the training data determined that generating the final value from an ensemble of the top-ranked adaptations gave the best results.

Table 2. MAE of EAR, k-NN, LWLR and LR for the sample domains

Method	Domains				
	Auto (A)	MPG (AM)	Housing (H)	Abalone (AB)	Hardware (CH)
EAR1: nearest, local-local	1.77	2.23	2.21	0.79	31.32
EAR2: nearest, global-local	1.66	2.22	2.2	0.82	31.04
EAR3: nearest, global-global	2.15	2.22	2.23	0.95	38.25
EAR4: local, local-local	1.38	1.90	2.04	0.60	28.74
EAR5: local, global-local	1.44	**1.71**	**1.90**	0.60	28.8
EAR6: local, global-global	**1.36**	1.74	2.04	0.60	28.76
EAR7: global, local-local	4.95	4.99	4.22	0.93	78.06
EAR8: global, global-local	4.30	3.73	4.46	0.91	63.98
EAR9: global, global-global	1.37	1.95	2.25	**0.59**	**28.18**
k-NN	1.61	2.00	2.74	0.61	29.12
Locally Weighted LR (LWLR)	1.61	2.02	2.54	0.68	30.82
Linear Regression (LR)	2.62	2.55	4.53	0.62	51.91

For example, EAR4 (local, local-local) yields its best results (in all domains except Abalone) when usually five to nine adaptations are combined. There were some exceptions to the general pattern in favor of using ensembles of adaptations. For EAR9 (global, global-global) in most cases using one adaptation per case in the Auto, MPG and Housing domains yields best results (for the Hardware domain, often two cases are used). For the Abalone domain the optimal number of adaptations based on the training data is on the order of 20, but the difference between using one adaptation rule and greater numbers is minimal (1%).

Effect of Domain Characteristics on EAR's Performance: We observed that EAR showed less benefit for the Abalone data set than for the other data sets, with performance of EAR often comparable to k-NN. We hypothesize that the level of improvement from EAR over k-NN could be related to the diversity of case solutions in the case base.

If a relatively large number of cases share identical solutions in a domain, and the standard deviation of solutions is low, using an appropriate similarity measure in a retrieve-only system (e.g. k-NN) may be sufficient to generate good solutions with simple averaging combination, while with more diversity, more adaptation may be needed. Table 1, shows the average number of cases sharing the same solution and the standard deviation of the solutions in the sample domains, which shows that these characteristics of the Abalone data set are substantially different from the other data sets. However, more examination is needed.

Local vs. Global Knowledge for Generating Adaptations: Table 2 shows that in most domains, the performance of EAR4 (local, local-local) is competitive with the other versions of EAR, and is superior to the baseline methods, despite the fact that it uses limited information. For example, comparing EAR4 to the most global method, EAR9 (global, global-global), MAE's are 1.38 vs. 1.37, 1.9 vs. 1.95, 2.04 vs. 2.25, 0.60 vs. 0.59, and 28.74 vs. 28.18. Because it uses limited information, it is computationally much less expensive than the global methods. Thus the local method's performance at worst has a minimal accuracy penalty, and sometimes is substantially better. Also, it has the benefit of reducing computational cost and permitting a lazy approach to adaptation rule generation).

EAR7 (global, local-local) and EAR8 (global, global-local) usually yield the worst results. In those two methods all cases are considered as base cases for estimating the target value, so adaptation generated from neighbor cases may not be appropriate for addressing the differences between the input problem and the base cases.

EAR vs. LWLR and k-NN. In all domains, the performance of EAR4 surpasses or equals that of the baseline methods, sometimes substantially so. EAR4 has almost the same performance as k-NN in Abalone and Hardware domains. In all domains, one of the nine versions of EAR has the best performance.

In Auto, MPG and Housing domains that EAR4 shows higher accuracies compared to the other baseline methods, one side paired t-test is used to assess the significance of those results. The null hypothesis is always MAE of EAR4 being less than that of k-NN

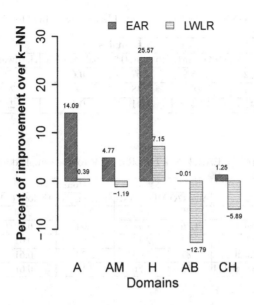

Fig. 2. Percent of improvement in MAE by EAR and LWLR over k-NN

and LWLR. For the comparison of EAR4 to k-NN in the Auto domain, $p<.007$, in MPG, $p<.062$ (so not significant), and in Housing, $p<.001$. Same values for comparing EAR4 versus LWLR are $p<.051$ (not significant), $p<0.05$ and $p<.001$ in the same order.

Figure 2 contrasts the relative improvement of EAR4 over k-NN (14%, 5%, 26%, 0% and 1%) with the relative improvement of LWLR over k-NN (0%, -1%, 7%, -13% and -6%) in the the Auto, MPG, Housing, Abalone and Hardware domains respectively.

EAR vs. Feature Subset Ensemble. As another baseline, we also compared EAR4's performance to a previously used approach for applying ensembles to CBR, feature subset ensembles (FSE). FSE uses a combination of k-NN predictors, each of which predicts based on a different subsets of case features (all subsets are of fixed size) [8]. The feature subsets are selected randomly with replacement (each subset includes at least two features), with each ensemble containing predictors based on 100 different subset of features, with evaluation by ten-fold cross validation. Both EAR4 and the feature subset ensemble methods were compared with their best parameter settings, as determined by hill climbing and leave-one-out testing on the training data for each fold. For feature subset ensembles, this determined the k value to use, and the number of features to use. For EAR4, this determined the number of base cases and adaptation rules to be used. For each domain, the local neighborhoods were set to contain the top 5% nearest neighbors of the input query. Learning was disabled for both methods. Table 3, shows Mean Absolute Error for k-NN, Feature Subset Ensemble (FSE) and EAR4 (local, local-local) on the test domains.

Table 3. MAE of EAR4, k-NN and the Feature Subset Ensemble method for the sample domains

Method	Domains				
	Auto (A)	MPG (AM)	Housing (H)	Abalone (AB)	Hardware (CH)
k-NN	1.62	2.06	2.67	**0.61**	30.3
FSE	1.51	2.28	2.48	0.7	27.51
EAR4	**1.42**	**1.84**	**2.01**	0.63	**25.79**

Table 4. MAE of EAR, k-NN, LWLR and LR for the sample domains

Method	Domains				
	Auto (A)	MPG (AM)	Housing (H)	Abalone (AB)	Hardware (CH)
EAR4: local, local-local	1.38	1.90	**2.04**	**0.60**	**28.74**
EAR6: local, global-global	**1.36**	**1.74**	**2.04**	**0.60**	28.74
Random: local, local-local	2.54	2.11	3.04	0.61	38.95
Random: local, global-global	3.87	2.43	3.29	0.61	72.86
distance only: local, global-global	1.55	1.86	2.11	0.61	30.68

The results in Table 3 show that EAR outperforms FSE in all test domains. For the Abalone domain, k-NN slightly outperforms both ensemble methods, which we hypothesize to be due to lack of domain diversity. Figure 3, shows the percent of improvement of EAR4 (local, local-local) and SFE over k-NN in the test domains.

4.4   Effect of Context-Based Rule Ranking

A final question is how much EAR's context-based adaptation rule ranking approach benefits performance. We tested this by an ablation study comparing EAR4 and EAR6's performance with three different ranking methods: (1) random ranking of adaptation rules, (2) rule ranking by case distance only, and (3) EAR's approach, balancing case similarity and adaptation context similarity.

As Table 4 shows, random ranking has the worst performance among other methods, with especially bad performance for the global-global methods, which generate more rules. The comparative difference appears to increase for domains with higher standard deviation (e.g. Hardware), and is lowest for Abalone, which has the largest average number of cases per unique solution and the lowest solution standard deviation. There the random method shows same performance as the distance only method.

Expanding the pool of adaptations with global methods decreases accuracy for distance-only method in nearly all domains, while EAR is more robust. This provides some support for the contextual information in EAR enabling it to select more appropriate adaptations from the global pool.

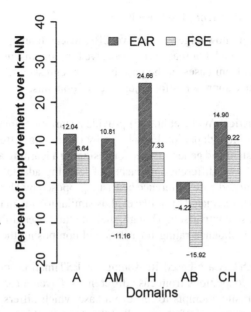

Fig. 3. Percent of improvement in MAE by EAR4 and LWLR over k-NN

5 Comparison to Previous Work

The EAR approach relates both to research on ensemble methods in CBR and on automatic adaptation rule generation for case-based regression.

5.1 Ensemble Methods in CBR

Ensemble methods aggregate results from a set of models. A number of general-purpose approaches have been proposed, such as Bagging [9], boosting [10] and random forests [11]. In CBR research, ensemble methods have been applied to improve accuracy by combining solutions from multiple subsets of a case base or from multiple case bases. For example, Cunningham and Zenobi [12] propose improving accuracy of nearest neighbor classifiers by using an ensemble of classifiers, each based on different feature subsets. Arshadi and Jurisica [13] present an ensemble method for combining predictions of a set of classifiers built based on disjoint subsets of cases from the original case base, for which the case features are selected locally by using logistic regression. Li and Sun [14] propose using an ensemble of CBR systems, with randomly generated feature subsets used for similarity assessment in each individual CBR system, and forming the final solution by combining the results of those individual systems. However, to our knowledge, previous CBR research has not considered the use of ensembles of case adaptation rules.

5.2   Learning Adaptations from the Case Base

Learning case adaptation knowledge is an active CBR research area, for which many approaches have been pursued. For reasons of space, we limit our discussion to methods which learn adaptations from cases in the case base for regression tasks, rather than more knowledge-intensive approaches for other types of domains.

Case Difference Heuristics. Wilke et al. [15] provide a starting point for knowledge-light approaches to learning adaptation knowledge by discussing different sources of knowledge in a CBR system and general issues for designing a learning algorithm. They use their framework for two different approaches of learning adaptation knowledge: weighted majority voting and case difference heuristic proposed by Hanney and Keane [3]. The latter approach investigated by Wilke et al. is similar to ours in that it generates adaptations based on case comparison. Though, their method uses different strategies for ranking rules (e.g. confidence rating for rules) and composing the final solutions compared to ours.

McSherry's [4] CREST (Case-based Reasoning for ESTimation) provides another approach to generating adaptations from case differences. Given a case to adapt, McSherry's difference heuristic attempts to retrieve a case which differs from the input query only in the value of a single feature, called the distinguishing attribute. Next, a pair of cases with the same values for the distinguishing attributes as the query and (respectively) the case to be adapted are retrieved, and the solution of the retrieved case is adjusted based on their difference. Because more than one similar case may be retrieved for an input query, the final estimation of the target value can be calculated by averaging different estimations, generated by the same method. McSherry's method is similar to EAR's local approach, in generating adaptations based on neighbors to the input query. However, CREST adjusts the solutions of each base case by applying a single adaptation, while EAR uses an ensemble of adaptations.

McDonnell and Cunningham [5] refine the case difference heuristic to address two problems. The first is that the effect of variations in feature values on the solution may differ according to the feature considered. The second is that the effect of variations in a feature value on the solution may depend upon the values of other case features. Their method generates adaptations by comparing the input query to nearby cases, selecting cases for which the gradient is similar to the target case (using local linear regression to approximate the gradients), and deriving adaptations from those cases. This approach is in the spirit of EAR's context-based approach but not applied to ensembles of adaptations.

Learning Adaptation Rules from Linear Regression. Patterson et al. [16] propose a rule acquisition process based on k-NN and regression analysis. Given a new problem, the k nearest neighbors are retrieved and combined in a new generalized case in which features are the distance-weighted average of the individual case features. The k nearest neighbors are also used to train a linear regression model for predicting the difference between case solutions, which is applied to the generalized case to predict the target value for the input. Like EAR, this method uses a lazy approach for generating adaptations; it differs in that it relies on linear regression and single adaptations for generating and applying adaptations, instead of case differences and ensemble of adaptations, respectively.

Other Adaptation Learning Models for Case-Based Regression. Adaptation learning for regression also includes methods not based on direct case comparisons. Policastro et al. [17] propose a method for learning and applying adaptation knowledge from a case base by using two components, estimators and combiner. As estimators they use a multi-layer neural network, an M5 regression tree, and a support vector machine. As combiners, they consider the same three techniques, applied to combine the estimators' values.

Craw et al. [18], Jarmulak et al. [19], and Wiratunga et al. [20] propose automated acquisition of adaptation knowledge by repeatedly partitioning the case base to form a small set of probe cases, retrieving k similar cases for each probe case, and building adaptation rules based on pairs of probe cases and their top k neighbors. For each set, their method creates rule sets, each one containing adaptation cases that concentrate on differences for a single feature. From those, their method selects rules whose decision tree indexes have above-average predictive accuracy. An initial solution is generated by averaging, with possible refinement by adaptation rules each addressing differences in a single feature.

6   Conclusions and Future Research

This paper has introduced EAR, an approach for automatically generating sets of adaptation rules from a case base based on case differences and selecting ensembles of adaptations to apply. An experimental evaluation of nine variants of the EAR approach showed that EAR variants generally increased accuracy over baseline case-based regression and linear regression approaches, and that rule generation based on local information was sufficient to obtain accuracy competitive with the best performance obtained. Likewise, an ablation study provided support for the benefit of EAR's context-based rule ranking approach.

Opportunities for future research include developing more sophisticated adaptation selection and combination techniques, exploring other ensemble methods for the generation and combination of adaptations, and examining how EAR could apply to knowledge-rich domains. Yet another direction for extending this work is considering the confidence of cases to adapt and the adaptation rules in EAR. That is to some extent explored in [21]. Also the question of comparative benefit of using local vs. global adaptations is an interesting one for future research.

References

1. Mantaras, R., McSherry, D., Bridge, D., Leake, D., Smyth, B., Craw, S., Faltings, B., Maher, M., Cox, M., Forbus, K., Keane, M., Aamodt, A., Watson, I.: Retrieval, reuse, revision, and retention in CBR. Knowledge Engineering Review 20(3) (2005)
2. Aha, D., Kibler, D., Albert, M.: Instance-based learning algorithms. Machine Learning 6(1), 37–66 (1991)

3. Hanney, K., Keane, M.: The adaptation knowledge bottleneck: How to ease it by learning from cases. In: Leake, D.B., Plaza, E. (eds.) ICCBR 1997. LNCS, vol. 1266, pp. 359–370. Springer, Heidelberg (1997)

4. McSherry, D.: An adaptation heuristic for case-based estimation. In: Smyth, B., Cunningham, P. (eds.) EWCBR 1998. LNCS (LNAI), vol. 1488, pp. 184–195. Springer, Heidelberg (1998)

5. McDonnell, N., Cunningham, P.: A knowledge-light approach to regression using case-based reasoning. In: Roth-Berghofer, T.R., Göker, M.H., Güvenir, H.A. (eds.) ECCBR 2006. LNCS (LNAI), vol. 4106, pp. 91–105. Springer, Heidelberg (2006)

6. Frank, A., Asuncion, A.: UCI machine learning repository (2010), http://archive.ics.uci.edu/ml

7. Hall, M., Frank, E., Holmes, G., Pfahringer, B., Reutemann, P., Witten, I.H.: The weka data mining software: an update. SIGKDD Explor. Newsl. 11(1), 10–18 (2009)

8. Bay, S.D.: Combining nearest neighbor classifiers through multiple feature subsets. In: Proc. 15th International Conf. on Machine Learning, pp. 37–45 (1998)

9. Breiman, L.: Bagging predictors. Mach. Learn. 24(2), 123–140 (1996)

10. Schapire, R.E.: The strength of weak learnability. Mach. Learn. 5(2), 197–227 (1990)

11. Breiman, L.: Random forests. Mach. Learn. 45(1), 5–32 (2001)

12. Cunningham, P., Zenobi, G.: Case representation issues for case-based reasoning from ensemble research. In: Aha, D.W., Watson, I. (eds.) ICCBR 2001. LNCS (LNAI), vol. 2080, pp. 146–157. Springer, Heidelberg (2001)

13. Arshadi, N., Jurisica, I.: An ensemble of case-based classifiers for high-dimensional biological domains. In: Muñoz-Ávila, H., Ricci, F. (eds.) ICCBR 2005. LNCS (LNAI), vol. 3620, pp. 21–34. Springer, Heidelberg (2005)

14. Li, H., Sun, J.: Case-based reasoning ensemble and business application: A computational approach from multiple case representations driven by randomness. Expert Systems with Applications 39(3), 3298–3310 (2012)

15. Wilke, W., Vollrath, I., Althoff, K.D., Bergmann, R.: A framework for learning adaptation knowledge based on knowledge light approaches. In: Proceedings of the Fifth German Workshop on Case-Based Reasoning, pp. 235–242 (1997)

16. Patterson, D., Rooney, N., Galushka, M.: A regression based adaptation strategy for case-based reasoning. In: Proceedings of the Eighteenth Annual National Conference on Artificial Intelligence, pp. 87–92. AAAI Press (2002)

17. Policastro, C.A., Carvalho, A.C., Delbem, A.C.: A hybrid case adaptation approach for case-based reasoning. Applied Intelligence 28(2), 101–119 (2008)

18. Craw, S., Jarmulak, J., Rowe, R.: Learning and applying case-based adaptation knowledge. In: Aha, D.W., Watson, I. (eds.) ICCBR 2001. LNCS (LNAI), vol. 2080, pp. 131–145. Springer, Heidelberg (2001)

19. Jarmulak, J., Craw, S., Rowe, R.: Using case-base data to learn adaptation knowledge for design. In: Proceedings of the 17th International Joint Conference on Artificial Intelligence, IJCAI 2001, vol. 2, pp. 1011–1016. Morgan Kaufmann Publishers Inc. (2001)

20. Wiratunga, N., Craw, S., Rowe, R.: Learning to adapt for case-based design. In: Craw, S., Preece, A.D. (eds.) ECCBR 2002. LNCS (LNAI), vol. 2416, pp. 421–435. Springer, Heidelberg (2002)

21. Jalali, V., Leake, D.: On deriving adaptation rule confidence from the rule generation process. In: Delany, S.J., Ontañón, S. (eds.) ICCBR 2013. LNCS, vol. 7969, pp. 179–187. Springer, Heidelberg (2013)

iCaseViz: Learning Case Similarities through Interaction with a Case Base Visualizer

Debarun Kar, Anand Kumar, Sutanu Chakraborti, and Balaraman Ravindran

Department of Computer Science and Engineering,
Indian Institute of Technology Madras, Chennai- 600036, India
{debarunk,anandkr,sutanuc,ravi}@cse.iitm.ac.in

Abstract. Since the principal assumption in case-based reasoning (CBR) is that *"similar problems have similar solutions"*, learning a suitable similarity measure is an important aspect in CBR. However, learning case-case similarities is often a non-trivial task and involves significant amount of domain expertise. Most techniques that arrive at a pertinent similarity measure are often incomprehensible to the domain experts. These techniques also rarely enable the user to provide expert feedback which can then be utilized to develop better similarity measures. Our work attempts to bridge this knowledge gap by developing an iterative and interactive visualization framework called iCaseViz which learns the domain experts' notion of similarity by utilizing the user feedback. This work is different from similar work in other communities in the sense that it is tailored to cater to the needs of a system built primarily based on the CBR hypothesis. The case base visualizer demonstrated in this paper is also very efficient as it has insignificant delay during real-time user interaction on large case bases. We provide preliminary results on the efficiency of the visualizer and the effectiveness of our similarity learning algorithm on UCI datasets and a real world high dimensional case base.

1 Introduction

In the past decade the emergence of vast collections of data from various sources has resulted in researchers exploring different avenues of data analysis. Most of the datasets, like the astronomical, textual and social networks data, are high-dimensional which makes interpretation and analysis difficult, even for the domain experts. Different research communities have thus focussed on developing techniques to explore the huge amount of hidden information in the data. Statistical and machine learning techniques are often employed to gain meaningful insights about the data. However, the knowledge possessed by domain experts are mostly left unutilized in these kinds of information extraction processes. This is because it is difficult for them to interpret the data while manipulating certain unknown parameters. Similar issues are encountered in case base reasoning (CBR) when trying to mine meaningful information, like the identification of important features, from large and noisy case bases, like textual case bases.

Case base visualization is therefore important to make sense of the underlying representation of the data and to facilitate analysis. A domain expert can use

S.J. Delany and S. Ontañón (Eds.): ICCBR 2013, LNAI 7969, pp. 203–217, 2013.

the knowledge acquired from the visualization to reform his understanding of the data and make informed decisions regarding certain aspects of the data, like the identification of noisy cases and attributes in a case base. The importance of case base visualization and several CBR visualization techniques has been the topic of discussion in the past. However, most of these visualizers do not support interaction with the user, in the sense that the user cannot manipulate elements in the visualization to obtain a refined representation. This is mostly due the fact that real-time interaction with a visualizer is either very slow or not supported when dealing with data sets with large number of cases and attributes. This prevents the user from further analysis and gaining deeper insights about the data. Also, it is not always clear how to elicit user feedback. In this paper we present an application that allows the user to visualize and iteratively interact with it and then utilize the user feedback for effective and efficient multi-variate data analysis.

CBR systems are primarily built on the hypothesis that "*similar problems have similar solutions*". However, identifying the most appropriate similarity measure for a given case base is often a non-trivial task and requires significant amount of domain knowledge. The failure to encode a suitable similarity measure results in incorrect retrieval of cases. This, in turn, leads to poor quality solutions and the failure of a CBR system as a whole. Therefore, it is sometimes important to involve the domain experts in the process of explicitly specifying the similarity between two cases in a case base. However, reviewing the similarity between all pairs of cases as computed by a pre-determined similarity measure and modifying the incorrect similarities is an incredibly arduous, if not infeasible, task especially if the case base consists of a large number of cases. Also, the notion of similarity between cases cannot always be explicitly stated. The qualitative notion of similarity can be captured by showing two cases to the users and allowing them to manipulate a slider, specifying the extent of similarity on a normalized scale, say 0 to 1. However, doing this for every pair in a large case base is also laborious and time consuming. Moreover, the case base may contain noisy samples due to erroneous recording of past experiences. It is therefore important to identify and eliminate such noisy cases so as to form a case base which is representative of the problem solving domain under consideration. With the help of our visualizer and the user feedback we will attempt to learn a suitable similarity measure and also recognize noisy cases and eliminate them as part of the case base maintenance procedure.

In this paper, in order to learn a new set of similarities between cases we will learn an appropriate set of feature weights. We will see in the subsequent sections how we perform feature weighting to obtain an estimate of the relative impact of each feature on a particular target variable and also to compute the suitable global similarity between cases. For large number of features, this automated learning of feature weights not only spares the domain expert from manually encoding the weight of each feature, but it is also applicable in situations where the importance of every feature is unknown even to the experts.

The main motivation for our work comes while trying to find an important set of features and a suitable similarity measure for the soil nutrients' prediction task using a CBR system called InfoChrom [1]. In our previous work [2], we attempted to solve the problem of feature weighting by using alignment as a guiding measure. However, we were still unable to exploit the domain experts' knowledge of similarity between the cases present in the case base of InfoChrom. Due to the presence of a large number of cases (15167) in a high dimensional (176) feature space, the problem of manually encoding the relative similarities between cases became an uphill task. In this paper we attempt to address this problem by focussing on two important issues. Firstly, we want to emphasize the need for an interactive case base visualization tool which is efficient for large case bases. Secondly, we want to present a way to utilize user feedback from the visualizer to learn the domain experts' notion of similarity by finding a suitable set of feature weights. In Section 2, we discuss related work regarding visualization and similarity learning. We present the algorithm used by iCaseViz to revise the similarity measure in Section 3. In Section 4, we describe our application framework, its various components and the process of interaction with the visualizer. This is followed by preliminary experimental results and analysis on UCI data sets and the case base of InfoChrom in Section 5. Finally, we conclude by discussing how our work is different from a similar existing work and by describing a few possible extensions of our work in Section 6.

2   Related Work

Encoding a suitable similarity measure for a given data set has been an active research topic in several AI communities, like machine learning, case-based reasoning and data mining. Some papers focus on learning similarity measures by computing a set of relevant feature weights and combining the weighted feature level similarities to arrive at a global similarity measure. Our approach to learning similarity is based on the above idea. Research on feature weight learning primarily focusses on eliminating the *"curse of dimensionality"* problem for k-nearest neighbor (k-NN) based approaches by finding features which are more important than others for a particular prediction task [3–7]. An informative survey of similarity mechanisms in case base reasoning can be found in [8].

Past work on information visualization focuses mainly on visualizing high dimensional data with the help of Parallel Coordinates (PC) and Multi-Dimensional Scaling (MDS) plots. PC [9] draws features or coordinates as lines, parallelly and equidistant to each other (Fig. 1). A point in the n-dimensional coordinate space is drawn as a polyline in the PC plot, where the polyline connects all the parallel axes. Therefore, this parallel coordinates representation in 2-d space enables us to view and find complex patterns in multivariate datasets. However, the biggest drawback of this technique is that the PC plot becomes cluttered as the number of data points and the number of coordinates increase. This hinders the user from gaining any insightful patterns about the data being viewed. PC has been used in case base visualization in [10]. This work also addressed the problem of axes reordering by utilizing feature similarities. Falkman [11] presents a visualization tool that projects cases onto a three dimensional PC plot.

Work involving lower dimensional projections in visual data analytics mostly involves projection using MDS [12]. This includes several spring layout based visualizations for clustering problems as well as applications to manipulate certain parameters to modify the visualization [13–18]. The task of feature selection and distance function learning using 2-d projection has been most recently addressed in [19]. Other past research on case base visualization which are closely aligned with our work include [20, 21], which uses a force-directed graph drawing algorithm for visualizing an evolving case base. Cbtv [22] allows the user to visualize the effect of several similarity measures on a case base. Other related work on the same topic include [23–25].

3   Our Approach to Learning Similarity

Learning similarity with the help of our visualizer is an iterative process and the steps are explained in detail in Sections 3.1, 3.2, 3.3 and 3.4. Let us assume that we are given a set of N cases $C_1, C_2,..., C_N$, with the problem space P consisting of the values of q features $F_1, F_2, ..., F_q$ and the solution space S consisting of the target variable T. Also, let the weight of feature F_i be w_i ($1 \le i \le q$), where $0 \le w_i \le 1$. Then we calculate the dissimilarity in the problem space between two cases C_a and C_b as the weighted sum of their feature space dissimilarities:

$$DisSimP(C_a, C_b, \boldsymbol{w}) = \sum_{i=1}^{q} w_i * DisSimF(C_{ai}, C_{bi}) \qquad (1)$$

where, $DisSimF(C_{ai}, C_{bi})$ denotes the dissimilarity between the corresponding values of feature F_i for the cases C_a and C_b, and is computed as follows:

$$DisSimF(C_{ai}, C_{bi}) = \frac{DistF(C_{ai}, C_{bi}) - Min(DistF_i)}{Max(DistF_i) - Min(DistF_i)} \qquad (2)$$

Here, $DistF(C_{ai}, C_{bi})$ denotes the city-block distance between the values C_{ai} and C_{bi}. $Max(DistF_i)$ and $Min(DistF_i)$ are the maximum and minimum distances between all pairs of values for feature F_i respectively. The target space dissimilarity between any two cases is calculated in the same way and is represented as $DisSimS(C_a, C_b)$. Please note that our formulation of feature space dissimilarity is such that the feature space similarity between two cases C_{ai} and C_{bi} for feature F_i will be $(1-DisSimF(C_{ai}, C_{bi}))$. The overall similarity between two cases can then be computed as the weighted sum of these feature space similarities. In our application, we work with dissimilarities between cases so that we can use in-built Matlab functions which takes a dissimilarity matrix as an argument, to perform the projection in Section 3.1.

3.1   Projecting the Case Base in Two Dimensions

We use classical multi-dimensional scaling (MDS) to project the high-dimensional case base onto two dimensions and show the output to the user in the form of a scatter plot. The MDS algorithm is provided with a dissimilarity matrix containing the dissimilarities between case pairs, computed in the original high dimensional space based on the current measure of dissimilarity. It then attempts to find a projection in lower dimensions that best preserves the relative dissimilarity between the cases. We use the MDS plot to help the user visualize the relative dissimilarities between the cases in the original high dimensional space. We also color the points in this MDS plot for effective visualization and analysis. The coloring scheme for the cases is discussed in Section 4.1.

3.2 User Action

The application provides the user with the capability to select any case from the scatter plot and move it, thus changing the relative similarities between the selected case and other cases in the case base. The facility to select multiple cases together and move them in the plot is also provided. This will affect the relative dissimilarities between the selected set of cases and all other cases in the case base, while the relative dissimilarities between the untouched cases remains the same. More details about the features that are incorporated in the application can be found in Section 4.1.

3.3 Capturing the Experts' Notion of Similarity

Once the user has changed the relative dissimilarities between the cases, we obtain the new locations of the cases in the 2-d scatter plot and compute the Euclidean distance between every pair of cases in the case base. This gives us an estimate of the domain experts' notion of similarity between the cases.

3.4 Feature Weighting and Similarity Learning

Once the user has modified the relative dissimilarities between the cases, the system solves an optimization problem to learn an appropriate feature weight vector w. These relative importance values for each feature denoted as w_i ($1 \leq i \leq q$) are then used to derive new dissimilarity values with the help of Eqn. 1.

In order to obtain the revised feature weight set w^t at time t, we solve the optimization problem as specified by Equation 3. For every pair of cases C_i and C_j, we take the squared difference between the user specified dissimilarities $DisSimP^{user}(C_i, C_j)$ as obtained by the users' modification of the scatter plot and the dissimilarities $DisSimP(C_i, C_j, w)$ in the high dimensional space obtained for a particular weight vector w. Alignment measures the extent to which the similarity hypothesis holds good in a particular case base. Local alignment measures this property of the case base in the neighbourhood of a particular case while global alignment is measured across the entire case base. Now, consider the scenario where the user moves all cases which are highly similar both in the problem and solution spaces to a single point. This is clearly not desirable even though this increases the local as well as the global alignment. Thus the need to preserve the structure in the original data in terms of inter-case similarities might pose a threat to the goal of improving the alignment. We devise the objective function to capture the tradeoff between these conflicting requirements. Now, consider another situation where the user moves two cases, which are highly similar both in the problem and solutions spaces, too far apart from each other. This will decrease the alignment and is also undesirable. To discourage the user movements as explained in the above two scenarios, we give weights to each of the squared difference terms in Equation 3. This is an important design step in our application. To do this, we first compute the complexity between two cases C_i and C_j as the product of their problem and solution side dissimilarities (Equation 6), normalize it (Equation 5) and replace the value thus obtained in Equation 4. The hyperbolic tangent function is used in Equation 4 as a smoothing function. The trade-off term will ensure that the squared difference term in Equation 3 receives less importance if the value of complexity between two cases is low and vice versa. Thus, if the user moves two cases, which are already highly similar both in the problem and solutions spaces, too close or too far apart from each other, the tradeoff term (Eqn. 4) will ensure that the movement receives less weight. Therefore, by incorporating the tradeoff term, random movements in the case space is prevented and the global structure of the case base is maintained in accordance with the CBR hypothesis.

$$\boldsymbol{w^t} = \underset{\boldsymbol{w}}{argmin}(\sum_{i \leq j \leq N}(tanh(Comp_{ij}^{norm})*(DisSimP^{user}(C_i, C_j) - DisSimP(C_i, C_j, \boldsymbol{w}))^2))$$

(3)

subject to $w_i \geq 0, (1 \leq i \leq q)$

$$tanh(Comp_{ij}^{norm}) = (2/(1 + \exp(-2 * Comp_{ij}^{norm}))) - 1$$

(4)

$$Comp_{ij}^{norm} = \frac{Comp_{ij} - Min(Comp_{ij})}{Max(Comp_{ij}) - Min(Comp_{ij})}$$

(5)

$$Comp_{ij} = DisSimP(C_i, C_j) * DisSimS(C_i, C_j)$$

(6)

4   The iCaseViz Application Framework and Its Components

This section describes the different components of our application and the features provided to the user. We also discuss the implementation details and how we obtain fast responses while loading and interacting with the data as compared to similar existing tools. It is divided into two sections: the visualizer (Sec. 4.1) and the analyzer (Sec. 4.2).

4.1   The Visualizer

The visualizer portion of the application is written in C++ using the Qt open source library. It has two main modules: the Parallel Coordinates (PC) visualizer

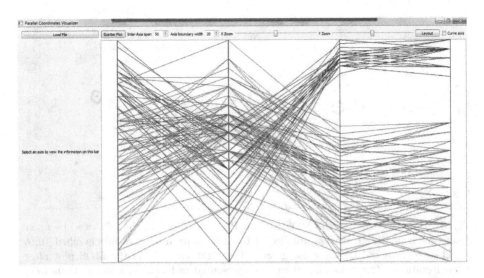

Fig. 1. The Parallel Coordinates visualizer

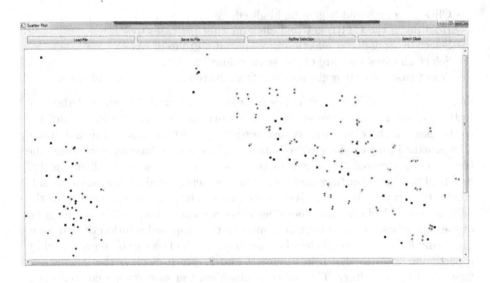

Fig. 2. The MDS Scatter Plot visualizer

(Figure 1) and the Scatter Plot visualizer (Figure 2). The PC module allows one
to reposition axis, change the spacing between axes, zoom along the X and Y
axes independently and scroll horizontally and vertically. It is also capable of
displaying axis specific information like the axis name when one selects an axis.
The Scatter Plot module displays an interactive two dimensional plot. One can
move the cases around and save the resulting plot data to a file. The Scatter
Plot gives the user multiple options to select the cases:

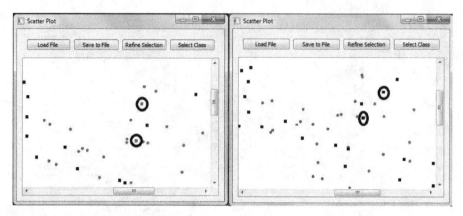

Fig. 3. A zoomed view of the MDS Scatter Plot visualizer. In the left view, two test cases (green in the visualizer) are circled to show that these were misclassified (into the class represented by the color green). The right view shows the MDS plot after a few iterations. The same circled cases (represented in black color in the visualizer) have now been correctly classified.

- Click on a case and it gets highlighted.
- Hold down the CTRL key and click on multiple cases to select.
- Select by dragging a selection box around a group of cases.
- Select all cases that are of the same colour.
- Select cases of only a given colour from currently selected set of cases.

Our 2-dimensional Scatter Plot has an additional feature in terms of the projection, which is missing from the earlier work on the same topic. In addition to the cases in the case base, we also project the test cases onto this 2-d space. The Scatter Plot visualizer represents cases in the case base as circles and the test cases as squares. The cases in the case base are colored both in the PC and Scatter Plot visualizer and their color is determined by the solution side similarity between the cases. Test cases whose solutions are satisfactory to the user are colored black while the other cases are colored according to the color of the case whose solution is most similar to the proposed solution of that case. For example, in the case of classification tasks, each class can be represented by a particular color as shown in Figures 1 and 2, which displays the Iris dataset from the UCI repository. The correctly classified test samples are colored black while the incorrectly classified test cases has the same color as the class it has been classified into. This enhances the visualization by providing additional information to the domain expert in the form of relative dissimilarities between the cases in the case base and the test cases. He gets an idea about the quality of the current similarity metric by not only viewing the relative positions of the cases but also the quality of the solutions produced for the test cases as can be seen by the color of a particular case. However, the test cases are only for viewing purpose and cannot be moved around in the scatter plot. Note that even if the expert is aware of the quality of solutions of the test cases, improving it cannot be enforced by arbitrarily moving the cases around. This is because of the tradeoff term in Section 3.4. So the cases in the case base have to be moved around so that some latent notion of similarity is captured. Hence this is a valuable technique even though the test data is shown to the user. Two zoomed views of the MDS Scatter Plot are given in Figure 3. The view on the left highlights two misclassified test cases and the right view highlights the same cases after a few iterations, when they have been correctly classified. Note that one iteration corresponds to a single user interaction with the MDS plot. This figure also displays the various components as tabs which the user can use during the interaction. The two visualizer modules are linked to each other. So when the user selects case(s) on the Scatter Plot, the corresponding polyline(s) on the PC plot is highlighted. Our PC visualizer thus provides a consolidated view of the entire case base and cases of interest across multiple dimensions and aids the domain expert in his interaction with the MDS Scatter Plot. Discussion of the implementation details of the two visualization components is provided below.

Parallel Coordinates Visualizer: The PC Visualizer reads data in CSV format. The first row is expected to be the headings row, for example the names of the features. Each subsequent row is expected to have an extra column that gives the quality of the solution for the corresponding case, represented by a particular colour. Figure 4 shows the sequence of activities that lead to the image getting drawn on screen. Once the data is loaded, a layout is created. The height of the layout is determined by the axis that has the largest range. The inter-axis interval and the number of axis determine the length of the canvas. This canvas is logically split into canonical rectangles the size of the viewing area (dimensions of the viewing windows scaled according to zoom factors). This means the visible rectangle on screen (determined by zoom setting and position of the scroll bars) can be made of at most four adjacent canonical rectangles. The canonical rectangles' dimensions change when the view is zoomed in or out. For large datasets, to make the application responsive, these canonical rectangles are cached. This allows the application to save time spent on repetitive drawing.

MDS Plot Visualizer: The scatter plot is realized using the capabilities of QGraphicsScene and QGraphicsView. The Qt Graphics View framework allows

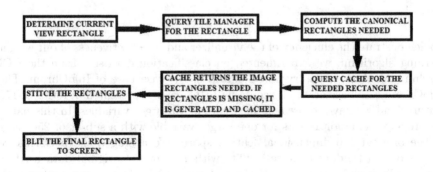

Fig. 4. Sequence of activities for the Parallel Coordinates visualizer

the developer to describe a scene in terms of its constituent items. The framework allows for easy implementation of user interaction with object like selection and moving objects through Qt's event system.

In our implementation, the Scatter Plot modules load data from the files into a QGraphicsScene. This scene is attached to a view that displays the contents of the scene. The Scatter Plot is linked to the PC visualizer through the event system. Whenever the user selects cases in the scatter plot, a signal is passed to the PC module indicating the change in status. This in turn marks the appropriate polyline as selected, resulting in a change in its colour. The image cache is purged and the view is redrawn to show the change.

4.2   The Analyzer

The analyzer module uses Matlab to perform the data analysis task. This involves, among other things, computing the dissimilarity between pairs of cases and finding the most similar cases to suggest a solution. We have used in-built Matlab functions to perform MDS. To solve the optimization problem in Equation 3, we use the default solver (SDPT3) in CVX, a Matlab-based convex optimization tool for specifying and solving convex programs [26, 27]. We have found that SDPT3 is slow when we are working with large case bases and we are currently looking for fast solvers to enhance the performance of our application.

5   Evaluation

All our experiments have been conducted on an Intel Core i5 processor (M450 2.40 GHz) with 4GB RAM and 64-bit Windows 7 operating system. In our experiments we use k-NN based retrieval strategy to propose a solution for a test case. We calculate its dissimilarity with all the other cases in the case base using Equation 1. Then we obtain the solution for the test sample by taking a weighted average of the solutions of its k nearest neighbors, with the problem side dissimilarities acting as weights. Please note that we have considered w_i = 1 $(1 \leq i \leq q)$ when no feature weighting is employed while calculating the dissimilarity between the cases.

5.1   Datasets

To demonstrate the efficiency of the visualizer and the effectiveness of our weight learning algorithm, we experimented on classification data sets from the UCI machine learning repository and a subset of the case base of InfoChrom. The InfoChrom case base originally contains 15167 cases represented in terms of 176 features and we have to predict the values of 15 target variables. In this paper we provide prediction results for one target variable with a subset of 250 cases in the original 176 dimensional feature space. We averaged the performance results over 5 random train-test splits with 70% of the original data used as the case base and the rest form the test cases. All results are reported with the value of k set to 3. For the data sets referred to in this paper, a brief

Table 1. Characteristics of the UCI data sets used for evaluation purposes

Data set	Kind of Prediction Task	Number of Features (all continuous)	Number of Classes (for Classification) / Range of the target variable (for Regression)
Iris	Classification	4	3
Glass Identification	Classification	9	7
Waveform-21	Classification	21	3
InfoChrom Case Base	Regression	176	26-246

description about the number and types of features and target variables are shown in Table 1. We measured performance in terms of classification accuracy and percentage error for classification and regression tasks respectively. Due to the lack of domain knowledge for the UCI data sets, the user interacted with the application based on the class information of each case, as indicated by the colour. For the InfoChrom case base, the users utilized their domain expertise to move the cases in the MDS plot.

5.2 Experimental Results and Observations

Figures 5 and 6 show the change in performance of iCaseViz on the Iris and Glass Identification data and the InfoChrom case base over several iterations. It is evident from the figures that as the domain expert interacts more with the system and the notion of similarity as captured by the system evolves, there is a noticeable improvement in performance of the system. For the case bases used to demonstrate the effectiveness of our similarity learning algorithm, we show the correlation between the dissimilarities in the original and MDS spaces in

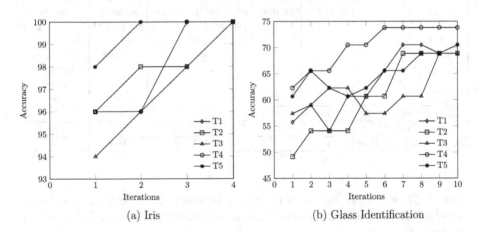

(a) Iris

(b) Glass Identification

Fig. 5. Performance of iCaseViz on (a)Iris and (b)Glass Identification data sets. T1-T5 indicate 5 different random test sets. Note that in (a), the performance curve for T1 is not clearly visible as portions of it have merged with those of T3 and T4.

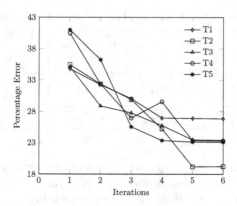

Fig. 6. Performance of iCaseViz on a subset of the InfoChrom case base for predicting the target variable 'potassium'. T1-T5 indicate 5 different random test sets.

Table 2. Correlation between dissimilarities calculated in original and MDS spaces

Data Set	Correlation
Iris	0.9905
Glass Identification	0.9188
InfoChrom Case Base	0.8563

Table 3. The time taken (in milliseconds) by various components of the visualizer. ϵ indicates that the time taken is negligible (of the order of nanoseconds).

Data set	Parallel Coordinates Plot					Scatter Plot
	Reading from file	Layout the data	Decide what to put on display	Filter the polylines	Draw onto memory image and display on screen	Reading from file, creating object & adding to QGraphics Scene
Iris	ϵ	ϵ	ϵ	ϵ	5	56
Glass Identification	ϵ	ϵ	ϵ	ϵ	11	4
Waveform-21	ϵ	ϵ	ϵ	6	202	126
InfoChrom Case Base	ϵ	ϵ	ϵ	ϵ	109	94

Table 2. These correlations show the extent to which the MDS algorithm maintains the relative dissimilarities of the original space, when projecting it to two dimensions.

We also report the time taken, in milliseconds, by the various components of the visualizers to display on a screen with a resolution of 1366 x 768, when the display window is maximized. We can see from Table 3 that the time taken are negligible and therefore the system is well-suited for real-time interaction, even on large case bases. Also, we noted that the response time of our system for user movements on the PC as well as the MDS plot is of the order of nanoseconds.

6   Conclusions and Future Work

In this paper, we have introduced iCaseViz, an interactive visualization framework that not only shows the relative positions of the cases in the original high dimensional problem space by projecting them onto a lower dimensional space but also lets the user explore this space by allowing them to change the relative similarities between the cases. Both the notions of problem side similarity and solution side similarity are captured by this visualization technique. This is coupled with a CBR centric optimization function that learns the domain experts' notion of similarity by arriving at an optimal set of feature weights. This work attempts to bridge the gap between the experts' knowledge about the problem domain and the case-based reasoning methodology by providing an iterative and interactive visualization application. The weighting term in the optimization function ensures that even though the domain expert is given the freedom to change the relative similarities between cases, attempts to modify the similarities in a way which destroys the inherent structure of the case base to a large extent, are strongly discouraged by the system. This contributes towards making the system robust to any unwarranted changes. We also show that the system is able to learn a suitable set of feature weights very fast over a small number of iterations, not only for UCI datasets but also for high dimensional real world data, as is evident from the performance graphs. The visualizer in its current format can handle case bases with more than 20,000 cases and over 175 features very easily and with minimal delay in response during interactions. This is a significant development over previous applications which became unresponsive and most often even failed to load case bases with around 10,000 cases represented in terms of around 15 features.

Our work is most similar to [19] as we have a common objective of interacting with an MDS plot to learn a dissimilarity measure by finding an appropriate set of feature weights. However, our work is different from [19] in a lot of aspects. Firstly, our work is more CBR-specific in the sense that we discourage the users from making any updates that can possibly lead to a configuration where the CBR hypothesis is violated both locally and across the entire case base. We do this by introducing a tradeoff term that prevents a potential decrease in the alignment of the case base. Dis-Function [19], on the other hand, provides an inertia against any updates made by the expert to the MDS plot. Secondly, in [19], a user is only allowed to select two sets of points and change the relative distances between them, while in our work we allow the expert to select any number of cases and modify the relative dissimilarities between those and the remaining cases in the case base. So, once the cases are moved, the user has an idea about the relative positions of all the cases in the case base which he is going to see in the next update. Since we also show the test cases, colored according to the quality of their solutions, along with the cases in the case base, this gives the domain expert a global picture at any point of time and thus helps him make informed decisions.

In the future we would like to perform more experiments on large case bases with various fast optimization problem solvers. This is because the current solver we are using is slow when working with large datasets and therefore the application takes time to compute the revised similarity values once the expert is done with the modifications. Also, currently the expert can only use the PC visualizer for viewing the data at several zoom levels and can re-position the axes in any order and with arbitrary gaps between them. We are looking to develop on this by providing an innovative visualization scheme based on parallel coordinates that will enable further interaction with the PC plot. Also, other modules are being developed and integrated with the existing tool which will provide additional information about the cases selected in the MDS plot. For example, a data viewer module will show the cases in terms of the features and their corresponding values. Selected cases will be highlighted in the data viewer. For the InfoChrom case base, where each case is a chromatogram image, an additional module to show the chromatogram for a selected case is being developed. We would also like to find out the utility of the PC visualizer in terms of the extent to which it aids the user in making decisions. This can be done by comparing the performance of the system when the PC plot is shown to the user as compared to when it is not. An integrated visualization and analysis application can then be built with further options to delete noisy cases and attributes, with the changes reflecting instantaneously in the PC and MDS visualizers. We are also interested in exploring various ways to learn a suitable kernel for a case base by using the expert information obtained through interaction with the scatter plot. We believe that encoding domain knowledge in the kernel function will be the key towards developing more accurate similarity measures for a particular prediction task.

References

1. Khemani, D., Joseph, M.M., Variganti, S.: Case based interpretation of soil chromatograms. In: Althoff, K.-D., Bergmann, R., Minor, M., Hanft, A. (eds.) ECCBR 2008. LNCS (LNAI), vol. 5239, pp. 587–599. Springer, Heidelberg (2008)
2. Kar, D., Chakraborti, S., Ravindran, B.: Feature weighting and confidence based prediction for case based reasoning systems. In: Agudo, B.D., Watson, I. (eds.) ICCBR 2012. LNCS, vol. 7466, pp. 211–225. Springer, Heidelberg (2012)
3. Aha, D.W.: Tolerating noisy, irrelevant and novel attributes in instance-based learning algorithms. Int. J. Man-Mach. Stud. 36(2), 267–287 (1992)
4. Wettschereck, D.: A study of distance-based machine learning algorithms. Ph.D. dissertation, Department of Computer Science, Oregon State University (1994)
5. Wettschereck, D., Aha, D.W.: Weighting features. In: Aamodt, A., Veloso, M.M. (eds.) ICCBR 1995. LNCS, vol. 1010, pp. 347–358. Springer, Heidelberg (1995)
6. Wettschereck, D., Aha, D.W., Mohri, T.: A review and empirical evaluation of feature weighting methods for a class of lazy learning algorithms. Artif. Intell. Rev. 11(1-5), 273–314 (1997)
7. Stahl, A.: Learning feature weights from case order feedback. In: Aha, D.W., Watson, I. (eds.) ICCBR 2001. LNCS (LNAI), vol. 2080, pp. 502–516. Springer, Heidelberg (2001)

8. Cunningham, P.: A taxonomy of similarity mechanisms for case-based reasoning. IEEE Transactions on Knowledge and Data Engineering 21(11), 1532–1543 (2009)
9. Inselberg, A., Dimsdale, B.: Parallel coordinates: a tool for visualizing multi-dimensional geometry. In: Proceedings of the 1st Conference on Visualization 1990, VIS 1990, pp. 361–378. IEEE Computer Society Press (1990)
10. Massie, S., Craw, S., Wiratunga, N.: Visualisation of case-based reasoning for explanation. In: Proceedings of ECCBR Workshop, Madrid, pp. 135–144 (2004)
11. Falkman, G.: The use of a uniform declarative model in 3D visualisation for case-based reasoning. In: Craw, S., Preece, A.D. (eds.) ECCBR 2002. LNCS (LNAI), vol. 2416, pp. 103–117. Springer, Heidelberg (2002)
12. Borg, I., Groenen, P.: Modern Multidimensional Scaling: theory and applications. Springer (2005)
13. Broekens, J., Cocx, T., Kosters, W.A.: Object-centered interactive multi-dimensional scaling: Ask the expert. In: Proceedings of the 18th Belgium-Netherlands Conference on Artificial Intelligencem, BNAIC (2006)
14. Buja, A., Swayne, D.F., Littman, M.L., Dean, N., Hofmann, H.: Xgvis: Interactive data visualization with multidimensional scaling. Technical report (2001)
15. desJardins, M., MacGlashan, J., Ferraioli, J.: Interactive visual clustering. In: Proceedings of the 12th International Conference on Intelligent User Interfaces, IUI 2007, pp. 361–364. ACM, New York (2007)
16. May, T., Bannach, A., Davey, J., Ruppert, T., Kohlhammer, J.: Guiding feature subset selection with an interactive visualization. In: IEEE VAST, pp. 111–120 (2011)
17. Endert, A., Han, C., Maiti, D., House, L., Leman, S., North, C.: Observation-level interaction with statistical models for visual analytics. In: IEEE VAST, pp. 121–130 (2011)
18. Okabe, M., Yamada, S.: An interactive tool for human active learning in constrained clustering. Journal: Emerging Technologies in Web Intelligence 3(1) (2011)
19. Brown, E.T., Liu, J., Brodley, C.E., Chang, R.: Dis-function: Learning distance functions interactively. In: IEEE VAST, pp. 83–92 (2012)
20. Smyth, B., Mullins, M., McKenna, E.: Picture perfect: Visualisation techniques for case-based reasoning. In: ECAI, pp. 65–72 (2000)
21. McArdle, G., Wilson, D.: Visualising case-base usage. In: Workshop Proceedings ICCBR, pp. 105–114 (2003)
22. Namee, B.M., Delany, S.J.: Cbtv: Visualising case bases for similarity measure design and selection. In: Bichindaritz, I., Montani, S. (eds.) ICCBR 2010. LNCS, vol. 6176, pp. 213–227. Springer, Heidelberg (2010)
23. McKenna, E., Smyth, B.: An interactive visualisation tool for case-based reasoners. Appl. Intell. 14(1), 95–114 (2001)
24. Chakraborti, S., Cerviño Beresi, U., Wiratunga, N., Massie, S., Lothian, R., Khemani, D.: Visualizing and evaluating complexity of textual case bases. In: Althoff, K.-D., Bergmann, R., Minor, M., Hanft, A. (eds.) ECCBR 2008. LNCS (LNAI), vol. 5239, pp. 104–119. Springer, Heidelberg (2008)
25. Freyne, J., Smyth, B.: Creating visualizations: A case-based reasoning perspective. In: Coyle, L., Freyne, J. (eds.) AICS 2009. LNCS, vol. 6206, pp. 82–91. Springer, Heidelberg (2010)
26. CVX Research Inc.: CVX: Matlab software for disciplined convex programming, version 2.0 beta (September 2012)
27. Grant, M., Boyd, S.: Graph implementations for nonsmooth convex programs. In: Blondel, V.D., Boyd, S.P., Kimura, H. (eds.) Recent Advances in Learning and Control. LNCIS, vol. 371, pp. 95–110. Springer, Heidelberg (2008)

A Multi-Objective Evolutionary Algorithm Fitness Function for Case-Base Maintenance

Eduardo Lupiani[1], Susan Craw[2], Stewart Massie[2],
Jose M. Juarez[1], and Jose T. Palma[1]

[1] University of Murcia, Spain
{elupiani,jmjuarez,jtpalma}@um.es
[2] The Robert Gordon University, Scotland, UK
{s.craw,s.massie}@rgu.ac.uk

Abstract. Case-Base Maintenance (CBM) has two important goals. On the one hand, it aims to reduce the size of the case-base. On the other hand, it has to improve the accuracy of the CBR system. CBM can be represented as a multi-objective optimization problem to achieve both goals. Multi-Objective Evolutionary Algorithms (MOEAs) have been recognised as appropriate techniques for multi-objective optimisation because they perform a search for multiple solutions in parallel. In the present paper we introduce a fitness function based on the Complexity Profiling model to perform CBM with MOEA, and we compare its results against other known CBM approaches. From the experimental results, CBM with MOEA shows regularly good results in many case-bases, despite the amount of redundant and noisy cases, and with a significant potential for improvement.

1 Introduction

Case-Base Maintenance (CBM) has as its main goals control of the number of cases within the case-base and maintaining the accuracy of the CBR system to resolve problems [14]. Redundant cases have a negative impact on the performance of the system, and noisy cases adversely affect the accuracy of the proposed solutions. Therefore, CBM algorithms usually try to remove both redundant cases and noisy cases.

There is a wealth of approaches to perform CBM in the literature [1,7,16,18–22]. The CNN algorithm only deletes redundant cases, focusing on retrieval efficiency [8]. The RNN algorithm extends CNN to consider noise cases as well [7]. The ENN algorithm only removes noisy cases, and RENN consists of multiple iterations of ENN, taking the output of one repetition as an input for the following iteration [21]. The family of algorithms DROP1, DROP2 and DROP3 were introduced to reduce redundancy and noisy cases [22]. The DROP family introduces the concept of *associate*, in an attempt to classify cases as redundant or noisy. COV-FP algorithms [20] and ICF [1] exploit the concepts *coverage* and *reachability* in order to reduce both the redundancy and noise levels [19].

S.J. Delany and S. Ontañón (Eds.): ICCBR 2013, LNAI 7969, pp. 218–232, 2013.

The main disadvantage of the aforementioned algorithms, with the exception of the COV-FP series, is their sensitivity to the order in which cases are examined. That is, given the same case-base, these CBM algorithms could provide different outcomes depending on the order of the cases in the case-base. Additionally, another commonplace feature of all these algorithms is their greedy approach to CBM goals, and the use of a lazy learning approach, such as k-nearest neighbour [3]. Furthermore, since each algorithm has a fixed deletion policy, the suitability of the algorithm to perform CBM is directly related to the redundancy and noise levels. For instance, those algorithms focused on case reduction underperform in cases-bases with few representative cases.

Consequently, CBM may be understood as a multi-objective optimization problem, minimising the case-base size and error rate at the same time. However, other objectives must be taken into consideration. In particular, it would be useful to estimate the optimum number of cases to resolve the entire problem domain [13], and to select a set of cases from the original case-base that maintains or improves accuracy.

In the last decades, Multi-Objective Evolutionary Algorithms (MOEAs) have been applied successfully in multi-objective optimization problems [2]. Therefore, CBM could be approached as a MOEA. To this end, considering that CBM algorithms should generate a case-base without redundant cases or noisy cases, and that is as small as possible to resolve the entire problem domain, three objectives based on Complexity Profiling [15] can be considered. Hence, MOEAs should get a good well-maintained case-base irrespective of the redundancy and noise levels of the original case-base.

In this work we propose to represent CBM as a 3-objective optimization problem, and we present a CBM algorithm based on MOEA using a novel fitness function.

The remainder of this work is as follows: in the next section we review the background of Complexity Profiling and the basic principles of MOEA. In section 3 we propose a fitness function for MOEA to perform CBM. In section 4, we evaluate the MOEA with different case-bases, and other CBM algorithms. Finally, in section 5 we present our conclusions and future work.

2   Background

2.1   Complexity Profiling

Massie *et al.* [16] introduced Complexity Profiling to estimate the proportion of redundant and noisy cases, as well as the existing error rate in the case-base. The foundation of this approach is a local complexity, which is an approximation to find the proportion of cases with the same solution in the nearest neighbour set of the case. Expression 1 describes the complexity function for a case:

$$complexity(c, k) = 1 - \frac{1}{k} \sum_{i=1}^{k} p(c, i), \qquad (1)$$

where k is the number of nearest neighbours to consider and $p(c, i)$ is the proportion of cases within the case's i-nearest neighbours that belong to the same solution as c. The codomain for *complexity* function is $[0, 1]$. The more the *complexity* of a case is, the more likely the case would be noisy.

Complexity Profiling is a global measure of the case-base, and it is composed by three different indicators:

1. the *error rate* is the average of all the local complexities measures;
2. the *noise* is the proportion of all the complexity measures with values greater than ϵ; and
3. the *redundancy* is the proportion of all the complexity measures with values equal to ρ.

The error, noise and redundancy are defined formally as follow:

$$error(M, k) = \frac{1}{|M|} \sum_{c \in M} complexity(c, k). \tag{2}$$

$$noise(M, k) = \frac{|\{c \in M | complexity(c, k) \geq \epsilon\}|}{|M|}. \tag{3}$$

$$redundancy(M, k) = \frac{|\{c \in M | complexity(c, k) = \rho\}|}{|M|}, \tag{4}$$

where M is a case-base, $c \in M$ is a case within M, and k is the number of neighbours of c. Experiments with $\epsilon = 0.5$ and $\rho = 0$ confirm that Complexity Profiling is a good predictor of accuracy and noise [16].

2.2   Multi-Objective Evolutionary Algorithms

Evolutionary Algorithms (EAs) are inspired in biological evolution [9], since they simulate biological processes to search for a solution to an optimization problem. EAs represent the problem with a string of binary values. The string is known as an individual, and each of its binary values as genes. For each individual the EA applies a function known as the fitness function, which indicates the suitability of the individual to resolve the optimization problem. The search for the best individual is an iterative process. Starting with a set of individuals known as the population, an EA uses three operations on it to create the next generation of individuals: reproduction, crossover and mutation. The reproduction operation aims to select the better individuals according to their fitness values. Crossover is applied only to selected individuals to create new individuals, usually exchanging their genes. Mutation flips randomly the genes of the individual to increase the diversity of individuals. At the end of the iteration process, the individuals within the final population are potential solutions to the optimization problem. Hence, a strategy is needed to choose the final solution as well.

Multi-Objective Evolutionary Algorithm (MOEA) is an EA that searches for a solution to a problem according to two or more optimization objectives. Unlike

EA, MOEA fitness function returns a value per each objective [23]. Expression 5 defines formally the optimization problem to minimize n objectives:

$$minimize(\Phi(x)) = minimize(\phi_1(x), \phi_2(x), \ldots, \phi_n(x)), \tag{5}$$

where x is an individual, Φ is the fitness function, and each ϕ_n is the fitness function associated to an objective. Given the fitness values of two individuals, it is possible to define a relation of dominance between them [5]. This dominance determines which individual is closer to the optimization objectives. Expression 6 defines formally the relation:

$$x \prec y \iff \tag{6}$$
$$\forall \phi_i(x), \phi_i(y) \in \Phi(x) : \phi_i(x) \leq \phi_i(y) \land$$
$$\exists \phi_j(x), \phi_j(y) \in \Phi(x) : \phi_j(x) < \phi_j(y),$$

where x and y are the individuals, $x \prec y$ expresses that x dominates y, and n is the number of objectives. MOEA generates generations of individuals, where non dominated individuals have higher odds of survival.

3   Multi-objective Optimization Fitness Function for Case-Base Maintenance

So to perform CBM with a MOEA, we need to set up the representation of the problem. On the one hand, we need to represent a case-base as an individual of the population. On the other hand, we need to define a fitness function to evaluate the suitability of the individual.

3.1   Case-Base Representation

The case-base is a string of binary values that creates an individual. The length of the string is the cardinality of the case-base. That is, each gene (binary value) of the individual (string) represents the presence of the case in the case-base.

Let M be the original case-base, denoted by $M = \{c_1, c_2, \ldots, c_n\}$, where c_i the i-th case of M ($|M| = n$). The space of all possible individuals of M is denoted by X. An individual $x \in X$ is formally defined as $x = x_1 x_2 \ldots x_{n-1} x_n$, where x_i is the i-th gene of the individual with values of $x_i \in \{true, false\}$.

In order to map the cases from the original case-base (M) to the elements of the individual, we introduce the following function:

$$\mathcal{M} : X \to \wp(M)$$
$$\mathcal{M}(x) = \overline{x} = \{c_i \in M | x_i = true\}. \tag{7}$$

For example, given the individual x with all elements set to true, $\mathcal{M}(x) = M$, otherwise if all elements are set to false then $\mathcal{M}(x) = \emptyset$. For the sake of clarity, we use the notation $\overline{x}$ as the case-base equivalent to the individual x.

3.2 Fitness Function to Perform CBM

We propose a fitness function based on Complexity Profiling to solve an optimization problem with three objectives:

1. to minimize the difference between the current number of cases in the solution and the estimated number of non redundant cases;
2. to minimize the number of redundant cases; and
3. to minimize the error rate level.

The first objective aims to estimate the minimum number of cases, the second is focused on avoiding case-bases with redundant cases, and the third leads the search to find a case-base with the minimum error rate. According to these objectives, the resulting case-base is expected to have smoother frontiers between the clusters of cases of different solutions, and with few cases within the clusters.

The formal description of the fitness function is shown as follows:

$$\Phi : X, \mathbb{N} \to \mathbb{R}^3 \tag{8}$$
$$\Phi(x, k) = (f_{size}(\overline{x}, k), redundancy(\overline{x}, k), error(\overline{x}, k)).$$

Note that the domain of the fitness function is an individual x, and a natural number k that sets the number of neighbours to consider in all the functions.

Function f_{size} is defined as follows:

$$f_{size} : X, \mathbb{N} \to \mathbb{R} \tag{9}$$
$$f_{size}(x, k) = ((|M| * (1 - redundancy(\overline{x}, k))) - length(x))^2,$$

where $length(x)$ is the number of elements of x set to true and $(|M| * (1 - redundancy(M, k))) - length(x)$ is the distance between the current number of cases in the solution and the estimated number of non redundant cases that the case-base should contain. This objective is squared to penalize those individuals with a greater number of cases.

The values returned by functions $redundancy(\overline{x}, k)$ and $error(\overline{x}, k)$ in the fitness function (expression 8) oppose each other since a lower error rate means a higher redundancy and vice versa.

3.3 NSGA-II

In this work we consider the well-known NSGA-II [5], a non-dominated sorting based MOEA. Given two individuals x and y representing two case-bases, and the fitness function $\Phi(x, k)$, the dominance relation for NSGA-II is defined as:

$$(f_{size}(x) \leq f_{size}(y) \wedge redundancy(x) \leq redundancy(y) \wedge error(x) \leq error(y)) \wedge$$
$$(f_{size}(x) < f_{size}(y) \vee redundancy(x) < redundancy(y) \vee error(x) < error(y)) . \tag{10}$$

For the sake of clarity we have omitted the parameter k of each function.

The main contributions of NSGA-II are a fast non-dominated sorting function and two operators to sort the individuals: a density estimation of the individuals in the population covering the same solution and a crowded comparison operator.

The *fast-nondominated-sort* algorithm details are shown in Alg.1. This function given a population P returns a list of the non-dominated fronts $\mathcal{F}$, where the individuals in front $\mathcal{F}_i$ dominates those individuals in front $\mathcal{F}_{i+1}$. That is, the first front contains the non-dominated individuals, the second front has those individuals dominated only once, the third contains individuals dominated up to twice, and so on. The individuals in the same front could have similar case-base representations; to avoid this situation NSGA-II uses the crowded comparison operator $\geq_n$, because individuals with lower density are preferred. To define formally the operator $\geq_n$, let x, y be two individuals, then $x \geq_n y$ if $(x_{rank} < y_{rank})$ or $((i_{rank} = j_{rank}) \wedge (i_{density} > j_{density}))$, where x_{rank} represents the front where the individual belongs. The *crowding-distance-assignment* procedure calculates the density per each individual (Alg.2).

Parameters are set up at the beginning, such as the number of generations and number of individuals N for population. Each generation t implies an iteration of the algorithm, where two populations P_t and Q_t of N individuals are used. When NSGA-II starts, the initial population P_0 is generated randomly. Furthermore, binary tournament selection, recombination, and mutation operators are used with individuals from P_0 to create a child population Q_0. Once P_0 and Q_0 are initialized, NSGA-II runs its main loop, which we can see in Alg.3. In each iteration, population P_t and Q_t are joined to create the population R_t, whose number of individuals is $2N$. After that, the individuals in R_t are sorted according to their dominance and crowding distances. The sorted individuals are added to population P_{t+1}. At the end of each iteration P_{t+1} is truncated to N individuals, and Q_{t+1} is generated using binary tournament selection, recombination, and mutation operators.

Once NSGA-II finishes, the final population P_t will contain as much individuals as potential solutions, and the non-dominated individuals are mapped to their corresponding case-bases. The case-base with the minimum error rate is chosen as the solution of the CBM algorithm. If two or more case-bases have the same error rate, then the algorithm chooses the first case-base found.

For further details of NSGA-II algorithms see [5].

3.4   Interpreting the MOEA Approach

A MOEA using our proposed fitness function tends to search for the minimum error rate and to delete the maximum number of cases, without exceeding a threshold of number of non-redundant cases that corresponds to $|M| * (1 - redundancy(M, k))$ (expression 8). Figure 2 depicts the target cases-bases of the fitness function for Iris dataset. That is, case-bases with a lower number of cases and with a similar error rate to the original case-base. To build the figure, we have created 1000 case-bases selecting from 5 to 70 random cases from Iris. Therefore, we have 1000 case-bases of 5 cases, 1000 cases-bases of 6 cases, and

Alg.1 fast-nondominated-sort(P)

Require: A population P
Ensure: list of the non-dominated fronts $\mathcal{F}$

```
 1: for p ∈ P do
 2:    for q ∈ P do
 3:       if p ≺ q then
 4:          S_p ← S_p ⋃{q}
 5:       else
 6:          if q ≺ p then
 7:             n_p ← n_p + 1
 8:          end if
 9:       end if
10:    end for
11:    if n_p = 0 then
12:       F_1 ← F_1 ⋃{p}
13:    end if
14: end for
15: i = 1
16: while F_i ≠ ∅ do
17:    H ← ∅
18:    for p ∈ F_i do
19:       for q ∈ S_p do
20:          n_q ← n_q − 1
21:          if n_q = 0 then
22:             H ← H ⋃{q}
23:          end if
24:       end for
25:    end for
26:    i = i + 1
27:    F_i ← H
28: end while
29: return F
```

Alg.2 crowding-distance-assignment($\mathcal{I}$)

Require: A set of individuals $\mathcal{I}$
Ensure: Each individual within $\mathcal{I}$ with a density measure.

```
 1: l ← |I|
 2: for i ∈ [1, N] do
 3:    I[i] ← 0
 4: end for
 5: for each objective m do
 6:    I ← sort(I, m)
 7:    I[1]_density ← ∞
 8:    I[l]_density ← ∞
 9:    for i ∈ [2, (l − 1)] do
10:       I[i]_density ← I[i]_density + (I[i +
          1].m − I[i − 1].m)
11:    end for
12: end for
```

Alg.3 NSGA-II main loop

Require: A fitness function Φ.
Ensure: a population P_t of potential solution

```
 1: R_t ← P_t ⋃ Q_t
 2: t ← 0, i ← 1
 3: F ← fast-nondominated-sort(P)
 4: while |P_{t+1}| < N do
 5:    crowding-distance-assignment(F_i)
 6:    P_{t+1} ← P_{t+1} ⋃ F_i
 7:    i ← i + 1
 8: end while
 9: sort(P_{t+1}, ≥_n)
10: P_{t+1} ← P_{t+1}[0 : N]
11: Q_{t+1} ← make-new-pop(P_{t+1})
12: t ← t + 1
```

Fig. 1. NSGA-II algorithm and main functions [5]

so on. Finally, a Hold-Out evaluation is used to measure the error rate of each
case-base, using 60% of the cases as the training set and 40% as the test set. For
each set of 1000 cases-bases the error rate given by the Hold-Out evaluation is
averaged. The plot shows for each case-base size the average error rate, and the
maximum and minimum observed values for each case-base size.

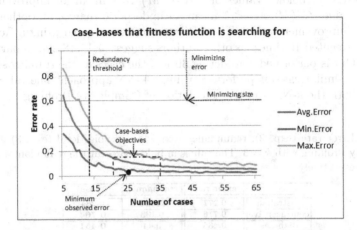

Fig. 2. Evolution of the error rate for Iris dataset. A lower number of cases is correlated with higher error rates. The fitness function goal is to find a case-base located
in between the redundancy threshold and the minimum observed value for this experimentation.

4   Experimental Evaluation

4.1   Experiments and Results

We have considered two measurements to study the suitability of our fitness
functions to perform CBM: the Reduction Rate, and the Competence Improvement [18]. The reason to use these two measures is because the suitability of
CBM algorithms is strongly related to the number of cases deleted by the CBM
process, and to the accuracy of the CBR system that will use the maintained
case-base. The measurements are defined as follows:

1. The Reduction Rate is the average number of cases removed, that is:

$$reduction(M', M) = \frac{|M'|}{|M|}. \tag{11}$$

2. The Competence Improvement, which quantifies the proportional improvement in accuracy of the CBR system. Note that this error is not related to
   the Complexity Profiling error measure. This measure is formally defined as
   follows:

$$CI(M', M) = \frac{eval\_error(M)}{eval\_error(M')},$$ (12)

where M is the initial case-base and M' is the case-base after the maintenance, with $M' \subseteq M$, and $eval\_error(M)$ is the proportion of times that the CBR system returns a wrong solution to the input problems using the Hold-Out approach described below. Values of $CI(M', M) > 1$ mean an improvement in accuracy, values $CI(M', M) < 1$ mean an underperformance, and otherwise it means no improvement at all. The CBR system is evaluated using a Hold-Out approach executed 10 times, as other authors suggest [4,17,18,20]. In particular, the Hold-Out is performed considering 30% of the cases as the training set. The retrieval of similar cases is performed using a k-NN approach. The value $k = 3$ is set for both the k-NN and the calculation of Complexity Profiling.

Table 1. Error rate (exp. 2), redundancy (exp. 4) and noise level (exp. 3) given by Complexity Profiling with $k = 3$. The values in bold represent the redundant and noise datasets, respectively.

	error rate	redundancy	noise
australian	0.277	**0.636**	0.284
contraceptive	0.716	0.133	**0.764**
diabetes	0.435	0.44	**0.451**
flags	0.6	0.289	**0.655**
glass	0.436	0.444	**0.453**
ionosphere	0.18	**0.772**	0.18
iris	0.064	**0.913**	0.06
liver-bupa	0.586	0.215	**0.597**
lymph	0.367	0.487	**0.392**
segment	0.056	**0.917**	0.058
sonar	0.218	**0.716**	0.245
vehicle	0.447	0.43	**0.468**
vowel	0.044	**0.9**	0.028
wine	0.069	**0.899**	0.056
zoo	0.083	**0.881**	0.089

In order to test the suitability of our proposal, we evaluate NSGA-II with our fitness function for CBM, using different standard datasets, and we do a comparative analysis considering some representative CBM algorithms from the literature. The results of each evaluation are the Reduction Rate and Competence Improvement measurements. In particular, our experiments consider:

- 15 datasets from the UCI repository [6]: australian, contraceptive, diabetes, flags, glass, ionosphere, iris, liver-bupa, lymph, segment, sonar, vehicle, vowel, wine and zoo. Each dataset has no missing values, and the nominal or string values in the datasets have been replaced by equivalent integer values. Finally, each record of the dataset is considered as a case, and the last attribute makes up the solution. Table 1 shows the levels of error rate, redundancy and noise given by Complexity Profiling using $k = 3$, $\epsilon = 0.5$ and $\rho = 0$ for expressions 2, 3 and 4. We consider a dataset as noisy where its noise level is higher than 0.4 and redundant when its redundancy level is higher than 0.5. Thus, there are eight redundant datasets and seven noisy datasets.

- 7 CBM algorithms: CNN, RNN, RENN, DROP1, DROP2, DROP3 and ICF.
- NSGA-II as MOEA using our fitness function. The number of individuals is 100, the number of generations 250, the mutation probability is 0.05 and the crossover probability is 0.9.

Figures 3, 4, 5 depict the result of Competence Improvement for each CBM algorithm and case-base. Each column represents the result CNN, RNN, RENN, DROP1, DROP2, DROP3 and ICF. NSGA-II returns a final population with individuals representing case-bases. In this experimentation we only consider three of them: the *minimum* as the solution case-base with the maximum error rate, the *maximum* as the case-base with the minimum error rate, and the case base with the minimum Complexity Profiling error, which is returned by NSGA-II. The lines in figures 3, 4, 5 represent the results of the maximum, minimum and the output case-bases obtained by NSGA-II using our fitness function.

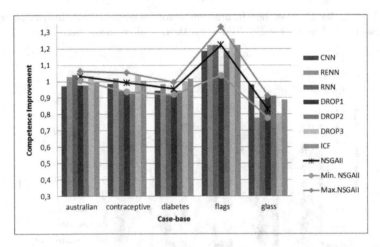

Fig. 3. Results for Competence Improvement for australian, contraceptive, diabetes, flags and glass case-bases for each CBM algorithm. Higher means better.

Table 2 shows the reduction rate results for each case-base and CBM algorithm. The highest reduction is highlighted in bold. For the sake of clarity, table 2 only shows the reduction given by the output case-base in NSGA-II, because maximum and minimum are very similar.

Concerning the duration of a CBM execution, we must take into consideration the size of the case-base, the number of features to describe the problem and solution, and the complexity to compute the similarity between cases. NSGA-II has the longest runtime among all the algorithms because other factors are implicated, such the crossover and mutation operator, the assigned probability to apply, and especially the amount of individuals in the population and the number of generations.

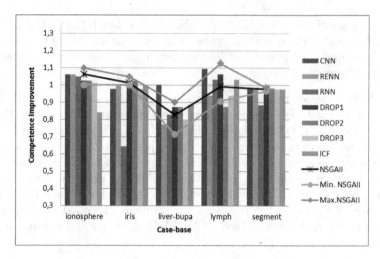

Fig. 4. Results for Competence Improvement for ionosphere, iris, liver-bupa, lymph and segment case-bases for each CBM algorithm. Higher means better.

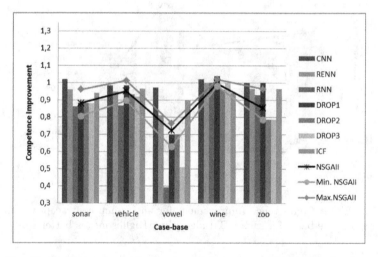

Fig. 5. Results for Competence Improvement for sonar, vehicle, vowel, wine and zoo case-bases for each CBM algorithm. Higher means better.

4.2   Discussion

According to the experiments, the case-base returned by the NSGA-II algorithm achieves the best improvements in competence (australian, ionosphere, and segment), or reaches a solution very close to the best observed among the other CBM algorithms. When the best case-bases of NSGA-II are considered, this algorithm achieves the best results in many of the datasets (australian, contraceptive, flags, ionosphere, iris, lymph, segment and vehicle). Additionally, the case-base with the worst Competence Improvement with NSGA-II only returns

Table 2. Reduction rate for each dataset and CBM algorithm

Datasets	CNN	RENN	RNN	DROP1	DROP2	DROP3	ICF	NSGAII
australian	0.595	0.162	0.598	0.5	0.5	0.581	0.415	**0.616**
contraceptive	0.246	0.552	0.266	0.5	0.5	**0.778**	0.686	0.47
diabetes	0.445	0.304	0.477	0.497	0.501	**0.652**	0.48	0.59
flags	0.259	0.533	0.37	0.496	0.504	**0.77**	0.637	0.705
glass	0.409	0.383	0.436	0.497	0.503	0.691	0.557	**0.713**
ionosphere	0.69	0.163	**0.763**	0.5	0.5	0.5834	0.425	0.698
iris	0.827	0.019	**0.856**	0.49	0.5	0.51	0.327	0.215
liver-bupa	0.271	0.408	0.383	0.5	0.5	**0.704**	0.588	0.642
lymph	0.534	0.223	0.612	0.495	0.505	0.612	0.466	**0.766**
segment	0.844	0.054	**0.881**	0.499	0.5	0.527	0.413	0.479
sonar	0.579	0.214	0.593	0.503	0.503	0.607	0.4	**0.651**
vehicle	0.413	0.325	0.455	0.499	0.501	**0.663**	0.498	0.588
vowel	0.471	0.108	**0.704**	0.499	0.5	0.555	0.327	0.372
wine	0.782	0.065	**0.815**	0.492	0.5	0.532	0.315	0.299
zoo	0.714	0.071	**0.771**	0.443	0.5	0.543	0.529	0.184
average	0.539	0.239	0.599	0.494	0.5012	0.621	0.471	0.533

the worst results in liver and sonar datasets among the CBM algorithm considered. In some datasets with high levels of redundancy, the best competence improvement results of NSGA-II are beaten by algorithms specialized in removing redundant cases, such as CNN. Similarity, NSGA-II is beaten in some noisy datasets by those algorithms specialized in deleting noisy cases, such as ICF in diabetes. However, overall the average Competence Improvement achieved by NSGA-II is consistent in all the experiments.

It is also worth mentioning that the worst case-base in the final population of NSGA-II are often very close to the competence improvement of the rest of the algorithms. However, our eager approach to choosing the final case-base seems insufficient for picking the best maintained case-base, suggesting that it is not enough to consider only the minimum error rate.

Figure 6 plots one point for each CBM algorithm, and each point corresponds to the Reduction Rates and Competence Improvement resulting from the average of all the Reduction Rates and Competence Improvement from the experiments. The figure depicts how difficult it is to achieve both great reductions and accuracy improvement at the same time, because a larger reduction results in worsening accuracy. The only exceptions to this tendency are CNN and the best results given by NSGA-II.

Finally, to identify whether a CBM algorithm deletes noisy or redundant cases, the Pearson product-moment correlation coefficient is computed between the error, redundancy and noise measure, which are returned by Complexity Profiling, and the accuracy and reduction rate given by the evaluation process. This correlation ranges from -1 to $+1$. Values in the interval $(-1, 0)$ indicates a negative correlation, values in the interval $(0, 1)$ note a positive correlation. That is, values close to -1 means the CBM algorithm does not delete that kind of cases, and values close to 1 point out that CBM deletes aggressively that type of case. Table 3 shows the coefficient values for each pair of results.

The NSGA-II correlation coefficients (table 3) indicate that the number of deleted cases is correlated with both noisy and redundancy levels. Thus, it seems

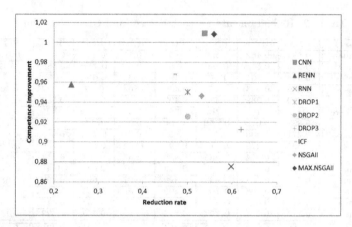

Fig. 6. Distribution resulting of averaging the datasets results for each CBM algorithm according to the reduction rate and competence improvement

that the fitness function aims the search of the maintained case-base deleting redundant cases and smoothing the frontiers between clusters of cases. Noisy cases are deleted more aggressively than the redundant cases though. In particular, NSGA-II achieves lower reduction rates in datasets with many redundant cases and few noise cases such as iris, vowel, wine and zoo. On the contrary, the reduction rate is greater in noisy datasets, such as contraceptive, diabetes, flags, glass, liver-bupa, lymph and vehicle. The coefficients also show that RENN, DROP3 and ICF are focused on deleting noisy cases, albeit DROP3 and ICF remove redundant cases as well. Moreover, CNN and RNN delete mainly redundant cases. DROP1 and DROP2 are focused on deleting cases near the borders. The rest of the CBM algorithms delete both types of cases equally.

Table 3. Correlation between Complexity Profiling error and accuracy, Complexity Profiling redundant level and reduction rate, and Complexity Profiling noise level and reduction rate

Pearson correlation	CNN	RENN	RNN	DROP1	DROP2	DROP3	ICF	NSGAII
Error & accuracy	-0,97	-0,91	-0,48	-0,86	-0,83	-0,64	-0,90	-0,85
Redundant & reduct. rate	0,91	-0,96	0,96	-0,33	-0,28	-0,95	-0,86	-0,60
Noise & reduction rate	-0,89	0,97	-0,96	0,31	0,30	0,97	0,88	0,59

5   Conclusions and Future Work

In this work we propose a multi-objective evolutionary approach to solve some tasks of Case-Base Maintenance. In particular, we present a novel fitness function based on Complexity Profiling [15]. We test the suitability of the approach on different datasets and compare the performance achieved to that of existing CBM algorithms from the literature.

Previous works are mainly focused on reducing either the number of redundant cases or noisy cases [1, 7, 8, 20–22], or aimed at selecting attributes [10, 12] or to both enhance the accuracy and reduce the size of the case-base [11]. However, the fitness function proposed in this work measures the redundancy of the case-base, the number of noisy cases and the error rate of the system. Therefore, this function aims to maintain the case-base following three objectives. The experiments show that the fitness function aims the search of the maintained case-base, to those case-bases with less redundant cases and smoother frontiers between clusters of cases.

The results obtained in the experiments show that the evolutionary approach outperforms general CBM approaches in many datasets, obtaining promising results even in worst cases. However, in our opinion, the most remarkable result of our proposal is the regularity of the behaviour with most datasets.

The runtime could be a limitation, in particular where CBM can not be performed off-line and the CBR system is stopped until the CBM process finishes. For this reason MOEA are not suitable in all scenarios. Nevertheless, using MOEA could be suitable when the case-base is built for the first time from a raw set of data, and where time is not the most important restriction. In this scenario, selection of an individual case-base from the final population could be done through an evaluation process using Cross-Validation or Hold-Out.

The use of genetic operators is limited in this work. Therefore, the next step will focus on the definition of specific crossover and mutation operators based on coverage and reachability.

Acknowledgements. This work was partially funded by the Seneca Research Foundation of the Region of Murcia under project 15277/PI/10, and by the Spanish Ministry of Science and Innovation+European FEDER+PlanE funds under the project TIN2009-14372-C03-01.

References

1. Brighton, H., Mellish, C.: On the consistency of information filters for lazy learning algorithms. In: Żytkow, J.M., Rauch, J. (eds.) PKDD 1999. LNCS (LNAI), vol. 1704, pp. 283–288. Springer, Heidelberg (1999)
2. Coello, C.C., Lamont, G., van Veldhuizen, D.: Evolutionary Algorithms for Solving Multi-Objective Problems. Genetic and Evolutionary Computation (2007)
3. Cover, T., Hart, P.: Nearest neighbor pattern classification. IEEE Transactions on Information Theory 13(1), 21–27 (1967)
4. Cummins, L., Bridge, D.: Maintenance by a committee of experts: The MACE approach to case-base maintenance. In: McGinty, L., Wilson, D.C. (eds.) ICCBR 2009. LNCS, vol. 5650, pp. 120–134. Springer, Heidelberg (2009)
5. Deb, K., Pratap, A., Agarwal, S., Meyarivan, T.: A fast and elitist multiobjective genetic algorithm: NSGA-II. IEEE Transactions on Evolutionary Computation 6(2) (2002)
6. Frank, A., Asuncion, A.: UCI machine learning repository (2010), http://archive.ics.uci.edu/ml

7. Gates, G.: Reduced nearest neighbor rule. IEEE Transactions on Information Theory 18(3), 431 (1972)
8. Hart, P.: Condensed nearest neighbor rule. IEEE Transactions on Information Theory 14(3), 515+ (1968)
9. Holland, J.H.: Adaptation in Natural And Artificial Systems. MIT Press (1975)
10. Ishibuchi, H., Nakashima, T., Nii, M.: Genetic-algorithm-based instance and feature selection. In: Frasson, C., McCalla, G.I., Gauthier, G. (eds.) ITS 1992. LNCS, vol. 608, pp. 95–112. Springer, Heidelberg (1992)
11. Cano, J.R., Herrera, F., Lozano, M.: Evolutionary stratified training set selection for extracting classification rules with trade off precision-interpretability. Data & Knowledge Engineering 60(1), 90–108 (2007)
12. Kim, K., Han, I.: Maintaining case-based reasoning systems using a genetic algorithms approach. Expert Systems with Applications 21(3), 139–145 (2001)
13. Leake, D., Wilson, M.: How many cases do you need? Assessing and predicting case-base coverage. In: Ram, A., Wiratunga, N. (eds.) ICCBR 2011. LNCS, vol. 6880, pp. 92–106. Springer, Heidelberg (2011)
14. Leake, D.B., Wilson, D.C.: Categorizing case-base maintenance: Dimensions and directions. In: Smyth, B., Cunningham, P. (eds.) EWCBR 1998. LNCS (LNAI), vol. 1488, pp. 196–207. Springer, Heidelberg (1998)
15. Massie, S., Craw, S., Wiratunga, N.: Complexity-guided case discovery for case based reasoning. In: 20th National Conference on Artificial Intelligence, AAAI 2005, vol. 1, pp. 216–221 (2005)
16. Massie, S., Craw, S., Wiratunga, N.: Complexity profiling for informed case-base editing. In: Roth-Berghofer, T.R., Göker, M.H., Güvenir, H.A. (eds.) ECCBR 2006. LNCS (LNAI), vol. 4106, pp. 325–339. Springer, Heidelberg (2006)
17. McKenna, E., Smyth, B.: Competence-guided case-base editing techniques. In: Blanzieri, E., Portinale, L. (eds.) EWCBR 2000. LNCS (LNAI), vol. 1898, pp. 186–197. Springer, Heidelberg (2000)
18. Pan, R., Yang, Q., Pan, S.: Mining competent case bases for case-based reasoning. Artificial Intelligence 171(16-17), 1039–1068 (2007)
19. Smyth, B., Keane, M.: Remembering to forget - a competence-preserving case deletion policy for case-based reasoning systems. In: International Joint Conference on Artificial Intelligence, IJCAI 1995, pp. 377–382 (1995)
20. Smyth, B., McKenna, E.: Competence guided incremental footprint-based retrieval. Knowledge-Based Systems 14(3-4), 155–161 (2001)
21. Wilson, D.: Asymptotic properties of nearest neighbor rules using edited data. IEEE Transactions on Systems Man and Cybernetics SMC 2(3), 408 (1972)
22. Wilson, D., Martinez, T.: Reduction techniques for instance-based learning algorithms. Machine Learning 38(3), 257–286 (2000)
23. Zitzler, E., Thiele, L.: Multiobjective evolutionary algorithms: A comparative case study and the strength pareto approach. IEEE Transactions on Evolutionary Computation 3(4), 257–271 (1999)

Mining and Retrieving Medical Processes to Assess the Quality of Care

Stefania Montani[1], Giorgio Leonardi[1,2], Silvana Quaglini[2],
Anna Cavallini[3], and Giuseppe Micieli[3]

[1] DISIT, Computer Science Institute, Università del Piemonte Orientale,
Alessandria, Italy
[2] Dipartimento di Informatica e Sistemistica, Università di Pavia, Italy
[3] IRCCS Fondazione "C. Mondino", Pavia, Italy - on behalf of the Stroke Unit
Network (SUN) collaborating centers

Abstract. In a competitive healthcare market, hospitals have to focus
on ways to deliver high quality care while at the same time reducing costs.
To accomplish this goal, hospital managers need a thorough understanding of the actual processes. Process mining can be used to extract process
related information (e.g., process models) from data. This process information can be exploited to understand and redesign processes to become
efficient high quality processes. Process analysis and redesign can take
advantage of Case Based Reasoning techniques.

In this paper, we present a framework that applies *process mining*
and *case retrieval* techniques, relying on a novel distance measure, to
stroke management processes. Specifically, the goal of the framework is
the one of analyzing the quality of stroke management processes, in order
to verify: (i) whether different patient categories are differently treated
(as expected), and (ii) whether hospitals of different levels (defined by
the absence/presence of specific resources) actually implement different
processes (as they auto-declare). Some first experimental results are presented and discussed.

1   Introduction

Healthcare institutions are increasingly facing pressure to reduce costs, while
at the same time improving the quality of care. In order to reach such a goal,
healthcare administrators and expert physicians need to evaluate the services the
institution provides. Service evaluation requires to analyze medical processes,
which are often automated and logged by means of the workflow technology.

Process analysis (PA) covers functions of simulation and diagnosis of processes. While simulation can support performance issues evaluation, diagnosis can highlight e.g., similarities, differences, and adaptation/redesign needs.
Indeed, the existence of different patients categories, or of local resource constraints, can make differences between process instances necessary, and
process adaptation compulsory (even when the medical process implements a
well-accepted clinical guideline). Proper PA techniques are strongly needed when

S.J. Delany and S. Ontañón (Eds.): ICCBR 2013, LNAI 7969, pp. 233–240, 2013.

a given process model does not exist, e.g., because a full clinical guideline has not been provided, and only some recommendations are implemented. In this case, *process mining* techniques [4] can be exploited, to extract process related information (e.g., process models) from log data. It is worth noting, however, that the mined process can also be compared to the existing guideline (if any), e.g., to check conformance, or to understand the required level of adaptation to local constraints. Thus, the mined process information can always be used to understand, adapt and redesign processes to become efficient high quality processes.

The *agile workflow* technology [15] is the technical solution which has been invoked to deal with process adaptation/redesign. In order to provide an effective and quick adaptation support, many agile workflow systems share the idea of recalling and reusing concrete *examples of changes* adopted in the past. To this end, *Case Based Reasoning (CBR)* [1] has been proposed as a natural methodological solution (see e.g, [10,11,7]). In particular, the *case retrieval* step has been extensively studied in PA applications, since the nature of processes can make distance calculation and retrieval optimization non-trivial [12,13,2,8].

In this paper, we propose a framework for medical process analysis and adaptation, which relies on **process mining** and **case retrieval** techniques.

Specifically, our goal is the one of analyzing the quality of stroke management processes, in order to verify: (i) whether different patient categories are differently treated (as expected), and (ii) whether hospitals of different levels (defined by the absence/presence of specific resources for stroke management) actually implement different processes (as they auto-declare).

First, our system extracts process models from a database of real world process logs. In particular, we learn different models for every patient category, and/or for every hospital. Given one of the models as an input, we then retrieve and order the most similar models we have learned. An examination of the distance among the models, to be conducted by a medical expert, can provide information about the quality of the processes, by verifying and quantifying issues (i) and (ii) above. To this end, we have introduced a proper *distance definition*, that extends previous literature contributions [5,3,2] by considering the available information, learned through process mining.

Experimental results (related to issue (ii)) and future research directions are discussed in the paper as well.

2   Methods

2.1   Process Mining and the ProM Tool

Process mining describes a family of a-posteriori analysis techniques exploiting the information recorded in logs, to extract process related information (e.g., process models).

Traditionally, process mining has been focusing on discovery, i.e., deriving process models and execution properties from enactment logs. It is important to mention that there is no a-priori model, but, based on process logs, some model, e.g., a Petri net, is constructed. However, process mining is not limited to process models (i.e., control flow), and recent process mining techniques are more and more focusing on other perspectives, e.g., the organization perspective, the performance perspective or the data perspective. Moreover, as well stated in [6], process mining also supports conformance analysis and process enhancement.

To be able to understand whether the healthcare organizations under study achieve their goals of providing timely and high quality medical services, we conducted several experiments (see also [9]) using the process mining tool called ProM, extensively described in [14]. ProM is a platform independent open source framework which supports a wide variety of process mining and data mining techniques, and can be extended by adding new functionalities in the form of plug-ins.

In particular, we relied on ProM's Heuristic miner [16] for mining the process models, and on a performance analysis plug-in which projects information of the mined process on places and transitions in a Petri net.

2.2   Distance Definition for Case Retrieval

In order to retrieve process models and order them on the basis of their distance with respect to a given query model, we have introduced a distance definition that extends previous literature contributions [5,3,2] by properly considering the available information, learned through process mining.

In particular, since mined process models are represented in the form of graphs (where nodes represent activities and edges provide information about the control flow), we define a distance based on the notion of graph edit distance [3]. Such a notion calculates the minimal cost of transforming one graph into another by applying insertions/deletions and substitutions of nodes, and insertions/deletions of edges.

As in [5], we provide a normalized version of the approach in [3], and as in [5,2], we calculate a *mapping* between the two graphs to be compared, so that edit operations only refer to mapped nodes (and to the edges connecting them).

Moreover, with respect to all the previous approaches, we introduce two novel contributions:

1. we calculate the cost of node substitution $fsubn$ (see Definition 2 below) by applying **taxonomic distance** [13,12] (see Definition 1), and not string edit distance on node names as in [5]. Indeed, we organize the various activities executable in our domain in a taxonomy, where activities of the same type (e.g., Computer Assisted Tomography (CAT) *with* or *without* contrast) are connected as close relatives. The use of this definition allows us to explicitly take into account this form of domain knowledge: the closer two activities are in the taxonomy, the less penalty has to be introduced for substitution;

2. we add a cost contributions related to edge substitution ($fsube$ in Definition 2 below), that incorporates information learned through process mining, namely (i) the percentage of patients that have followed a given edge, and

(ii) the reliability of a given edge, i.e., of the control flow relationship between two activities. The percentage of patients that followed an edge is calculated as the fraction over all the traces in the database in which the activities connected by the edge at hand take place in sequence. The reliability of a relationship (e.g., activity x follows activity y) is not only influenced by the number of occurrences of this pattern in the logs, but is also (negatively) determined by the number of occurrences of the opposite pattern (y follows x). Both items (i) and (ii) are outputs of Heuristic miner [16].

Formally, the following definitions apply:

Definition 1: Taxonomic Distance

Let α and β be two activities in the taxonomy t, and let γ be the closest common ancestor of α and β. The *Taxonomic Distance* $dt(\alpha, \beta)$ between α and β is defined as:

$$dt(\alpha, \beta) = \frac{N_1 + N_2}{N_1 + N_2 + 2 * N_3}$$

where N_1 is the number of arcs in the path from α and γ in t, N_2 is the number of arcs in the path from β and γ, and N_3 is the number of arcs in the path from the taxonomy root and γ.

Definition 2: Extended Graph Edit Distance. Let $G1 = (N1, E1)$ and $G2 = (N2, E2)$ be two graphs, where Ei and Ni represent the sets of edges and nodes of graph Gi. Let M be a partial injective mapping [5] that maps nodes in $N1$ to nodes in $N2$ and let *subn*, *sube*, *skipn* and *skipe* be the sets of substituted nodes, substituted edges, inserted or deleted nodes and inserted or deleted edges with respect to M. In particular, a substituted edge connects a pair of substituted nodes in M. The fraction of inserted or deleted nodes, denoted *fskipn*, the fraction of inserted or deleted edges, denoted *fskipe*, and the average distance of substituted nodes, denoted *fsubn*, are defined as follows:

$$fskipn = \frac{|skipn|}{|N1| + |N2|}$$

$$fskipe = \frac{|skipe|}{|E1| + |E2|}$$

$$fsubn = \frac{\sum_{n,m \in M} dt(n, m)}{|subn|}$$

where n and m are two mapped nodes in M.

The average distance of substituted edges $fsube$ is defined as follows:

$$fsube = \frac{\sum_{(n1,n2),(m1,m2)\in M}(|rel(e1) - rel(e2)| + |pat(e1) - pat(e2)|)}{|2 * sube|}$$

where edge $e1$ (connecting node $n1$ to node $m1$) and edge $e2$ (connecting node $n2$ to node $m2$) are two substituted edges in M, $rel(ei)$ is the reliability $\in [0,1]$ of edge ei as extracted by Heuristic miner [16], and $pat(ei)$ is the percentage of patients that crossed edge ei.

The extended graph edit distance induced by the mapping M is:

$$ext_{edit} = \frac{wskipn * fskipn + wskipe * fskipe + wsubn * fsubn + wsube * fsube}{wskipn + wskipe + wsubn + wsube}$$

where $wsubn$, $wsube$, $wskipn$ and $wskipe$ are proper weights $\in [0,1]$.

The extended graph edit distance of two graphs is the minimal possible distance induced by a mapping between these graphs. To find the mapping that leads to the minimal distance we resort to the greedy algorithm described in [5].

3 Experimental Results

In clinical practice, no support is available to physicians/administrators to verify whether hospitals of different levels actually implement different processes when caring a specific pathology (see issue (ii) described in the Introduction). In a previous version of this work [9], process mining was relied upon to provide physicians with a graphical view of the mined processes. A visual inspection of those figures was a first help towards the fulfillment of the tasks related to issue (ii). However, mined processes can be huge and very complex, so that an automated comparison among them, like the one we are providing in this framework, can truly be an added value for quality evaluation.

In the rest of this section, we discuss our experimental results, related to issue (ii). In particular, we wished to test whether the level of 37 hospitals located in the Lombardia Region (Northern Italy) could be verified (or corrected) through our framework, when referring to stroke care. Hospital levels (i.e., 1, 2, 3) have to be defined in Lombardia Region according to the available human and instrumental resources. Every hospital auto-declares its own level. Specifically, we mined the stroke management processes implemented in all 37 hospitals. We then chose one level-2 hospital as a query, and we retrieved and ordered the mined processes of the 36 others (21 of which were declared as level-2 hospitals as well). We performed retrieval and ordering both resorting to the distance defined in [5], and to the novel one introduced in section 2.2. Results are reported in figure 1.

First, we can observe that our distance is able to discriminate among every single mined processes, while the one in [5] only identifies some macro-classes, composed by several processes, whose distance from the query does not change

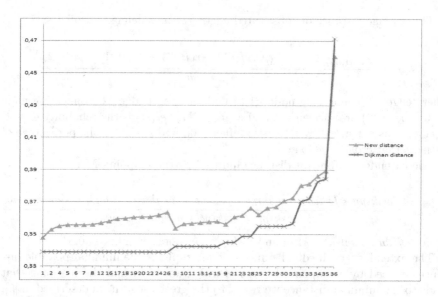

Fig. 1. Retrieval and ordering of 36 mined processes, implemented in 36 different hospitals in the Lombardia region, with respect to the selected query process (on the x-axis: process number; on the y-axis: distance value from the query). Results are shown in two different framework settings: when relying on the metric in [5] (Dijkman distance), and when relying on the metric defined in section 2.2

(see horizontal segments in figure 1). We believe that the finer distinction we could obtain is due to the use of taxonomic distance, and of edge information, which are disregarded by [5]. This additional information can be very significant from a medical viewpoint. For instance, hospitals 2 and 20 are not distinguishable according to [5], but in hospital 20 more than 70% of the patients undergo ECG immediately after CAT, while in hospital 2 this occurs for only 10% of the patients. Almost all patients undergo these tests in the two hospitals indeed, but within different control flow patterns. In hospital 20 there seems to be a behavioral rule pushing for the pattern CAT *immediately followed by* ECG, while in the other hospital this direct sequential pattern does not exist. This is an edge-related information extracted by Heuristic miner, and properly used by our metric for providing its finer ordering.

As for the declared hospital levels, we considered the 22 closest processes (i.e., hospitals) with respect to the query. This number was chosen because it is the sum of the number of processes in the two closest macro-classes when resorting to [5] (16 processes belong to the first macro-class, 6 to the second), and with [5] it is not possible to further refine the ordering among these examples. If the auto-declared level of these examples was correct (and confirmed by the mined processes), we should find 21 level-2 hospitals in this set. However, this did not happen. When resorting to [5], we found only 13 level-2 hospitals in these nearest neighbors. Of them, only 9 were listed in the closest 16 (i.e., the first macro-class). When exploiting our distance, we still found 13 level-2 hospitals in the first 22, but 11 of them were in the first 16. Our results were thus closer to the expected ones.

We analyzed the situation of the remaining 8 level-2 hospitals, that were not found in the nearest neighbors. Very interestingly, 7 of these missing examples are the very same when resorting to the two different metrics. Indeed, the visual examination of the graphs highlight important differences with respect to the query hospital. For example, one of them does not perform the thrombolisys treatment, even if typical of level 2 stroke units. We have to say that some local conditions (e.g., specific resources availability) may have recently changed, altering the real level of some hospitals with respect to the originally declared one. This conclusion thus supports the quality of the implemented metrics, and of our novel contribution in particular.

As a final consideration, we can quickly comment on 4 cases, that were differently ordered by the two metrics. According to the auto-declared levels, our ordering is closer to reality in 3 of them (no. 9, 22 and 24), while in the fourth case (no. 26) our metric overestimates the distance between the hospital and the query. Despite the overall positive outcome, this motivates further improvements, like the ones we will discuss in section 4.

4   Discussion, Conclusions and Future Work

This work showed that process mining and case retrieval techniques can be applied successfully to clinical data to gain a better understanding of different medical processes adopted by different hospitals (and for different groups of patients). It is interesting to analyze the differences, to establish whether they concern only the scheduling of the various tasks or also the tasks themselves. In this way, not only different practices may be discovered that are used to treat similar patients, but also unexpected behavior may be highlighted.

In this paper we have shown some first experimental results. More tests are obviously needed, including leave-one-out style experiments and comparisons with other metrics, and are planned for the next months.

In the future we also wish to extend our contribution, by including the treatment of time in $fsube$ (see Definition 2 in section 2.2). Indeed, by projecting the mined process on a Petri Net (see section 2.1), we can obtain information about delays between activities, possible overlaps and synchronizations . We would like to explicitly compare this information between mapped processes. We believe that, since in emergency medicine the role of time is clearly central, this enhancement could represent a relevant added value in our framework, and make it even more reliable and useful in practice.

Acknowledgments. This research is partially supported by the GINSENG Project, Compagnia di San Paolo.

References

1. Aamodt, A., Plaza, E.: Case-based reasoning: foundational issues, methodological variations and systems approaches. AI Communications 7, 39–59 (1994)
2. Bergmann, R., Gil, Y.: Retrieval of semantic workflows with knowledge intensive similarity measures. In: Ram, A., Wiratunga, N. (eds.) ICCBR 2011. LNCS, vol. 6880, pp. 17–31. Springer, Heidelberg (2011)
3. Bunke, H.: On a relation between graph edit distance and maximum common subgraph. Pattern Recognition Letters 18(8), 689–694 (1997)
4. Van der Aalst, W., van Dongen, B., Herbst, J., Maruster, L., Schimm, G., Weijters, A.: Workflow mining: a survey of issues and approaches. Data and Knowledge Engineering 47, 237–267 (2003)
5. Dijkman, R., Dumas, M., Garca-Banuelos, R.: Graph matching algorithms for business process model similarity search. In: Proc. International Conference on Business Process Management, pp. 48–63 (2009)
6. IEEE Taskforce on Process Mining: Process Mining Manifesto, http://www.win.tue.nl/ieeetfpm
7. Kapetanakis, S., Petridis, M., Knight, B., Ma, J., Bacon, L.: A case based reasoning approach for the monitoring of business workflows. In: Bichindaritz, I., Montani, S. (eds.) ICCBR 2010. LNCS, vol. 6176, pp. 390–405. Springer, Heidelberg (2010)
8. Kendall-Morwick, J., Leake, D.: On tuning two-phase retrieval for structured cases. In: Lamontagne, L., Recio-García, J.A. (eds.) Proc. ICCBR 2012 Workshops, pp. 25–334 (2012)
9. Mans, R., Schonenberg, H., Leonardi, G., Panzarasa, S., Cavallini, A., Quaglini, S., Van der Aalst, W.: Aprocess mining techniques: an application to stroke care. In: Proc. Medical Informatics Europe (MIE), pp. 573–578 (2008)
10. Minor, M., Tartakovski, A., Schmalen, D., Bergmann, R.: Agile workflow technology and case-based change reuse for long-term processes. International Journal of Intelligent Information Technologies 4(1), 80–98 (2008)
11. Montani, S.: Prototype-based management of business process exception cases. Applied Intelligence 33, 278–290 (2010)
12. Montani, S., Leonardi, G.: Retrieval and clustering for supporting business process adjustment and analysis. Information Systems, doi: http://dx.doi.org/10.1016/j.is.2012.11.006
13. Montani, S., Leonardi, G.: Retrieval and clustering for business process monitoring: results and improvements. In: Agudo, B.D., Watson, I. (eds.) ICCBR 2012. LNCS, vol. 7466, pp. 269–283. Springer, Heidelberg (2012)
14. van Dongen, B.F., de Medeiros, A.K.A., Verbeek, H.M.W(E.), Weijters, A.J.M.M.T., van der Aalst, W.M.P.: The proM framework: a new era in process mining tool support. In: Ciardo, G., Darondeau, P. (eds.) ICATPN 2005. LNCS, vol. 3536, pp. 444–454. Springer, Heidelberg (2005)
15. Weber, B., Wild, W.: Towards the agile management of business processes. In: Althoff, K.-D., Dengel, A.R., Bergmann, R., Nick, M., Roth-Berghofer, T.R. (eds.) WM 2005. LNCS (LNAI), vol. 3782, pp. 409–419. Springer, Heidelberg (2005)
16. Weijters, A., Van der Aalst, W., Alves de Medeiros, A.: Process Mining with the Heuristic Miner Algorithm, BETA Working Paper Series, WP 166. Eindhoven University of Technology, Eindhoven (2006)

Leveraging Historical Experience to Evaluate and Adapt Courses of Action

Alice M. Mulvehill, Brett Benyo, and Fusun Yaman

Raytheon/BBN Technologies
Cambridge, MA 02138, USA
{amm,bbenyo,fusun}@bbn.com

Abstract. In many planning domains there may be multiple potential solutions to a given problem. Each solution may require different resources, involve more or less risk, and result in desirable or undesirable effects. Reuse of historical plans is a strategy that can be employed to solve planning problems. While the retrieval of similar historical plans can be facilitated with sophisticated annotation and search engines, evaluating the usefulness of historical plans tends to be subjective, is context sensitive, and difficult when no single historical plan can be used to develop a new plan. Course of action (COA) evaluation is a method that can be used to compare a set of alternative solutions. An agent-based tool called MICCA (Mixed-Initiative Course of Action Critic Advisors) can aid human operators or software agents in evaluating and adapting historical plans for use in achieving one or more objectives in some current or future hypothetical world state. In this paper we introduce MICCA and describe how case base reasoning (CBR) and generative planning techniques are utilized to support COA evaluation and adaptation.

Keywords: course of action planning, case base reasoning, plan adaptation, agent based systems, blackboard technology.

1   Introduction

Reuse of historical plans is a strategy that can be employed to solve planning problems when access to past experience is available, time is limited, there is uncertainty about the current and/or future state of the world, and/or the decision maker lacks sufficient domain expertise to solve the current problem. While the retrieval of similar historical plans can be facilitated with sophisticated annotation and search engines, evaluating the usefulness of historical plans tends to be subjective, is context sensitive, and difficult when no single historical plan can be used to develop a new plan. The process is further complicated when the historical plan is very old, was created for a different problem domain and/or by a different user or software application.

In many planning domains there may be multiple potential solutions to a given problem. Each solution may require different resources, involve more or less risk, and result in desirable or undesirable effects. Course of action (COA) evaluation is a method that can be used to compare a set of alternative solutions.

S.J. Delany and S. Ontañón (Eds.): ICCBR 2013, LNAI 7969, pp. 241–254, 2013.
© Springer-Verlag Berlin Heidelberg 2013

MICCA (Mixed-Initiative Course of Action Critic Advisors) is an agent-based mixed-initiative tool that has been developed to aid human operators or software agents in evaluating and adapting historical plans for use in achieving one or more objectives in some current or future hypothetical world state. MICCA has been designed as a generalized approach that, in theory, can be utilized in any planning domain that leverages historical experience. The system utilizes two cycles: evaluation and adaptation. During evaluation, potentially useful historical plans are critiqued by both general and domain specific evaluation agents that evaluate the risk, adaptability and cost of each plan. During adaptation, selected candidate plans are refined to operate in a current problem environment. The revised candidates are then reevaluated along similar or other (user-defined) dimensions.

To date MICCA has been applied in two domains: the Rovers domain which was a benchmark domain in the Third International Planning Competition [1] and the Joint Air/Ground Operations Unified Adaptive Replanning (JAGUAR) domain [2] which was the focus of research for the Defense Advanced Research Projects Agency (DARPA) sponsored JAGUAR program. In the Rovers domain one or more land Rovers are tasked to collect rock samples or images from a planet surface. In the JAGUAR air mission planning domain specialized aircraft and weapon combinations are tasked to satisfy a military air mission objective, e.g., strike a certain target, take photos of a particular geographic region, provide command and control, etc.

In this paper we introduce MICCA and describe how case base reasoning (CBR) and generative planning techniques are utilized to support COA evaluation and adaptation. The paper includes a description of how the case base technology is used, a discussion of case base experimentation and some performance results.

2   Overview

At a very general level a course of action (COA) describes how a problem can be solved or a goal can be achieved. Representationally, a COA is similar to a plan in that it is generally comprised of a set of activities that can be performed in some order to achieve a particular goal. Each activity may include a recommendation for resource capabilities and possibly time sequencing. In certain domains such as military air mission planning, multiple COAs are traditionally evaluated to determine how to best achieve a goal or objective [3]. During the evaluation process the COAs are compared with each other to determine the benefits, risks and effects (both positive and negative) that are associated with each approach.

MICCA has been specifically developed to support the COA evaluation process. The MICCA system consists of a set of agents, a blackboard (BB), and a suite of specialized user interfaces (UI) that display COA options, evaluation and adaptation information, and provide detailed justification and rationale to the user [4]. Communication between MICCA agents with external databases, historical case base repositories, domain specific models, knowledge bases, and with the human operator is supported by the blackboard [5]. Figure 1 displays an overview of the MICCA evaluation and adaptation agent framework. In the next section an overview of the MICCA process is provided.

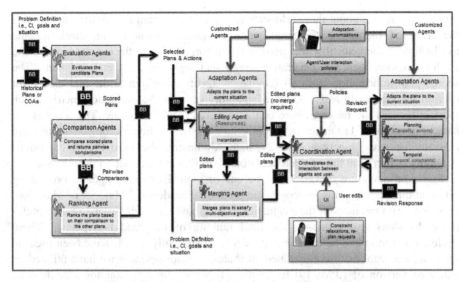

Fig. 1. Evaluation and Adaptation Agent Framework

2.1   The MICCA Process

Although MICCA is not a CBR system, the MICCA process emulates the basic CBR cycle: retrieve, reuse, revise, and retain [6]. The MICCA process starts with the receipt of one or more objectives. The objectives may have preferences for resources and some temporal requirements. Candidate COAs/plans, obtained from an external COA development system or as the result of a query from a historical repository, are provided to MICCA for evaluation and alignment with the current world state. For the research reported in this paper, COA candidates are retrieved from a historical case base repository. The case base technology used in MICCA was originally developed to store executed air mission data for the JAGUAR domain [7]. In order to support MICCA experimentation, a small subset of the JAGUAR mission data was used. In order to support experimentation with the Rovers domain, Rover plans were created with JSHOP, a Hierarchical Task Network (HTN) planner [8]. To align the two domains, the JAGUAR mission data was also converted into an HTN plan format.

Historical plans are stored in the case base in an XML data format that consists of five pieces of information: a unique string plan ID, the HTN plan, a set of case features, the world state that the plan was created for, and the set of objectives the plan was created to satisfy. The case features are a set of feature names and values that serve as descriptive meta data about each case and facilitate search and similarity matching. A Case Base Retrieval Agent was created in MICCA to use the published objectives as the basis for case base queries. Once retrieved plans are provided to MICCA they are processed by a set of Evaluation agents. The Evaluation agents can utilize information about the world state and relevant domain information to evaluate the retrieved plans. For example, one Evaluation agent might check for the availability of the actor that was used in the historical plan in the current world state. Each Evaluation agent will generate a scored plan object for each retrieved plan. The scored plan information from all Evaluation agents is written to the blackboard for each of the plans. In MICCA, the adaptability score is the evaluation metric which most influences the ranking score of the retrieved candidate plans. The score is the sum of two ratios: 1) Objective Alignment: the ratio of the case objectives that are relevant for new objectives, and 2) State Alignment: the ratio of the conditions necessary to carry out the relevant case objectives that are satisfied in the new state.

Retrieved plans are ranked based by one or more Evaluation agent scores along different dimensions using lexicographic preference models (LPMs). A LPM defines an order of importance on the evaluation criteria and uses this order to make preference decisions and influence the final ranking of the candidate plans. These evaluation criteria can be domain dependent. Historically, LPMs have been used on numeric, Boolean, or discrete valued attributes. In our research we have utilized an extended version of LPMs [9] to handle attributes within a monotonic continuous domain (such as the scores generated by Evaluation agents).

Figure 2 displays the list of ranked plans that have been retrieved from the case base for a problem in the JAGUAR domain. The UI displays some high level information about each of the retrieved plans such how many of the objectives each retrieved plan can satisfy in the "objectives" column. Scores from each of the Evaluation agents are also presented (the last four columns display this data in Figure 2). The user can use the "choose columns" option to display additional data. The data presented is derived from the case features obtained from the case base. The human operator can select one or more plans to publish to the blackboard for continued adaptation and re-evaluation. Although the highest ranked retrieved plan in Figure 2 satisfies all three of the objectives in this example, the operator can select several retrieved plans in order to generate multiple COAs/plans. In addition, if few or no retrieved plans satisfy all of the objectives, the user can request that certain retrieved plans be merged to create a plan that can satisfy all of the objectives.

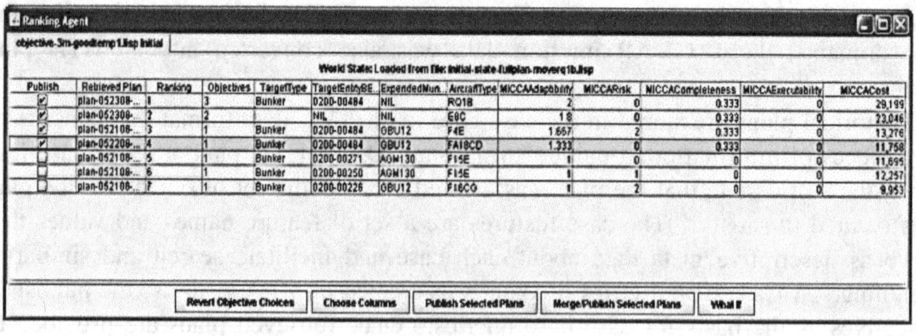

Publish	Retrieved Plan	Ranking	Objectives	TargetType	TargetEntityBE...	ExpendedMun...	AircraftType	MICCAAdaptability	MICCARisk	MICCACompleteness	MICCAExecutability	MICCACost
✓	plan-052308-...	1	3	Bunker	0200-00484	NIL	RQ1B	2	0	0.333	0	29,199
✓	plan-052308-...	2	2		NIL	NIL	E8C	1.8	0	0.333	0	23,046
✓	plan-052108-...	3	1	Bunker	0200-00484	GBU12	F4E	1.667	2	0.333	0	13,276
✓	plan-052208-...	4	1	Bunker	0200-00484	GBU12	FA18CD	1.333	0	0.333	0	11,758
	plan-052108-...	5	1	Bunker	0200-00271	AGM130	F15E	1	0	0	0	11,695
	plan-052108-...	6	1	Bunker	0200-00250	AGM130	F15E	1	1	0	0	12,257
	plan-052108-...	7	1	Bunker	0200-00226	GBU12	F16CG	1	2	0	0	9,953

Fig. 2. Ranking Agent UI

3   Case Revision – Preprocessing

Because the past and the present are rarely equivalent, a retrieved case must be revised in order for it to be used in a different context. Because the complexity of revision can vary, MICCA provides a variety of revision capabilities in multiple steps. The first step is associated with the instantiation of a historical plan within the current world state and is viewed as a preprocessing step for the more sophisticated MICCA Adaptation agents. This process is similar to what Mitra and Basak [10] call "substitutional adaptation" and involves the substitution of resource instances and revisions to task time values so that they are aligned with the current time.

If no single historical retrieved plan can be used to satisfy all of the objectives of the current problem, the revision will require that parts of several retrieved cases be merged to form a single coherent solution. In MICCA this first step is performed by two specialized Instantiation agents: an Editing agent and a Merge agent. Similar to a case base planner, these agents (1) edit the historical plan so that it conforms to the existing constraints and resource availabilities in the current state, and (2) merge elements from multiple historical plans together to produce a candidate plan that covers as many of the current objectives as possible.

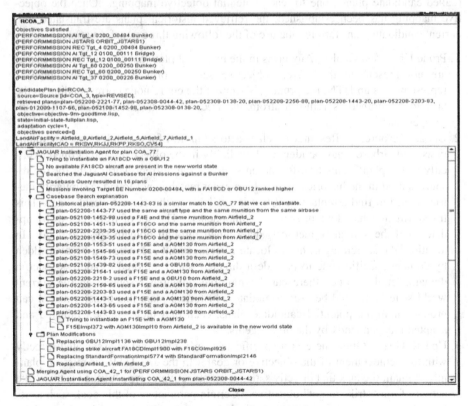

Fig. 3. Instantiation Agent Activity

Figure 3 is an example showing how these agents are used to create a candidate plan. In the example provided, the case base has been queried by the agents through the case base web interface for missions of type "AI" and a target of type "Bunker". 16 cases have been retrieved. The agents evaluate each retrieved case and select the most similar case for revision, where similarity is based on a set of additional case features including the Takeoff airbase and the specific target location. Plan modifications on this case are then made by the Editing agents. The Merge agent is used to combine edited plans to satisfy as many of the objectives in the new problem as possible.

3.1   The Editing Agent

This agent maps the current objectives to the goals of each retrieved historical plan. It assumes that each retrieved historical plan (which can consist of multiple tasks) can be mapped to one or more of the current objectives. This mapping can be automatically computed by objective similarity matching, or manually defined by a user through the Ranking Agent UI. If a single task in a retrieved historical plan can be mapped to multiple current objectives, the Editing agent will create multiple copies of the plan (called candidate plans), one for each potential objective mapping. Once the objective mapping has been established, the retrieved historical plans are translated into current candidate plans through the use of the following three transformations:

1. Prune Unneeded Goals: Any tasks in the historical plan that satisfy objectives that are not present in the current objective set are pruned.   Since the plan is represented as an HTN, the pruning is simple; the entire node corresponding to the task to be pruned is removed from the tree.   This transformation is domain-independent.

2. Update Resources: Resources such as actors (aircraft for example), sensors, munitions, and airbases may be identified explicitly in the objectives and defined explicitly in the plan.   This transformation searches the current world state for the resources used in the historical plan.   If an exact match is not found, a similarity metric is used to find a replacement that is available in the current world state.   This transformation has domain specific components.   The set of resources to be updated and the similarity metric is determined through domain specific code.   In addition, the search algorithm to locate a replacement resource can be overridden by domain specific code to provide a directed search tailored to address specific domain details.   Since there can be multiple potential resources in the current world state that could be used to update a historical plan, this transformation can produce multiple potential candidate plans, each of which can be evaluated and adapted independently by the other agents.

3. Update Goal: Often there is a specific task in a plan that is associated directly with the achievement of the objective or goal.   In the JAGUAR air mission planning domain the task that directly satisfies an objective of type "strike" is the strike activity for a "strike target" objective.   Certain parameters of this goal, such as its location, can be modified from historical values to match the requirements of the

current world state. Details about which properties of the goal can be edited are defined in domain specific code. In the JAGUAR domain, for example, we allow the Editing agent to modify the location of the goal in the historical candidate plan, thus allowing MICCA to use a historical plan that has a goal of the same type as the current objective, but at a different location. The Adaptation agents will later modify the plan to route the actor to the new location.

The output of the Editing agent is a candidate plan (or a set of candidate plans). These candidate plans may not be executable yet, and may violate constraints such as temporal or spatial constraints. For example if the current objective is "Travel from A to C" and the historical objective was "Travel from A to B", the Editing agent will replace B's in the plan with C's, provided that the domain specific Update Goal procedure allows this modification. Note that not every instance of B has to be replaced by C. Hierarchical plan structure, namely the decomposition tree, is used to identify the sub-tree related to the historical goal "Travel from A to B" and to replace all instances of B in the sub-tree with C. Details about how to get from A to C are not handled by the Editing agent; it simply substitutes C for B. An Adaptation agent will re-compute the route from A to C.

3.2   The Merger Agent

When no single retrieved plan achieves all of the current objectives, but a collection of retrieved plans can satisfy a subset of objectives, the Merger agent is used to combine several retrieved plans in order to produce a full candidate plan that achieves more of the desired objectives. Merging plans in an effective way generally requires a complex reasoner, such as a case based planner. MICCA approximates this CBR process by using a mixed initiative approach. We assume that MICCA may revise candidate plans that can achieve only a subset of the current objectives. The user can choose multiple retrieved plans to serve as the basis for a single objective, which would give the Merger agent multiple options for merging, and thus a decision to make. If no choice is made by the user, the Merger agent will consider' all possible retrieved plans for a specific objective. The Merger agent will produce multiple merged candidate plans, up to a threshold number, if there are multiple options to cover a specific objective. In order to produce candidate plans that are as different as possible, the merge algorithm used by the Merger agent will examine how often each chosen retrieved plan has been used to cover an objective in the set of output merged candidate plans, and choose the retrieved plan that has been used the least.

Although the plans that are being merged are revised to work in the current (not historic) state, the combined plan might require further revisions to ensure coherence of the pieces. For example, the temporal order on the objectives may need to be revised. Thus, after merging the plans, which in this case means concatenating them in the correct order, removing duplications and combining the objectives, another revision cycle will be triggered. In the case when an objective is still uncovered because none of the historical plans satisfy this objective, a message is sent to the user that there remains an unsatisfied objective.

4   MICCA Plan Adaptation

Once the retrieved plans are revised with current world state elements, specialized Adaptation agents begin to repair the plans to satisfy the objectives of the current problem solving situation. These agents are managed by a Coordination agent that keeps track of Adaptation agents and their capabilities, and keeps track of domains and their needs. The particular Adaptation agents used and their process sequence are defined in a policy. One of the prime functions of the Adaptation agents is to repair the causal and temporal problems in the edited candidate plans. In some cases, these agents will remove parts of the historical plans that are irrelevant to the current state or objective. When a current objective still cannot be satisfied by any of the published candidate plans, the JSHOP plan generator will be used to generate the missing plan elements. Adapted plans are re-evaluated by the Evaluation agents and presented to the user. The process can be repeated if the user is not satisfied with the results or if the user wants to experiment with hypothetical future situations.

5   Experimentation Process

Two different planning domains were used to test the generality of the MICCA design, architecture, agent types, and the evaluation and adaptation strategies. The experimentation results presented in this paper describe how CBR and generative techniques were used by the MICCA agents.

The MICCA system was first used to evaluate and adapt previously generated plans for the Rovers domain. The general problem in this domain is to find one or more Rovers to navigate a planet surface, find samples, take pictures, and communicate them back to a lander. An example of a Rover plan is provided in Figure 4.

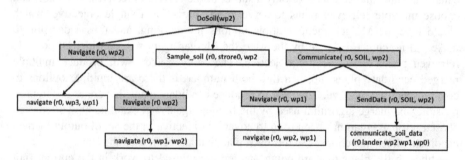

Fig. 4. Example Hierarchical Rover Plan where the Blue Nodes are Methods for Composite Actions and the Yellow Nodes are Simple Actions

The JSHOP planning software was used to generate 14 Rover plans, each with associated initial world states. Case features were created to support search by goal type and objective and to support plan evaluation. Two case base repositories were created to store the Rover plans, one to store the tasks and objectives and a second case repository to store the actual plan data. To start the process, a human operator uses information about a current goal/objective to query the Rover Task case base. The query returns a list of Rovers (with PlanIDs) who are capable of satisfying the objective. The user then selects a PlanID and queries for the HTN/Plan from the Rover Plan case base with the index PlanID. The retrieved plan(s) are published onto the blackboard – triggering the MICCA evaluation and adaptation cycles.

The second domain used was the JAGUAR air mission planning domain. Like the Rovers domain, there is an actor, in this case an aircraft with particular functional capabilities to perform certain activities. The aircraft is located at some origin and has to travel to another location in order to accomplish an objective – either strike a target with a weapon or take a photograph of the target location. The JAGUAR domain also contains detailed specifications of tasks and causal models. Unlike the Rovers domain, each JAGUAR air mission tends to satisfy one objective. Multiple objectives as well as supporting tasks are enumerated as a set of missions in a larger plan structure called the Air Tasking Order (ATO). A specialized plan object was created in order to support the evaluation and adaptation of multiple objectives for this domain.

To utilize JAGUAR data in MICCA we developed JAGUAR HTN models and modified the existing JAGUAR executed plan case base to include the HTN plan data. This involved domain engineering, plan extraction and finally case base creation. To scope our effort, we initially focused on one of the ATO strike mission types – Air Interdiction (AI), which is a strike mission that can be employed to destroy, neutralize or delay the enemy's military capability and two other mission types, reconnaissance (REC) and JSTARS. An experimental case base was created that contained 2 JSTARS and 3 REC missions and 20 strike missions. In this case base each case includes HTN formatted plans, world states and generated objectives. The structures of HTN Methods and operators were extracted from the JAGUAR models. This included 24 Methods and 14 Operators. We manually selected the predicates that appear in the modeled process constraints and represented most of these constraints as the preconditions and effects of the methods and operators. 19 axioms were manually defined to support the complex evaluation of certain conditions; such as what does it mean to be in battle space. We verified the validity of the domain with the JSHOP planner, which produced plans for simple problems. The JAGUAR domain also supports very complex routes and routing procedures, which were beyond the scope and the interests our research, so we simplified the routing in the JAGUAR HTNs.

6   Test Cases and Experimentation Results

Case base historical data, test scenarios and problem sets were established for both domains to support testing. Our objective was to determine if basic functions of MICCA could be used to evaluate and adapt plans in different problem domains with increased problem complexity. To support experimentation in the Rovers domain 16 random problems were generated with JSHOP. A typical problem was composed of one or more objectives where the objective is of the form: do-rock, do-soil, or do-image at a particular location (waypoint) in a current world state.

MICCA was tested on two levels of complexity during the Rover experiments. The first level of complexity involved a simple test for MICCA to leverage a historical plan to support a single new objective, e.g., (do-rock at waypoint12). The second level of complexity presented a situation with more complex objectives: two objectives and two goal types and/or three objectives and three goal types and a case base where no single historical plan could support a set of objectives.

Experimentation results demonstrate that the MICCA agents were able to successfully produce improved plans for single objectives and for the composite objectives that were causally sound (i.e., Rovers continuously move in the space and do not jump from one location to another without any proper navigation action) and complete (i.e., achieving all goals).

For the air mission planning domain, the goal of experimentation was again to measure MICCA performance given varying levels of complexity. We were also interested in two other characteristics: (1) the amount of time required by MICCA to generate one or more useful COAs, and (2) the influence that a human operator has on the generation of revised COAs. Several tests were conducted to measure MICCA performance on single objective plans that included 3 mission types: AI, REC and JSTARS. MICCA was able to successfully and quickly generate one or more COAs in all tests with a single objective.

Tests were then conducted to measure MICCA performance on building COAs to satisfy an increasingly complex set of multiple objectives. The following multiple objective problem situations were tested:

- 3m simple – 3 objectives, no temporal conflicts
- 3m   – 3 objectives, 3 mission types
- 4m   – 5 objectives, 3 mission types
- 9m   – 9 objectives, 3 mission types
- 50m   – 50 objectives, 3 mission types

Table 1 contains the MICCA performance results for the multiple objective test cases. All of these tests were run on a single laptop (Windows 7, i7 2.2 GHz processor, 3 GB RAM available to MICCA). Note that while MICCA agents can be run on different machines and parts of the problem can be worked on in parallel we did not perform any multi machine scalability experiments in this research effort. The 50m scenario was designed as a stress test to see if the system could operate with a larger and more complex problem.

For each multiple objective test case, MICCA was run once in the mixed-initiative (MI) mode and once in the auto-mode (auto). In the MI mode the user decides which plans to adapt and merge to form the final COA(s). In the auto mode the Merger agent automatically picks candidates to form the final COAs. The number of "Candidate Plans" listed in Table 1 is the number of plans that were retrieved initially from the case base. For all of the tests, the case base consisted of 25 cases. The results also indicate that when there is a difference in the "number of COAs generated", the difference is due to the interaction by the user. The results demonstrate that fewer COAs are generated when the Mode is MI even when the Auto mode is bound by a configuration parameter to control the number of COAs produced.

Table 1. Air Mission Planning Performance Results

Scenario Name	Interaction Mode	Eval. Agents	Candidate Plans	Objectives	Published	Conflicts	# COAs generated	Uncovered Objectives	Time to Run
3m simple	MI	4	7	3	6	0	4	0	~2 minutes
3m simple	Auto	4	7	3	7	0	4	0	~2 minutes
3m	MI	4	7	3	4	3	2	0	~2 minutes
3m	Auto	4	7	3	7	3	4	0	~2 minutes
4m	MI	4	10	5	7	4	4	0	~2 minutes
4m	Auto	5	10	5	10	4	4	0	~2 minutes
9m	MI	4	25	9	11	2	2	2^1	~10 minutes
9m	Auto	4	25	9	23	5	4	$2 - 3^2$	~10 minutes
50	Auto	4	25	50	21	100	1*	21^3	~97 minutes

Statistics in this table reflect the usage of MICCA at the end of the project and for each of the air mission planning scenarios. Auto means that all of the candidate plans were published. MI means that the user interacted and published only the candidate plans of interest.

*One COA was generated because MICCA encounter insufficient resources. The algorithm will only produce one COA if there is no way to produce a complete plan.
1 : (insufficient REC a/c)
2 : (insufficient REC a/c)
3: 21 RECs (there are only 3 REC a/c available); 1 JSTARs (only one available), all 24 AI objectives were satisfied.

On average, the agents completed the evaluation and adaptation cycles in no more than 10 minutes. The increase in processing time for some of the scenario tests is directly related to the time required to perform temporal reasoning during plan adaptation. This is most apparent in the 50m stress test scenario where MICCA took 97 minutes to run. The system did complete all processing cycles for the stress test scenario and was able to satisfy most of the strike (AI) objectives. The system was not able to generate missions for some of the REC and JSTARS objectives. This is due to the low number of REC and JSTARS aircraft available in the current world state.

7   Discussion

For both domains, the case base was used to provide historical plans to MICCA for use as the basis for candidate plans to satisfy current objectives, and to find alternative actor types for use by the Editing agent if the historical actors were not available in the current world state. In the Rovers domain, the Instantiation agents demonstrated CBR-like revision behavior. The Editing agent mapped the current objective to the goals of the historical plan and pruned unrelated ones; provided simple resource mapping; updated the Rovers used in the historical case with appropriate Rovers that existed in the current world state; and replaced other Rover components (camera) as necessary to support the current objective. In cases where no historical plan was able to cover all of the objectives in the problem set, the Merger agent merged multiple retrieved plans (N) that handled a subset of the current objectives (M) into a single plan to cover more objectives. The maximum possible number of merged plans is N * M. In general, the Merge agent produced up to R merge candidates; where R was set by a human administrator. Partial candidates were generated with the data from each individual retrieved plan. Ties were resolved by sorting by order from the Ranking agent. In the Rovers domain, the Adaptation agents were used to fix causal relationships in the plan, e.g., action B depends on A. The case base was not used by the Adaptation agents to repair broken dependencies; instead the Adaptation agents used the JSHOP HTN planner to support repair.

For the air mission planning domain, the case base was regularly referenced during the editing operations. The historical plan case base data used by the Editing agent to determine appropriate resource substitutions also constrained resource choices. This proved to be a useful strategy for operating in the JAGUAR domain where specific resource dependencies exist and where a typical world state can contain multitude resource options that are suitable for a specific task and objective. For example, aircraft Type/weapon combinations cannot be easily determined. Instead of reasoning with domain specific rules to determine a specific aircraft Type/weapon combination, the Editing agents queried the historical plans for historically used legal combinations. When none of the aircraft Type/weapon combinations from the historical pairs matched the availability in the current world state, the Editing agent did not produce a candidate plan. Figure 5 contains some revision statistics. Here, the "Parameter Edits" value highlights how many changes were made by the Editing agent for each RCOA (Revised COAs) produced during the MICCA session.

COA Comparison Summary				
Adaptation	RCOA 1	RCOA 2	RCOA 3	RCOA 4
Satisfied Temporal Constraints	24.0	34.0	34.0	34.0
Unsatisfied Temporal Constraints	1.0	0.0	0.0	0.0
Unverified Temporal Constraints	9.0	0.0	0.0	0.0
Parameter Edits	27.0	26.0	21.0	27.0
Unsatisfied Tasks	1.0	1.0	1.0	1.0
Failed Replans	0.0	0.0	0.0	0.0
Task Edits	0.0	1.0	2.0	2.0
Open Causal Goals	9.0	9.0	9.0	9.0
Base Changes	0.0	1.0	2.0	2.0
Unsatisfied Resource Constraints	0.0	0.0	0.0	0.0
High Priority Temporal Constraint Edits	0.0	4126476.0	6493238.0	622476.0
Medium Priority Temporal Constraint Edits	9266619.0	2400000.0	4842619.0	2400000.0
Low Priority Temporal Constraint Edits	0.0	0.0	0.0	0.0

OK

Fig. 5. COA Comparison and Revision Data

Because the revised plans in the JAGUAR domain are aggregates of multiple missions, the merge capability was augmented to produce "aggregated case features". When two candidate plans are merged into one, the case features are also merged.

Some merged features are a set (this plan uses F16CG and E8C), while other features are a sum (2+2=4 anomalies). Figure 6 contains some example feature aggregates, e.g., AircraftType and TargetType that were generated for the revised (edited, merged and adapted) plans/RCOAs that were produced.

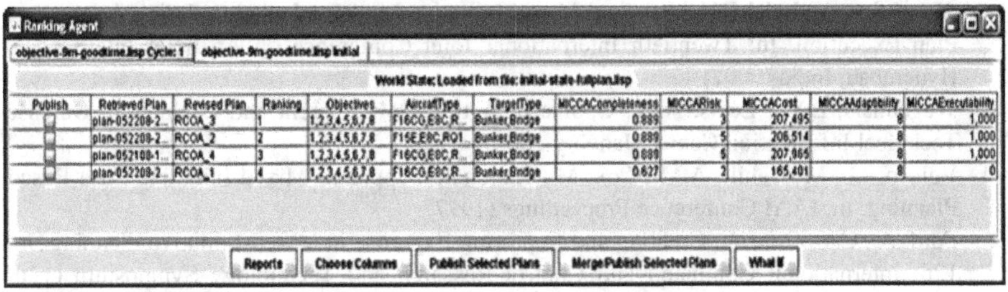

Fig. 6. Revised Plans in Ranked Order

8   Conclusion

Our work with the Rovers domain and the JAGUAR air mission planning domain showcases how the MICCA process emulates the basic CBR cycle: retrieve, reuse, revise, and retain. Our experimentation indicates that while COA development can be facilitated with the re-use of past experience, the complexity of adaptation is a function of the domain complexity. In our research this is evidenced by the improvements that were made to the Adaptation and Instantiation agents used in the Rovers domain so that they could more efficiently handle the revision of resource and task dependencies in the JAGUAR domain. In addition to improvements to the capabilities of these agents, the Adaptation agents were also modified to work more closely with the Instantiation agents. For example, if the Editing agent changes the target, a specialized Adaptation agent will modify the path.

Historical candidates were provided by a case base and the case base was used to support case revision in all of the experiments. When no single historical plan could be used to satisfy objectives, parts of several relevant historical plans were merged then edited to support the problem. When no historical plans could be used to satisfy an objective, the JSHOP generative planner was used to generate a solution.

Our results indicate that enabling the agents to access the case base during revision improved the revision scope and accuracy. This was particularly evident in the JAGUAR domain where the Instantiation agents were able to support the replacement of many interdependent elements of the JAGUAR plans, including: the actor, aircraft/weapon pair loading, formation, target, and starting airbase of candidate retrieved plans, leveraging information from case base queries.

Acknowledgements. This work is based upon work funded by the Air Force Research Laboratory (AFRL), Contract No. FA8750-10-C-0184.

References

1. Long, D., Fox, M.: The 3rd International Planning Competition: Results and Analysis. Artificial Intelligence Journal (AIJ) 20, 1–59 (2003)
2. Mulvehill, A.M., Benyo, B., Cox, M., Bostwick, R.: Expectation Failure as a Basis for Agent-Based Model Diagnosis and Mixed Initiative Model Adaptation during Anomalous Plan Execution. In: Twentieth International Joint Conference on Artificial Intelligence, Hyderabad, India (2007)
3. Wagenhals, L.W., Levis, A.H.: Course of Action Development and Evaluation, Defense Technical Information Center (January 2000)
4. Veloso, M., Mulvehill, A.M., Cox, M.: Rationale-Supported Mixed-Initiative Case-Based Planning. In: IAAI Conference Proceedings (1997)
5. Ford, A., Carozzoni, J.: Creating and Capturing Expertise in Mixed-Initiative Planning. In: 12th International Command and Control Research and Technology Symposium (12th ICCRTS), Newport, RI, June 19-21 (2007)
6. Aamodt, A., Plaza, E.: Case-based reasoning: Foundational issues, methodological variations, and system approaches. AI Com – Artificial Intelligence Communications 7(1), 39–59 (1994)
7. Mulvehill, A.M., Krisler, B., Bostwick, R.: Deriving Reliable Model Revisions from Executed Plan Data Analysis. In: 14th International Command and Control Research and Technology Symposium, Washington, D.C. (2009)
8. Nau, D.S., Au, T.C., Ilghami, O., Kuter, U., Muñoz-Avila, H., Murdock, J.W., Wu, D., Yaman, F.: Applications of SHOP and SHOP2. IEEE Intelligent Systems 20(2), 34–41 (2005)
9. Yaman, F., des Jardins, M.: More-or-Less CP-Networks. In: Uncertainty in Artificial Intelligence, Vancouver, Canada, July 20-22 (2007)
10. Mitra, R., Basak, J.: Methods of Case Adaptation: A Survey. International Journal of Intelligent Systems 20, 627–645 (2005)

The COLIBRI Open Platform
for the Reproducibility of CBR Applications*

Juan A. Recio-García, Belén Díaz-Agudo, and Pedro Antonio González-Calero

Department of Software Engineering and Artificial Intelligence
Universidad Complutense de Madrid, Spain
jareciog@fdi.ucm.es, {belend,pedro}@sip.ucm.es

Abstract. There is an increasing requirement in the scientific software development area of promoting the interchange of resources to ensure the reproducibility and validation of the results. This paper presents the COLIBRI STUDIO environment that supports researchers in the generation of Case-based Reasoning (CBR) systems by means of workflow-like representations with different degrees of abstraction. These workflows – called *templates*– can be shared with the community to promote their future reference and reproducibility.

1 Introduction

COLIBRI is a platform for developing Case-Based Reasoning (CBR) software. Its main goal is to provide the infrastructure required to develop new CBR systems and its associated software components. COLIBRI is designed to offer a collaborative environment where users could share their efforts in implementing CBR applications. It is an open platform where users can contribute with different designs or components that will be reused by other users. In general terms, this process -named the COLIBRI development process- proposes and promotes the collaboration among independent entities (research groups, educational institutions, companies) involved in the CBR field. To enable this collaboration, the development process defines several activities to interchange, publish, retrieve, instantiate and deploy workflows that conceptualize CBR systems.

The first advantage of our approach is the reduction of the development cost through the reuse of existing templates and components. This is one of the aspirations of the software industry: that software development advances, at least in part, through a process of reusing components. In this scenario, the problem consists of composing several software components to obtain a system with a certain behaviour. To perform this composition it is possible to take advantage of previously developed systems. This process has obvious parallels with the CBR cycle consisting of the steps retrieve, reuse, revise and retain. The expected benefits are improvements in programmer productivity and in software quality.

---

* Supported by Spanish Ministry of Science and Education (TIN2009-13692-C03-03).

S.J. Delany and S. Ontañón (Eds.): ICCBR 2013, LNAI 7969, pp. 255–269, 2013.

Our COLIBRI development process [1] has an additional advantage that is analysed in this paper: the collaboration among users promotes the repeatability of the results achieved by other researchers. In science the reliability of the experimental results must be backed up by the reproducibility of the experiments. We propose a development process that promotes the reproducibility of experiments for the Case-based Reasoning realm.

The paper runs as follows: Section 2 introduces reproducibility and why it is relevant in the CBR field. Section 3 presents the COLIBRI platform and its associated development process based on templates. Section 4 presents the tools that support reproducibility in COLIBRI and some issues associated to this process. Finally, Section 5 presents an experimental evaluation and Section 6 concludes the paper.

2   Reproducibility for the Case-Based Reasoning Field

Reproducibility is a key element in the scientific method and enables researchers to evaluate the validity of each other's results. In general terms, scientific experiments are often done manually and are prone to error, slowing the pace of discoveries. The work by Gil et al. [2,3] analyse this problem and propose workflow technology as a suitable solution to aid researchers in the publication, discovery and reuse of existing computational processes. Workflows capture processes in a declarative manner so that they can be reproduced by other groups or replicated on other datasets.

Reproducibility is a requirement for any scientific research domain, and CBR is not alien to it. Advances in CBR are performed by several research groups that are continuously proposing novel techniques. These techniques must be compared to the existing ones to validate their correctness. However, the development of CBR systems is not an obvious task and requires a significant knowledge engineering effort and different skills on software development. This is a limitation that slows down the development of the CBR field. Our working hypothesis is that Case-based Reasoning requires tools and procedures to support the scientific method like any other domain does. CBR methodologies and implementations should be appropriately indexed and made available for referencing and reuse. Benefits are manifold: automation of system generation, systematic exploration of the CBR design space, validation support, optimization and correct reproducibility are the most significant. Additionally, such capabilities have enormous advantages for educational purposes.

These goals have been achieved by existing computational workflow systems for generic computation [2]. Taverna is a workflow approach for the integration of components in the bioinformatics field [4]. It hides the complexity of the access to the processing services. Pegasus is another widely used workflow system that enables the composition of distributed resources [5]. However, to assist scientists in the composition, management and execution of these workflows, authors provide a complementary environment named Wings [6]. This tool starts with a high-level user description and uses knowledge about components, data and workflows to automatically generate the workflows for Pegasus [7]. The Wings system points out the central role of a knowledge rich representation of workflows and their components, that is provided by means of ontologies. These ontologies are formalized in the OWL language [8], and the associated Description Logics reasoners [9,10] provide the reasoning capabilities to ensure the correctness of the workflow and compatibility with the data being processed. Moreover, Wings defines different layers of abstraction in the specification of workflows. These different specifications must be taken into account in order to provide the appropriate tools for the users. The highest layers of abstractions define the overall behaviour of the system whereas the lower layers include computational and execution details.

These contrasted ideas for the automatic reuse of workflows in generic scientific domains have being applied to the CBR field in the COLIBRI platform. COLIBRI defines an architecture, development process and tools that provide the benefits of the workflows technologies previously described: reproducibility, automatic generation, and evaluation. We present this platform next.

3   The COLIBRI Platform

Addressing the task of developing a CBR system raises many design questions: how are cases represented? Where is the case base stored and how are the cases loaded? How should algorithms access the information inside cases? How is the background knowledge included? and so on. A successful approach to solve these issues is to turn to the expertise obtained from previous developments. Therefore, COLIBRI proposes an architecture that states how to design CBR systems and their composing elements. The definition of this architecture is the key element of the platform as it enables the compatibility and reuse of components and workflows created by independent sources.

The main items of the COLIBRI architecture are:

- Persistence: defines how to organize the storing and loading of cases from different media like data bases, textual files, etc. It proposes the use of specialized connectors that perform this task.
- Knowledge models: define a clear and common structure for the basic knowledge models found in a CBR application: cases, queries, connectors, similarity metrics, case-base organizations, etc.
- CBR system organization: COLIBRI organizes CBR systems into: precycle, where the required knowledge models (mostly cases) are initialized; cycle, which performs the four traditional CBR tasks; and postcycle, where resources are released.
- Methods: They are software components that implement the different algorithms involved in the retrieval/reuse/revise/retain cycle. The platform also includes methods for maintenance and evaluation.

Any architecture will not be useful without a reference implementation that enables users to create tangible applications. The main features of this implementation should be reusability and extensibility to let users adapt it to the target system. In the COLIBRI platform this building block is provided by the jCOLIBRI framework. jCOLIBRI is a mature software framework for developing CBR systems in Java that has evolved over time, building on several years of experience[1] [11].

After jCOLIBRI was sufficiently mature we continued with the next step in our platform: creating graphical development tools to aid users in the development of CBR systems through the reuse of existing designs. These tools are enclosed in the COLIBRI STUDIO IDE[2] and are described in detail in the next section.

The incorporation of such tools into the platform has required the definition of a software development process that identifies the task required to implement CBR systems and the different user roles involved in this process. The following section describes this software development process, its activities, the associated user roles and the tools that support it.

Summarizing, the COLIBRI platform comprises: (i) an architecture that states how to design CBR systems, (ii) an implementation of such architecture in the jCOLIBRI framework, and (iii) a development process that identifies the tasks required to implement CBR systems according to the architecture and reusing the software components provided by jCOLIBRI.

3.1   The COLIBRI Development Process

The main feature of the development process proposed in COLIBRI is reuse (somehow inherent to CBR). We propose the reuse of both system designs and their components. Reusable designs are called *templates* and comprise CBR system workflows which specify the behaviour of a set of CBR systems. In general terms, the platform provides a catalogue of templates and lets users select the most suitable one and adapt it to the concrete requirements of the target application. It is a kind of CBR process for developing CBR systems [12].

The tools provided by our platform COLIBRI are targeted to different user roles. We have identified several roles that address the development of CBR systems from different points of view: senior researchers design the behaviour of the application and define the algorithms that will be implemented to create software components that are composed in order to assemble the final system. On the other hand, developers will implement these systems/components. Furthermore, during the last few years there has been an increasing interest in using jCOLIBRI as a teaching tool, and consequently, our platform also supports this task.

User roles are associated to activities that comprise the COLIBRI development process. Next, we describe these activities and the user roles that perform them:

---

[1] Download, reference and academic publications can be found at the web site: www.jcolibri.net

[2] Available at: www.colibricbrstudio.net

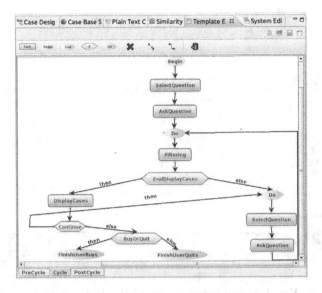

Fig. 1. Screenshot of the template generation tool

Template generation. A template is a workflow-based representation of a CBR system where several tasks are linked together to define the desired behaviour. They should be generated by 'expert' users although other users may create them. Here reputation will play a significant role as we will discuss in Section 4.1. This is the first activity in our software development process. Figure 1 shows a basic template in our specialized tool to design templates.

Template publishing. Templates can be shared with the community. Therefore there is a second tool that lets users publish a template in the COLIBRI central repository.

Template retrieval and adaptation. Although the publication of templates is a key element in the platform, the main use case consists of retrieving and adapting the template to generate a new CBR system. Here the actors are not only CBR experts: developers, teachers, students, or inexperienced researchers will perform these activities. Due to their importance, these activities are referred to as Template-Based Design (TBD). TBD begins with the retrieval of the template to be adapted from the central repository. This retrieval is performed by means of a recommender system proposes the most suitable template depending on the features of the target CBR system. It follows a "navigation by proposing" approach where templates are suggested to the user. Next, the adaptation of the template retrieved consists of assigning components that solve each task (see Figure 2). These components are the ones provided by the jCOLIBRI framework. For a detailed description of the TBD we point readers to [12] although we include a discussion about the implications of these activities from the reproducibility point of view in Section 4.1.

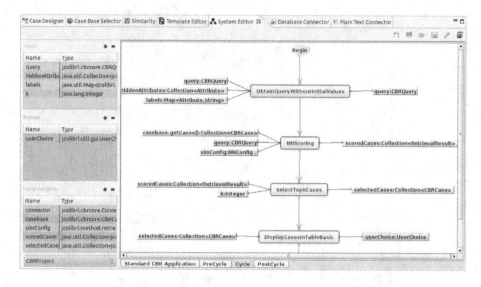

Fig. 2. Screenshot of the template adaptation tool

Component development. The design of components is closely related to the advance in CBR research as they implement the different algorithms being developed by the community. Therefore, this is the second main task of expert researchers. However, we do not expect that expert researchers will implement the components. This task will be delegated to users in a 'developer' role. We also contemplate the role of 'junior researcher' that could design and even implement his own experimental components. Again, these components could be shared with the community by means of the publication tool that uploads it to the COLIBRI repository.

System Evaluation. As we have mentioned, one of the most relevant benefits of COLIBRI is that it provides an easy to use experimental environment to test new templates and components. Consequently another activity in the development process is the evaluation of the generated systems. It enables the comparison of the performance of different CBR system implementations.

Up to now, we have referred to templates as a general term. Next we will provide more details about their formalization and conceptualization.

3.2   Template Categorization

Researchers may have different points of view regarding the data to be published and shared with the community. Some researchers may be keen on sharing all the details of the experiment, whereas others may desire to publish only an abstract description. This fact must be taken into account in a collaborative environment like COLIBRI STUDIO. The level of abstraction has an impact in the reproducibility of the CBR system because abstract templates do not provide enough details to generate the original system. However, the tools in COLIBRI STUDIO support the instantiation of these abstract representations to end up with executable systems.

Consequently, the COLIBRI development process defines the following categorization of templates according to their level of abstraction:

Abstract templates. Abstract templates are a high level representation of CBR systems. This kind of templates comprises CBR system designs which specify behaviour by means of tasks but they do not explicitly define functional details. A task defines, in an independent way, a piece of functionality that must be provided by the system. Tasks form a workflow that states the execution (control flow) of the application. Figure 1 shows an abstract template in the template editor tool in COLIBRI STUDIO.

Instantiated templates. The functionality specified by tasks can be provided by different software components. This way each task in a template can be *solved* by a component that implements the expected behaviour. When every task in a template has a component assigned –usually from the jCOLIBRI framework–, that template is considered *instantiated* with a concrete configuration of reasoning algorithms that implement a particular CBR application. The instantiation activity is supported by the adaptation tool of COLIBRI STUDIO. This activity implies the connection between components' inputs and outputs to define the dataflow of the system. Figure 2 shows an instantiated template generated from an abstract template through the template adaptation tool.

Executable Systems. An instantiated template encapsulates the algorithmic steps that perform the reasoning cycle of a CBR application. However, a fully functional CBR system requires additional knowledge models such as the case base or the retrieval (similarity) and reuse knowledge. These models can be defined by means of the tools in COLIBRI STUDIO. When an instantiated template is configured with these knowledge models, we obtain an executable system. COLIBRI STUDIO is able to generate executable systems automatically because the control flow of the application is defined by the initial abstract template, and the concrete algorithmic details are specified during the data flow configuration of the instantiation process.

Once we have identified the abstraction levels in the representation of templates, we can describe how to exploit these representations to support reproducibility of CBR applications.

4   Reproducibility in COLIBRI STUDIO

COLIBRI STUDIO supports the publishing of templates, components and data in a central repository that can later be explored by the user to retrieve and reuse these existing resources (i.e. the Template-based Design). TBD enables the reproduction of existing systems and their further extension or evaluation. However, the nature of the element being published has an impact in the reproducibility of the system. Next we analyse these dependencies.

Abstract templates. At the lowest reproducibility level we find abstract templates. Users reusing abstract templates have the support of COLIBRI STUDIO to instantiate them with the components from jCOLIBRI. However the executable system obtained may not be identical to the original one as the algorithms chosen to solve each task may be different. Furthermore, the knowledge containers (case base, similarity, etc.) are not provided and must be supplied by the user reusing the template.

Instantiated templates. Provide a higher reproducibility level than abstract ones. They specify the components used to solve each task but do not include the knowledge containers. Note that those components could be custom implementations created by the developers of the original system/template. If an instantiated template includes such components, they should also be published to ensure reproducibility.

Executable systems. They are systems completely reproducible by other users. They include all the required elements.

Components. System developers can also publish components that provide the behaviour defined by a task. These components may be just just new algorithms made available to the community to solve common tasks, or key components for the instantiation of custom templates published in the repository by the same authors.

Knowledge models. This is the last element of a CBR system that can be shared both as part of an executable system, or independently. COLIBRI STUDIO supports the publishing of four types of knowledge containers:

- Case Structure. Defines the typed attributes that form a case. They are created through one of the tools in COLIBRI STUDIO.
- Case Base. A case base is stored in a persistence media (data base, text file, ontology, ...) and loaded through a connector into the working memory of cases. Connectors are components provided by the jCOLIBRI framework or created ad-hoc by developers. They are configured with xml files that define how cases are mapped from the persistence media into a concrete case structure. These elements form the case base knowledge model: persistence, connector and configuration. They can be packaged together and shared with the community.
- In-memory organization. Once cases are loaded from persistence (through a connector) the are indexed for use by the components that instantiate a template. There are several in-memory case base organizations such as linear lists or k-d trees. Developers can also implement their own organizations.
- Similarity knowledge. Similarity is defined in COLIBRI by means of global and local similarity functions assigned to each attribute that forms the case structure. A tool in COLIBRI STUDIO allows users to define the configuration of the similarity metrics. It is the last knowledge model of a CBR system that our approach can share with the community.

Evaluation protocol. COLIBRI includes a complete framework for the evaluation of CBR systems. It includes cross-validation techniques and different performance metrics. Although it is an optional requirement, the evaluation configuration can be also published to support the reproducibility of an empirical study.

4.1   Reputation and Provenance

Once we have presented the elements to be published and retrieved, several questions arise: Who should publish these elements? What impact does the publisher have in the retrieval of templates? As we mentioned in Section 3.1, templates should be generated by expert users. And this expertise should be taken into account during the retrieval process. It implies that the authorship reputation within the community should have an impact in the recommendation process that aids users in the retrieval of templates. Reputation can be hard-coded into the system, listing well-known researchers as trustable authors, or may appear, as described below, through collaborative recommendation. As users evaluate the contributions from other users, reputation will grow for those authors contributing templates that others like. Templates are not the only element that is influenced by this provenance knowledge; the retrieval of components, case bases, similarity knowledge, etc. should be also weighted according to the reputation of the authors. Provenance is a raising topic associated to reproducibility [13,14]. This factor must be also considered during the publication process and, consequently, provide the mechanisms to support the inclusion of the provenance knowledge. Although we have not included provenance mechanisms in our platform yet, we plan to implement collaborative approaches.

The current implementation of the recommender system for the retrieval of templates follows a content-based approach. It means that the requisites of the user are compared to the description of the templates to find the most suitable ones. However, instead of using the details of the template to perform the recommendation, it could be possible to follow a collaborative approach. Collaborative recommenders consider the opinions or ratings of the users to select the most suitable items. This is an alternative approach that could be implemented in the repository of templates to allow users to rate or comment templates, components, etc. This way, ratings and comments become a measure for the reputation of the author. Elements with higher valuation will be proposed first to users, and these users may later rate their experience. This rating process can be considered as the revision step of our CBR approach. Template-based design consists in retrieving and adapting templates. We plan to include this further stage where users provide revisions in a collaborative way.

The descriptions of templates and the other elements in the platform use a semantic representation that enhances the capabilities of the tools in COLIBRI STUDIO. This representation includes functional descriptions and must also be capable to integrate the provenance knowledge that we have described. Next we present the details of this semantic representation.

4.2   Semantic Representation

As we have described previously, the COLIBRI development process involves several elements that must be integrated to obtain the target CBR system. Templates, components and knowledge models are combined by the user or the automatic tools provided by COLIBRI STUDIO. However, a major issue arises: how does COLIBRI STUDIO control the semantic coherence of the system being generated? Methods are only able to solve certain tasks; tasks are thought to be performed over concrete case structures; and depending on the case structure developers need specific persistence connectors and similarity metrics. Additionally, reputation and provenance knowledge should also integrated in the development process. Therefore, there is a semantic interdependency between the elements involved in the development of a CBR system that must be controlled by COLIBRI STUDIO to support not only the local development of systems but also the publishing and reuse from the central repository.

The answer to these questions in the COLIBRI platform is the *CBROnto* ontology [15]. This ontology guides the description of components, the design of templates and their associated retrieval and adaptation activities. There are many formalisms for representing the templates such as UML. However we have decided to use a representation based on ontologies to enhance the whole development process. The most significant benefit of this representation is the reasoning capabilities. Several ontologies have been proposed for modelling systems. Most of them come from the field of Semantic Web Services (SWS). This community explores -among other things- the methodologies for automatically composing web services to create executable software systems. Each service is described by means of ontologies that define its semantic behaviour. Therefore, in the SWS community there are different standards to represent the behaviour of software components and their composition. The most significant examples are OWL-S [16] and WSMO [17].

Therefore, we have created our *CBROnto* ontology using the OWL language and integrated the vocabulary needed to represent templates (in their different categories), components and knowledge models. This ontology is used to describe each element in the COLIBRI platform and ensures their correct integration into executable systems. Conceptualizations in *CBROnto* include semantic restrictions that define the interdependencies between each element. These restrictions are represented using the Description Logics capabilities of the OWL language. Semantic restrictions can be used during the retrieval and adaptation of templates to guide the users. It requires the application of knowledge intensive similarity and adaptation techniques that take advantage of the semantic descriptions. For example, the recommender system that supports the retrieval of templates exploits this knowledge to find the most suitable template according to the user requirements. The similarity metric that compares templates to the requirements of the user exploits the semantic knowledge in *CBROnto* that conceptualizes the structure of the template, its associated components and several other high level descriptions provided by the author of the template.

Furthermore, the instantiation process that adapts templates (Figure 2) is guided by the ontological knowledge. *CBROnto* dictates the possible assignments of components to tasks, because it represents the functional requisites of tasks and the capabilities of the components. For example, in the Java signature of a component we can state that it requires a list of cases. However, without an additional mechanism such as *CBROnto* we could not define the number or nature of the cases contained (structured, textual, etc.). The reasoning capabilities of OWL enables the definition of such restrictions for tasks and components. Consequently, our system is able to find the proper components that satisfy the semantic requirements of a task.

An example of template representation using *CBROnto* is provided in Figure 3. It corresponds to the template being instantiated in Figure 2 that is composed of the FormFilling, Scoring, Selection and Display tasks plus a conditional task to ask the user. *CBROnto* defines the concepts required to define such a workflow of tasks: Task, Sequence, If-Then-Else, ConditionalTask, etc. On the other hand, components able to solve a task in a template –called *methods* in the ontology– are described by their pre/post conditions. To illustrate such representation Figure 4 presents a graphical visualization of the semantic signature in *CBROnto* of the *Nearest Neighbor Scoring* method. As we can observe, pre/post conditions are defined by means of concepts in the ontology that enable the reasoning about its composition. Additionally, the solves property indicates the task(s) solved by the method. In this case, the NNScoring method can solve the Scoring task. Therefore, it can be assigned to that task in the template adaptation tool as shown in Figure 2 (see second task). We can also observe how its inputs (case base, query, similarity configuration) and outputs (retrieval collection) are layered out on the left and right sides of the task being solved to enable the definition of the dataflow of the template.

CBROnto shares many conceptual ideas with the Wings system described in Section 2. It includes an ontology that describes the elements in a workflow system that is exploited to validate the system being generated. Our approach differs in the way this ontology is exploited. In COLIBRI the knowledge rich representation of templates and components is used in a case-based fashion. Templates are retrieved and adapted following the typical CBR approach. This adaptation step is not supported in Wings. Therefore, we can define COLIBRI STUDIO as a knowledge-intensive CBR application where cases are knowledge-rich workflows: the templates [12].

The role of *CBROnto* in the COLIBRI development process is a large topic that cannot be detailed in this paper due to space restrictions. We point users to [18] as it provides a deeper description of the Template-based Design activity, leaving aside the reproducibility features of the COLIBRI platform. *CBROnto* still does not include the reputation and provenance knowledge required to implement a proper repository of templates. This is our on-going work.

```
<CBRApplicationTemplate rdf:about=" cbronto:Template1">
<hasPreCycle rdf:resource=" cbronto:Template1_PreCycle"/>
<hasCycle>
 <Cycle rdf:about=" cbronto:Template1_Cycle">
  <hasFlow>
   <Sequence rdf:about=" cbronto:Template1_Cycle_Sequence">
    <Components>
     <Task rdf:resource=" cbronto:Begin"/>
     <Task rdf:resource=" cbronto:FormFilling"/>
     <Task rdf:resource=" cbronto:Scoring"/>
     <Task rdf:resource=" cbronto:Selection"/>
     <Task rdf:resource=" cbronto:Display"/>
    </Components>
   </Sequence>
   <If-Then-Else rdf:about=" cbronto:Template1_Cycle_If_1">
    <ifCondition>
     <ConditionalTask rdf:resource=" cbronto:Continue"/>
    </ifCondition>
    <then rdf:resource=" cbronto:Template1_Cycle_If_1_Then"/>
    <else rdf:resource=" cbronto:Template1_Cycle_If_1_Else"/>
   </If-Then-Else>
  </hasFlow>
 </Cycle>
</hasCycle>
<hasPostCycle rdf:resource=" cbronto:Template1_PostCycle"/>
</CBRApplicationTemplate>
```

Fig. 3. Semantic representation of a template using the *CBROnto* ontology (simplified)

5   Experimental Evaluation

To evaluate the benefits of the COLIBRI development process we ran an experimental evaluation with 50 students from an Artificial Intelligence and Knowledge Based Systems course. Students had intermediate programming skills and no previous experience in this field. Several lessons introduced recommender systems and they were proposed to freely design and implement a recommender given several domains: movies, videogames, music, etc.

Initially, students specified the design of their recommender systems and afterwards COLIBRI STUDIO was introduced. COLIBRI STUDIO was provided with repository of 16 templates for building a wide range of recommender systems. Again, they could decide the implementation process: using the template-based approach or build the systems from scratch. By using several surveys we measured the suitability of our development process.

First, we were interested in the coverage of the repository of templates. Coverage is a key element in our platform as the COLIBRI development process is based on the idea of finding and reusing existing designs that are similar enough to the target system to reproduce. Therefore, students were asked to rank each template in the repository according to its similarity to the recommender designed previously. Figure 5(a) shows the maximum scoring assigned to templates, with an average similarity value of 8.25 in a scale from 0 to 10. From this data we can confirm that the coverage of the repository was good enough to perform the activities of retrieval and adaptation of templates.

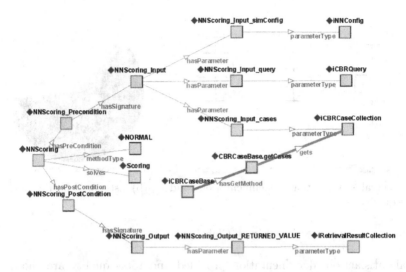

Fig. 4. Simplified visualization of the semantic representation of the Nearest Neighbor Scoring method using the *CBROnto* ontology

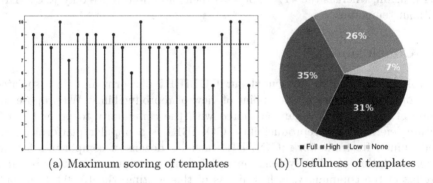

(a) Maximum scoring of templates   (b) Usefulness of templates

Fig. 5. Evaluation of the COLIBRI development process

Next, students implemented their previous designs and were asked about the usefulness of the template-based development process. Results are shown in Figure 5(b). 7% of the students decided not to use templates at all, and 26% used templates although they found the selected template much less useful than expected. This group of students retrieved a template supposed to be relevant to implement their design, but during its adaptation realized that it was not. On the other hand, 35% of the students found the template-based development process very useful to implement their designs and the remaining 31% could implement their recommenders almost directly from the template (with a very low adaptation effort).

Finally, we measured the global opinion of our users with COLIBRI STU-DIO and its associated development process. The results obtained regarding the

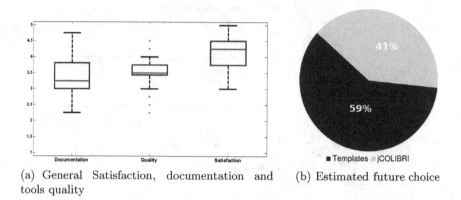

(a) General Satisfaction, documentation and tools quality

(b) Estimated future choice

Fig. 6. Global opinion

general satisfaction, documentation provided and tools quality are shown in Figure 6(a). When asked if they would use again the template-based design to implement future recommenders, the 59% of the students were keen on using it again, whereas the 41% expressed their intention to use only jCOLIBRI without templates.

6   Conclusions

In this paper we have presented our COLIBRI platform and its associated development process from the point of view of reproducibility. We propose the reuse of templates that can be shared with the community to promote their future reference and reproducibility. COLIBRI is a popular platform in the community. Its framework jCOLIBRI includes dozens of components for the implementation of CBR systems, many of them contributed by other research groups in the community. It has hit, as of this writing, the 15.000 downloads mark with users in 150 different countries. On the other hand, the configuration tools that supports the Template-based design have been recently published. It includes, by now, all the components from jCOLIBRI and up to 16 templates for the generation of recommender systems. We are working on the inclusion of templates for other CBR families such as textual CBR, web CBR or group recommendations. In case COLIBRI STUDIO would reach the same popularity than jCOLIBRI, reproducibility in CBR would be closer to become a reality.

References

1. Recio-García, J.A., Díaz-Agudo, B., González-Calero, P.A.: Template based design in colibri studio. In: Proceedings of the Process-oriented Case-Based Reasning Workshop at ICCBR 2011, pp. 101–110 (2011)
2. Gil, Y.: From data to knowledge to discoveries: Artificial intelligence and scientific workflows. Scientific Programming 17, 231–246 (2009)

3. Gil, Y., Deelman, E., Ellisman, M.H., Fahringer, T., Fox, G., Gannon, D., Goble, C.A., Livny, M., Moreau, L., Myers, J.: Examining the challenges of scientic workows. IEEE Computer 40, 24–32 (2007)
4. Oinn, T., Greenwood, M., Addis, M., Alpdemir, N., Ferris, J., Glover, K., Goble, C., Goderis, A., Hull, D., Marvin, D., Li, P., Lord, P., Pocock, M., Senger, M., Stevens, R., Wipat, A., Wroe, C.: Taverna: lessons in creating a workflow environment for the life sciences. Concurrency and Computation: Practice and Experience 18, 1067–1100 (2006)
5. Deelman, E., Singh, G., Su, M.H., Blythe, J., Gil, Y., Kesselman, C., Mehta, G., Vahi, K., Berriman, G.B., Good, J., Laity, A.C., Jacob, J.C., Katz, D.S.: Pegasus: A framework for mapping complex scientific workflows onto distributed systems. Scientific Programming 13, 219–237 (2005)
6. Gil, Y., Ratnakar, V., Kim, J., González-Calero, P.A., Groth, P.T., Moody, J., Deelman, E.: Wings: Intelligent workflow-based design of computational experiments. IEEE Intelligent Systems 26, 62–72 (2011)
7. Gil, Y., Ratnakar, V., Deelman, E., Mehta, G., Kim, J.: Wings for pegasus: Creating large-scale scientific applications using semantic representations of computational workflows. In: AAAI, pp. 1767–1774. AAAI Press (2007)
8. W3C: Owl web ontology language overview, World Wide Web Consortium (2004), http://www.w3.org/TR/owl-features/
9. Baader, F., Calvanese, D., McGuinness, D.L., Nardi, D., Patel-Schneider, P.F. (eds.): The Description Logic Handbook: Theory, Implementation, and Applications. Cambridge University Press (2003)
10. Sirin, E., Parsia, B.: Pellet: An OWL DL Reasoner. In: Haarslev, V., Möller, R. (eds.) Description Logics. CEUR Workshop Proceedings, vol. 104, CEUR-WS.org (2004)
11. Díaz-Agudo, B., González-Calero, P.A., Recio-García, J.A., Sánchez, A.: Building CBR systems with jCOLIBRI. Special Issue on Experimental Software and Toolkits of the Journal Science of Computer Programming 69, 68–75 (2007)
12. Recio-García, J.A., Bridge, D.G., Díaz-Agudo, B., González-Calero, P.A.: CBR for CBR: A Case-Based Template Recommender System for Building Case-Based Systems. In: Althoff, K.-D., Bergmann, R., Minor, M., Hanft, A. (eds.) ECCBR 2008. LNCS (LNAI), vol. 5239, pp. 459–473. Springer, Heidelberg (2008)
13. Gil, Y., Szekely, P., Villamizar, S., Harmon, T.C., Ratnakar, V., Gupta, S., Muslea, M., Silva, F., Knoblock, C.A.: Mind your metadata: Exploiting semantics for configuration, adaptation, and provenance in scientific workflows. In: Aroyo, L., Welty, C., Alani, H., Taylor, J., Bernstein, A., Kagal, L., Noy, N., Blomqvist, E. (eds.) ISWC 2011, Part II. LNCS, vol. 7032, pp. 65–80. Springer, Heidelberg (2011)
14. Moreau, L., Clifford, B., Freire, J., Futrelle, J., Gil, Y., Groth, P.T., Kwasnikowska, N., Miles, S., Missier, P., Myers, J., Plale, B., Simmhan, Y., Stephan, E.G., den Bussche, J.V.: The open provenance model core specification (v1.1). Future Generation Comp. Syst. 27, 743–756 (2011)
15. Díaz-Agudo, B., González-Calero, P.A.: CBROnto: a task/method ontology for CBR. In: Haller, S., Simmons, G. (eds.) Procs. of the 15th International FLAIRS 2002 Conference, pp. 101–105. AAAI Press (2002)
16. The OWL Services Coalition: OWL-S: Semantic Markup for Web Services, http://www.daml.org/services/owl-s/1.1/overview/ (2004)
17. Cristina Feier, J.D.: Wsmo primer (2005), http://www.wsmo.org/
18. Recio-García, J.A., González-Calero, P.A., Díaz-Agudo, B.: Template-based design in colibri studio. Journal on Information Systems (to appear)

Refinement-Based Similarity Measure over DL Conjunctive Queries[*]

Antonio A. Sánchez-Ruiz[1], Santiago Ontañón[2],
Pedro Antonio González-Calero[1], and Enric Plaza[3]

[1] Dep. Ingeniería del Software e Inteligencia Artificial
Universidad Complutense de Madrid, Spain
antsanch@fdi.ucm.es, pedro@sip.ucm.es
[2] Computer Science Department
Drexel University
Philadelphia, PA, USA 19104
santi@cs.drexel.edu
[3] IIIA, Artificial Intelligence Research Institute
CSIC, Spanish Council for Scientific Research
Campus UAB, 08193 Bellaterra, Catalonia (Spain)
enric@iiia.csic.es

Abstract. Similarity assessment is a key operation in case-based reasoning and other areas of artificial intelligence. This paper focuses on measuring similarity in the context of Description Logics (DL), and specifically on similarity between individuals. The main contribution of this paper is a novel approach based on measuring similarity in the space of *Conjunctive Queries*, rather than in the space of concepts. The advantage of this approach is two fold. On the one hand it is independent of the underlying DL, and thus, there is no need to design similarity measures for different DL, and on the other hand, the approach is computationally more efficient than searching in the space of concepts.

1 Introduction

Description Logics (DL) are one of the most widespread standards for knowledge representation in many application areas [3]. Gaining momentum through the Semantic Web initiative, DL popularity is also related to a number of tools for knowledge acquisition and representation, as well as inference engines, that have been made publicly available. For these reasons, DL has also become the technology of choice for representing knowledge in knowledge-intensive case-based reasoning systems [19,7,10].

In this paper, we focus on the problem of similarity assessment in DL, in order to enable general purpose case-based reasoning systems that use this formalism to represent domain knowledge. Specifically, we focus in the problem of measuring similarity between individuals. The similarity measure presented in

---

[*] Partially supported by Spanish Ministry of Economy and Competitiveness under grants TIN2009-13692-C03-01 and TIN2009-13692-C03-03 and by the Generalitat de Catalunya under the grant 2009-SGR-1434.

S.J. Delany and S. Ontañón (Eds.): ICCBR 2013, LNAI 7969, pp. 270–284, 2013.

this paper, S_Q, works as follows: 1) given two individuals, we convert them into *DL Conjunctive Queries*, 2) the similarity between the two queries is measured using a refinement-operator-based similarity measure [17,20]. The approach presented in this paper differs from previous work [8,20] in that the S_Q similarity is defined over the space of DL Conjunctive Queries, rather than in the space of DL concepts.

There are three main advantages in the S_Q similarity approach: 1) the conversion process from individuals to queries does not lose information (the conversion to concepts usually causes some loss of information), 2) the language used to represent conjunctive queries is independent of the particular DL being used (and thus our approach can be applied to any DL), and 3) assessing similarity in the space of queries is computationally more efficient than assessing similarity in the space of concepts, as we will show in the experimental evaluation section.

The rest of this paper is organized as follows. Section 2 introduces the necessary concepts of Description Logics, Conjunctive Queries and Refinement Operators respectively. Then, in Section 3, we introduce a new refinement operator for Conjunctive Queries. Section 4 presents the S_Q similarity measure between individuals, which is illustrated with an example in §4.3. Section 5 presents an experimental evaluation of our approach. The paper closes with related work, conclusions and directions for future research.

2   Background

Description Logics [3] are a family of knowledge representation formalisms, which can be used to represent the conceptual knowledge of an application domain in a structured and formally well-understood way.

Description Logics (DL) represent knowledge using three types of basic entities: *concepts, roles* and *individuals*. Concepts provide the domain vocabulary required to describe sets of individuals with common features, roles allow to describe relationships between individuals, and individuals represent concrete domain entities. DL expressions are built inductively starting from finite and disjoint sets of atomic concepts (N_C), atomic roles (N_R) and individual names (N_I).

The expressivity and the reasoning complexity of a particular DL depends on the available concept constructors in the language. Although the proposed similarity measure is independent of the description logic being used (the only effect being computation time), in this paper we will use the $\mathcal{EL}$ logic, a light-weight DL with good computational properties that serves as a basis for the OWL 2 EL profile[1]. $\mathcal{EL}$ is expressive enough to describe large biomedical ontologies, like SNOMED CT [6] or the Gene Ontology [2],while maintaining important properties such as concept subsumption being polynomial. The $\mathcal{EL}$ concept constructs are the top concept, intersection and existential restrictions (see Table 1).

A DL knowledge base (KB), $\mathcal{K} = (\mathcal{T}, \mathcal{A})$, consists of two different types of information: $\mathcal{T}$, the *TBox* or terminological component, which contains concept and role axioms and describes the domain vocabulary; and $\mathcal{A}$, the *ABox* or

---

[1] http://www.w3.org/TR/owl2-profiles/

Table 1. $\mathcal{EL}$ concepts and semantics

Concept	Syntax	Semantics
Top concept	$\top$	$\Delta^{\mathcal{I}}$
Atomic concept	A	$A^{\mathcal{I}}$
Conjunction	$C \sqcap D$	$C^{\mathcal{I}} \cap D^{\mathcal{I}}$
Existential restriction	$\exists R.C$	$\{x \in \Delta^{\mathcal{I}} \mid \exists y : (x, y) \in R^{\mathcal{I}} \wedge y \in C^{\mathcal{I}}\}$

Table 2. TBox axioms

Axiom	Syntax	Semantics
Concept inclusion	$C \sqsubseteq D$	$C^{\mathcal{I}} \subseteq D^{\mathcal{I}}$
Disjointness	$C \sqcap D \equiv \bot$	$C^{\mathcal{I}} \cap D^{\mathcal{I}} = \emptyset$
Role domain	$domain(R) = A$	$(x, y) \in R^{\mathcal{I}} \to x \in A^{\mathcal{I}}$
Role range	$range(R) = A$	$(x, y) \in R^{\mathcal{I}} \to y \in A^{\mathcal{I}}$

assertional component, which uses the domain vocabulary to assert facts about individuals. For the purposes of this paper, a TBox is a finite set of concept and role axioms of the type given in Table 2, and an ABox is a finite set of axioms about individuals of the type shown in Table 3.

Regarding semantics, an interpretation is a pair $\mathcal{I} = (\Delta^{\mathcal{I}}, \cdot^{\mathcal{I}})$, where $\Delta^{\mathcal{I}}$ is a non-empty set called the *interpretation domain*, and $\cdot^{\mathcal{I}}$ is the *interpretation function*. The interpretation function relates each atomic concept $A \in N_C$ with a subset of $\Delta^{\mathcal{I}}$, each atomic role $R \in N_R$ with a subset of $\Delta^{\mathcal{I}} \times \Delta^{\mathcal{I}}$ and each individual $a \in N_I$ with a single element of $\Delta^{\mathcal{I}}$. The interpretation function can be extended to complex concepts as shown in Table 1.

An interpretation $\mathcal{I}$ is a *model* of a knowledge base $\mathcal{K}$ iff the conditions described in Tables 2 and 3 are fulfilled for every axiom in $\mathcal{K}$. A concept C is *satisfiable* w.r.t. a knowledge base $\mathcal{K}$ iff there is a model $\mathcal{I}$ of $\mathcal{K}$ such that $C^{\mathcal{I}} \neq \emptyset$.

The basic reasoning operation in DL is *subsumption*, that induces a subconcept-superconcept hierarchy. We say that the concept C is subsumed by the concept D (C is more specific than D) if all the instances of C are also instances of D. Formally, C is subsumed by D w.r.t. the knowledge base $\mathcal{K}$ ($C \sqsubseteq_{\mathcal{K}} D$) iff $C^{\mathcal{I}} \subseteq D^{\mathcal{I}}$ for every model $\mathcal{I}$ of $\mathcal{K}$. When the knowledge base $\mathcal{K}$ is known we can simplify the notation and write $C \sqsubseteq D$. Finally, an *equivalence axiom* $C \equiv D$ is just an abbreviation for when both $C \sqsubseteq D$ and $D \sqsubseteq C$ hold, and a *strict subsumption axiom* $C \sqsubset D$ simply means that $C \sqsubseteq D$ and $C \not\equiv D$.

Table 3. ABox axioms

Axiom	Syntax	Semantics
Concept instance	$C(a)$	$a^{\mathcal{I}} \in C^{\mathcal{I}}$
Role assertion	$R(a, b)$	$(a^{\mathcal{I}}, b^{\mathcal{I}}) \in R^{\mathcal{I}}$
Same individual	$a = b$	$a^{\mathcal{I}} = b^{\mathcal{I}}$
Different individual	$a \neq b$	$a^{\mathcal{I}} \neq b^{\mathcal{I}}$

TBox

$$Topping \sqsubseteq \top$$
$$Cheese \sqsubseteq Topping$$
$$Mozzarella \sqsubseteq Cheese$$
$$\dots$$
$$Pizza \sqsubseteq \top$$
$$CheesyPizza \equiv Pizza \sqcap \exists hasToppi$$
$$MeatyPizza \equiv Pizza \sqcap \exists hasToppi$$
$$Margherita \sqsubseteq Pizza \sqcap \exists hasToppi$$
$$\exists hasTopping.Moz.$$
$$range(hasTopping) \sqsubseteq Topping$$

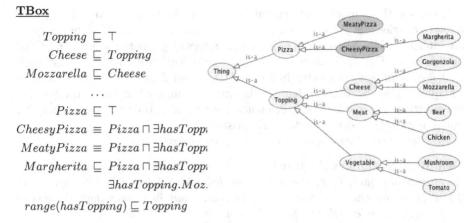

ABox

$Margherita(p1), Pizza(p2), hasTopping(p2, t1), Chicken(t1), hasTopping(p2, t2),$
$Vegetable(t2)$

Fig. 1. Example of knowledge base

Figure 1 shows an example knowledge base that we will use in the rest of the paper. The TBox contains axioms to define some vocabulary about pizzas and ingredients: *Mozzarella* is a type of *Cheese*; *Margherita* is a type of *Pizza* with *Tomato* and *Mozzarella*; a *CheesyPizza* is any *Pizza* with *Cheese*, etc. The ABox, in turn, contains axioms to describe two individuals: a margherita pizza and a pizza with chicken and vegetable toppings.

2.1 DL Conjunctive Queries

DL knowledge bases can be queried in order to retrieve individuals that meet certain conditions –in a way similar to that queries are used to retrieve data in databases. In order to define queries, along with the set of atomic concepts (N_C), atomic roles (N_R) and individual names (N_I) from knowledge bases, we need as well a disjoint set of variable names (N_V).

Definition 1. (Conjunctive Query)
A DL conjunctive query $Q(\boldsymbol{x}, \boldsymbol{y})$ is a logic formula $\exists \boldsymbol{y}.\psi(\boldsymbol{x}, \boldsymbol{y})$ where ψ is conjunction of terms of the form $A(x)$, $R(x, y)$, $x = y$ and $x \neq y$, in which $A \in N_C$ is an atomic concept, $R \in N_R$ is an atomic role, and x and y are either individual names from N_I or variable names taken from the sets $\boldsymbol{x}, \boldsymbol{y} \subset N_V$.

The sets $\boldsymbol{x}$ and $\boldsymbol{y}$ contain, respectively, all the *answer variables* and *quantified variables* of the query. A boolean conjunctive query $Q(\emptyset, \boldsymbol{y})$, or just $Q(\boldsymbol{y})$, is a query in which all the variables are quantified.

To define the semantics of general DL queries, let us begin considering only boolean queries. Let $VI(Q)$ be the set of variables and individuals in the query Q. An interpretation $\mathcal{I}$ is a model of a boolean query $Q(\boldsymbol{y})$, noted as $\mathcal{I} \models \exists \boldsymbol{y} : Q(\boldsymbol{y})$ or shortly as $\mathcal{I} \models Q$, if there is a variable substitution $\theta : VI(Q) \rightarrow \Delta^{\mathcal{I}}$ such that $\theta(a) = a^{\mathcal{I}}$ for each individual $a \in VI(Q)$, and $\mathcal{I} \models \alpha\theta$ for each term α in the query. The notation $\alpha\theta$ denotes the query atom α where the variables of α are substituted according to θ. A knowledge base $\mathcal{K}$ entails a boolean query Q, noted as $\mathcal{K} \models Q$, if every model of $\mathcal{K}$ satisfies Q.

Now let us consider queries with answer variables.

Definition 2. (Query Answer)
An answer to a query $Q(\boldsymbol{x}, \boldsymbol{y})$ w.r.t. a knowledge base $\mathcal{K}$ is a variable substitution θ that maps the answer variables in $\boldsymbol{x}$ to individuals in $\mathcal{K}$ such that the boolean query $Q(\boldsymbol{x}\theta, \boldsymbol{y})$ is entailed by $\mathcal{K}$ as defined above.

The notation $Q(\boldsymbol{x}\theta, \mathbf{y})$ represents the query where all the distinguished variables have been replaced according to θ. Note that, for interpreting boolean queries, we use a substitution that maps variables to arbitrary elements of the domain $\Delta^{\mathcal{I}}$ whereas for a query answer we require the answer variables to be mapped to named individuals in the ABox.

Definition 3. (Query Answer Set)
The answer set of a query $Q(\boldsymbol{x}, \boldsymbol{y})$ w.r.t. $\mathcal{K}$, noted as $Q(\mathcal{K})$, is the set containing all the answers to the query $Q(\boldsymbol{x}, \boldsymbol{y})$ w.r.t. $\mathcal{K}$.

For example, given the knowledge base in Figure 1, let us consider the queries:

$$Q_1(\{x_1\}, \{\}) = Pizza(x_1)$$
$$Q_2(\{x_1\}, \{y_1\}) = Pizza(x_1) \wedge hasTopping(x_1, y_1) \wedge Tomato(y_1)$$

Now, the query Q_1 below will retrieve all the existing pizzas ($Q_1(\mathcal{K}) = \{\{p1/x_1\}, \{p2/x_1\}\}$), while the query Q_2 will retrieve only those pizzas with tomato ($Q_2(\mathcal{K}) = \{\{p1/x1\}\}$). Notice that the reasoner infers that $p1$ has tomato because it is a marguerita pizza although there is no individual of type tomato explicitly asserted in the ABox.

2.2   Query Subsumption

We can define a subsumption relation between queries similar to the subsumption relation between concepts. In this way, queries can be organized into a hierarchy where the most general queries are above the most specific ones.

Definition 4. (Query Subsumption)
A query $Q(\mathbf{x}, \mathbf{y})$ is subsumed by another query $Q'(\mathbf{x}, \mathbf{y}')$ w.r.t. $\mathcal{K} = (\mathcal{T}, \mathcal{A})$ (noted as $\mathcal{K} \models Q \sqsubseteq Q'$) if, for every possible ABox $\mathcal{A}'$ and the knowledge base $\mathcal{K}' = (\mathcal{T}, \mathcal{A}')$ it holds that $Q(\mathcal{K}') \subseteq Q'(\mathcal{K}')$ (i.e. that the answer set of Q is contained in the answer set of Q').

Query containment is very closely related to query answering. The standard technique of *query freezing* [21] can be used to reduce query containment to query answering in DL [16]. To decide query subsumption, we build a canonical ABox $\mathcal{A}_Q$ from the query $Q(\mathbf{x}, \mathbf{y})$ by replacing each of the variables in $\mathbf{x}$ and $\mathbf{y}$ with fresh individual names not appearing in the KB. Let θ be the substitution denoting the mapping of variables $\mathbf{x}$ to the fresh individuals. Then, for $\mathcal{K} = (\mathcal{T}, \mathcal{A})$, $\mathcal{K} \models Q \sqsubseteq Q'$ iff θ is in the answer set of Q' w.r.t. to $\mathcal{K}_Q = (\mathcal{T}, \mathcal{A}_Q)$.

Note that Definition 4 assumes that both Q and Q' share the same set of answering variables, which is enough for the purposes of this paper. For example, considering the pizza knowledge base, query Q_3 below subsumes Q_4 because any margherita pizza is also a pizza with tomato and thus any answer to Q_4 is also an answer to Q_3.

$$Q_3(\{x_1\}, \{y_1\}) = Pizza(x_1) \wedge hasTopping(x_1, y_1) \wedge Tomato(y_1)$$
$$Q_4(\{x_1\}) = Margherita(x_1)$$

2.3   Refinement Operators

This section briefly summarizes the notion of *refinement operator* and the concepts relevant for this paper (see [13] for a more in-depth analysis of refinement operators). Refinement operators are defined over *quasi-ordered sets*. A *quasi-ordered set* is a pair $(S, \leq)$, where S is a set, and $\leq$ is a binary relation among elements of S that is *reflexive* and *transitive*. If $a \leq b$ and $b \leq a$, we say that $a \approx b$, or that they are *equivalent*. Refinement operators are defined as follows:

Definition 5. (Refinement Operator)
A refinement operator ρ over a quasi-ordered set $(S, \leq)$ is a function such that $\forall a \in S : \rho(a) \subseteq \{b \in S | b \leq a\}$.

In other words, refinement operators (sometimes called *downward refinement operators*) generate elements of S which are "smaller" (which in this paper means "more specific"). The complementary notion of an *upward refinement operator*, that generates elements that are "bigger", also exists, but is irrelevant for this paper. Typically, the following properties of operators are considered desirable:

- A refinement operator ρ is *locally finite* if $\forall a \in S : \rho(a)$ is finite.
- A downward refinement operator ρ is *complete* if $\forall a, b \in S | a \leq b : a \in \rho^*(b)$.
- A refinement operator ρ is *proper* if $\forall a, b \in S, b \in \rho(a) \Rightarrow a \not\approx b$.

where ρ^* means the *transitive closure* of a refinement operator. Intuitively, *locally finiteness* means that the refinement operator is computable, *completeness* means we can generate, by refinement of a, any element of S related to a given element a by the order relation $\leq$ (except maybe those which are equivalent to a), and *properness* means that a refinement operator does not generate elements which are equivalent to a given element a.

Regarding DL queries, the set of DL conjunctive queries and the subsumption relation between queries (Definition 4) form a quasi-ordered set. In this way, we only need to define a refinement operator for DL conjunctive queries to specialize or generalize them.

3 A Refinement Operator for DL Conjunctive Queries

The following rewriting rules define an downward refinement operator for DL
Conjunctive Queries. A rewriting rule is composed of three parts: the applicabil-
ity conditions of the rewriting rule (shown between square brackets), the original
DL query (above the line), and the refined DL query (below the line).

(R1) Concept Specialization

$$[A_2(x_1) \notin Q \wedge A_2 \sqsubset A_1 \wedge \not\exists A' : A_2 \sqsubset A' \sqsubset A_1]$$

$$\frac{Q(\boldsymbol{x}, \boldsymbol{y}) = A_1(x_1) \wedge \alpha_1 \wedge \ldots \wedge \alpha_n}{Q'(\boldsymbol{x}, \boldsymbol{y}) = A_1(x_1) \wedge A_2(x_1) \wedge \alpha_1 \wedge \ldots \wedge \alpha_n}$$

(R2) Concept Introduction

$$[x_1 \in V(Q) \wedge A_1 \in max\{A \in N_A \mid \forall A'(x_1) \in Q : A \not\sqsubseteq A' \wedge A' \not\sqsubseteq A\}]$$

$$\frac{Q(\boldsymbol{x}, \boldsymbol{y}) = \alpha_1 \wedge \ldots \wedge \alpha_n}{Q'(\boldsymbol{x}, \boldsymbol{y}) = A_1(x_1) \wedge \alpha_1 \wedge \ldots \wedge \alpha_n}$$

(R3) Role Introduction

$$[x_1, x_2 \in V(Q) \wedge R_1 \in max\{R \in N_R \mid \forall R'(x_1, x_2) \in Q : R \not\sqsubseteq R' \wedge R' \not\sqsubseteq R\}]$$

$$\frac{Q(\boldsymbol{x}, \boldsymbol{y}) = \alpha_1 \wedge \ldots \wedge \alpha_n}{Q'(\boldsymbol{x}, \boldsymbol{y}) = R_1(x_1, x_2) \wedge \alpha_1 \wedge \ldots \wedge \alpha_n}$$

(R4) Variable Introduction

$$[x_1 \in VI(Q), x_2 \in N_V \setminus V(Q)]$$

$$\frac{Q(\boldsymbol{x}, \boldsymbol{y}) = \alpha_1 \wedge \ldots \wedge \alpha_n}{Q'(\boldsymbol{x}, \boldsymbol{y} \cup \{x_2\}) = \alpha_1 \wedge \ldots \wedge \alpha_n \wedge \top(x_2)}$$

(R5) Variable Instantiation

$$[\theta : V(Q) \to N_I]$$

$$\frac{Q(\boldsymbol{x}, \boldsymbol{y}) = \alpha_1 \wedge \ldots \wedge \alpha_n}{Q'(\boldsymbol{x}, \boldsymbol{y}) = \alpha_1 \theta \wedge \ldots \wedge \alpha_n \theta}$$

Rules R1 and R2 refine a query either specializing an existing type or introducing
a new type that is neither more general nor more specific than the existing ones
(for example a sibling in the concept hierarchy). Rule R3 works analogously to
R2 but introducing roles instead of concepts. Note that we do not provide a rule
to specialize role assertions since the $\mathcal{EL}$ logic does not allow role hierarchies.
Rule R4 introduces a new quantified variable in the query, and R5 binds an
existing variable to a concrete individual in the knowledge base.

For the sake of space we do not provide proofs, but it is easy to verify that
the previous refinement operator is locally finite and not proper. The refinement
operator is also complete if we only consider the space of DL conjunctive queries

with a *fixed set of answer variables* (none of the above rules adds new answer variables) in which *all the variables represent distinct individuals.*

Although the assumption that all the variables are different restricts the set of queries that can be represented, it also simplifies to a large degree the similarity assessment process that we introduce in the following section, since it prevents refinement chains of infinite length in which all the queries are equivalent:

$$A(x_1) \rightarrow A(x_1) \wedge \top(y_1) \rightarrow A(x_1) \wedge \top(y_1) \wedge \top(y_2) \rightarrow \cdots$$

Dealing with these infinite chains using other approaches and exploring different refinement operators that offer a different trade-off of completeness and efficiency is part of our future work.

4   Similarity Based on Query Refinements

The similarity S_Q proposed in this paper consists of two main steps (described in the following two subsections). First, given individuals a and b, we transform them into conjunctive DL queries, Q_a and Q_b. Second, using the refinement operator presented above, we measure the similarity between Q_a and Q_b.

4.1   From Individuals to Queries

Given an individual a and an ABox, $\mathcal{A}$, we can define the *individual graph* of a as follows.

Definition 6. (Individual graph)
An individual graph $G_a \subseteq \mathcal{A}$ *is a set of ABox axioms with one distinguished individual* $a \in N_I$ *such that:*

- $\forall C \in N_C$, *if* $C(a) \in \mathcal{A}$ *then* $C(a) \in G_a$
- $\forall C \in N_C$, *if* $C(b) \in G_a$ *then* $\forall R \in N_R$, *if* $R(b,c) \in \mathcal{A}$ *then* $R(b,c) \in G_a$
- $\forall R \in N_R, C \in N_C$, *if* $C(b) \in G_a$ *and* $R(b,c) \in \mathcal{A}$ *then* $\forall D \in N_C$ *if* $D(c) \in \mathcal{A}$ *then* $D(c) \in G_a$

In other words, if we represent the ABox as a graph, where each individual is a node, and each role axiom is a directed edge, the individual graph of an individual a would be the connected graph resulting from all the nodes and edges reachable from a.

We can transform an individual graph to an *equivalent* conjunctive query applying a substitution that replaces the distinguished individual by a new answer variable, and the remaining individuals by new quantified variables. Note that the conversion is straightforward since ABox axioms and DL query terms are alike and, what is more important, no information is lost in the translation.

For example, next we show an individual graph and its equivalent DL query:

$$G_{p1} = Pizza(p1) \wedge hasTopping(p1, t1) \wedge Chicken(t1) \wedge$$
$$hasTopping(p1, t2) \wedge Vegetable(t2)$$
$$Q_{p1}(\{x_1\}, \{y_1, y_2\}) = Pizza(x_1) \wedge hasTopping(x_1, y_1) \wedge Chicken(y_1) \wedge$$
$$hasTopping(x_1, y_2) \wedge Vegetable(y_2)$$

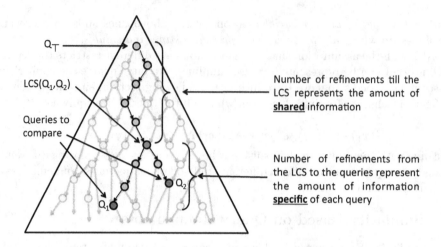

Fig. 2. Query distance based on refinements

4.2   Similarity over in the CQ Space

Our proposed similarity measure for conjunctive DL queries is based on the following intuitions (see Figure 2):

First, given two queries Q_1 and Q_2 such that $Q_2 \sqsubseteq Q_1$, it is possible to reach Q_2 from Q_1 by applying a complete downward refinement operator ρ to Q_1 a finite number of times, i.e. $Q_2 \in \rho^*(Q_1)$.

Second, the number of times a refinement operator needs to be applied to reach Q_2 from Q_1 is an indication of how much more specific Q_2 is than Q_1. Note, however, that since our refinement operator is not proper, some of the refinements in the refinement chain do not add new information to the previous query, and they should not be taken into account. The length of the chain of useful refinements (those that produce proper specializations) to reach Q_2 from Q_1, which will be noted as $\lambda(Q_1 \overset{\rho}{\to} Q_2)$, is an indicator of how much information Q_2 contains that was not contained in Q_1. In our experiments, we used a greedy search algorithm to compute this length, which does not ensure obtaining the shortest chain, but that is computationally efficient.

Third, given any two queries, their *least common subsumer* (LCS) is the most specific query which subsumes both. The LCS of two queries contains all that is shared between two queries, and the more they share the more similar they are. $\lambda(Q_T \overset{\rho}{\to} LCS)$ measures the distance from the most general query, Q_T, to the LCS, which is a measure of the amount of information shared by Q_1 and Q_2.

Finally, the similarity between two queries Q_1 and Q_2 can be measured as the ratio between the amount of information contain in their LCS and the total amount of information contained in Q_1 and Q_2. These ideas are collected in the following formula:

$$S_\rho(Q_1, Q_2) = \frac{\lambda_1}{\lambda_1 + \lambda_2 + \lambda_3}$$

where:

$$\lambda_1 = \lambda(Q_\top \xrightarrow{\rho} LCS(Q_1, Q_2))$$
$$\lambda_2 = \lambda(LCS(Q_1, Q_2) \xrightarrow{\rho} Q_1)$$
$$\lambda_3 = \lambda(LCS(Q_1, Q_2) \xrightarrow{\rho} Q_2)$$

Thus, the similarity between two individuals a and b, is defined as:

$$S_Q(a, b) = S_\rho(Q_a, Q_b)$$

where Q_a and Q_b are the queries corresponding to the individual graphs of a and b, respectively.

4.3   Example

In this section we show an example of the S_Q similarity works. Suppose we want to compute the similarity between a pizza margherita $p1$ and a pizza $p2$ with some vegetable and chicken. Figure 3 shows the queries representing both pizzas and the chain of refinements used to compute their similarity. Note that the refinements marked with an asterisk do not add new information and therefore they are not taken into account while computing the length of the refinement paths (steps 5 and 9 do not add new information because the role $hasTopping$ has range $Topping$ and thus we can infer that type for y_3 and y_4). Their LCS describes the common part of the pizzas: both have at least two ingredients and one of them is a vegetable. Their similarity is determined as follows:

$$S_Q(p1, p2) = S_\rho(Q_{p1}, Q_{p2}) = \frac{6}{6 + 3 + 5} = 0.43$$

5   Experiments

In order to evaluate the S_Q similarity measure, we used the trains data set shown in Figure 4 as presented by Michalski [14]. Like in our previous work on similarity assessment [20], we selected this dataset since it is available in many representation formalisms (Horn clauses, feature terms and description logic), and therefore, we can compare our similarity measure with existing similarity measures in the literature. The dataset consists of 10 trains, 5 of them labelled as "West", and 5 of them labelled as "East."

We compared our similarity measure against 7 others: $S_{DL\rho}$ [20], a similarity measure for the $\mathcal{EL}$ description logic; González et al. [10], a similarity measure for acyclic concepts in description logic; RIBL [9], which is a Horn clause similarity measure; SHAUD [1], which is a similarity measure for feature terms; and S_λ, S_π, and $S_{w\pi}$ [18], which are similarity measures for feature terms but also based on the idea of refinement operators. For RIBL, we used the original version of the trains dataset, for SHAUD, S_λ, S_π, and $S_{w\pi}$, we used the feature term version

Queries for G_{p1} and G_{p2}

$Q_{p1}(\{x_1\},\{\}) = Margherita(x_1)$

$Q_{p2}(\{x_2\},\{y_1,y_2\}) = Pizza(x_2) \wedge hasTopping(x_2,y_1) \wedge Chicken(y_1) \wedge$
$\qquad\qquad\qquad hasTopping(x_2,y_2) \wedge Vegetable(y_2)$

Path from Q_T to $LCS(Q_{p1},Q_{p2})$

$1 : \top(x_3)$

$2 : Pizza(x_3)$

$3 : Pizza(x_3) \wedge \top(y_3)$

$4 : Pizza(x_3) \wedge hasTopping(x_3,y_3) \wedge \top(y_3)$

$5^* : Pizza(x_3) \wedge hasTopping(x_3,y_3) \wedge Topping(y_3)$

$6 : Pizza(x_3) \wedge hasTopping(x_3,y_3) \wedge Vegetable(y_3)$

$7 : Pizza(x_3) \wedge hasTopping(x_3,y_3) \wedge Vegetable(y_3) \wedge \top(y_4)$

$8 : Pizza(x_3) \wedge hasTopping(x_3,y_3) \wedge Vegetable(y_3) \wedge hasTopping(x_3,y_4) \wedge \top(y_4)$

$9^* : Pizza(x_3) \wedge hasTopping(x_3,y_3) \wedge Vegetable(y_3) \wedge hasTopping(x_3,y_4) \wedge Topping(y_4)$

Path from $LCS(Q_{p1},Q_{p2})$ to Q_{p1}

$10 : Pizza(x_3) \wedge hasTopping(x_3,y_3) \wedge Tomato(y_3) \wedge hasTopping(x_3,y_4) \wedge Topping(y_4)$

$11 : Pizza(x_3) \wedge hasTopping(x_3,y_3) \wedge Tomato(y_3) \wedge hasTopping(x_3,y_4) \wedge Mozzarella(y_4)$

$12 : Pizza(p1) \wedge hasTopping(p1,y_3) \wedge Tomato(y_3) \wedge hasTopping(p1,y_4) \wedge Mozzarella(y_4)$

Path from $LCS(Q_{p1},Q_{p2})$ to Q_{p2}

$13 : Pizza(x_3) \wedge hasTopping(x_3,y_3) \wedge Vegetable(y_3) \wedge hasTopping(x_3,y_4) \wedge Meat(y_4)$

$14 : Pizza(x_3) \wedge hasTopping(x_3,y_3) \wedge Vegetable(y_3) \wedge hasTopping(x_3,y_4) \wedge Chicken(y_4)$

$15 : Pizza(p2) \wedge hasTopping(p2,y_3) \wedge Vegetable(y_3) \wedge hasTopping(p2,y_4) \wedge Chicken(y_4)$

$16 : Pizza(p2) \wedge hasTopping(p2,t1) \wedge Vegetable(t1) \wedge hasTopping(p2,y_4) \wedge Chicken(y_4)$

$17 : Pizza(p2) \wedge hasTopping(p2,t1) \wedge Vegetable(t1) \wedge hasTopping(p2,t2) \wedge Chicken(t2)$

Fig. 3. Refinement paths to compute $S_\rho(Q_{p1},Q_{p2})$

of the dataset used in [17], which is a direct conversion from the original Horn clause dataset without loss, and for the DL similarity measures, we used the version created by Lehmann and Hitzler [15].

We compared the similarity measures in five different ways:

- Classification accuracy of a nearest-neighbor algorithm.
- *Average best rank* of the first correct example: if we take one of the trains, and sort the rest of the trains according to their similarity with the selected train, which is the position in this list (rank) of the first train with the same solution as the selected train (West or East).
- Jaro-Winkler distance: the Jaro-Winkler measure [22] can be used to compare two orderings. We measure the similarity of the rankings generated by our similarity measure with the rankings generated with the others.
- Mean-Square Difference (MSD): the mean square difference with respect to our similarity measure, S_Q.
- Average time take to compute similarity between two individuals.

Table 4 shows the results we obtained by using a leave-one-out evaluation. Concerning classification accuracy, we can see that our similarity measure S_Q

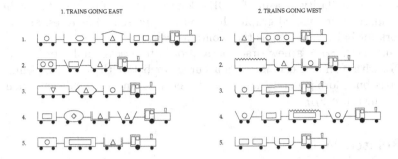

Fig. 4. Trains data set as introduced by Michalski [14]

Table 4. Comparison of several similarity metrics in the trains dataset (* These times do not take into account data preprocessing, required for these two techniques). Note that the Jaro-Winkler and MSD values are computed with respect to S_Q.

	S_Q	$S_{DL\rho}$	González et al.	RIBL	SHAUD	S_λ	S_π	$S_{w\pi}$
Accuracy 1-NN	70%	70%	50%	60%	50%	40%	50%	80%
Best Rank	1.4	1.4	1.5	2.0	2.0	2.3	2.1	1.7
Jaro-Winkler	-	0.75	0.68	0.79	0.73	0.76	0.76	0.77
MSD	-	0.03	0.21	0.07	0.06	0.07	0.11	0.16
Avg. Time	0.55s	175.74s	0.01s	0.01s	0.07s	0.04s	0.00s*	0.00s*

achieves a high classification accuracy, higher than most other similarity measures, except $S_{w\pi}$. We would like to emphasize that the trains data-set is only apparently simple, since the classification criteria is a complex pattern which involves several elements from different cars in a train. The only similarity measure that came close is $S_{w\pi}$, which achieved an 80% accuracy (it misclassified trains west 1 and west 3). Concerning the average best rank, our measure obtains the best score (tied with our previous measure $S_{DL\rho}$). Concerning the Jaro-Winkler and MSD results, we can see that in the trains data set S_Q produces similarities similar to $S_{DL\rho}$, RIBL, and S_λ.

Where our new similarity measure stands out is in terms of time. We can see that, compared to our previous $S_{DL\rho}$ similarity measure, the S_Q similarity is very fast. This is because the space of queries is narrower than the space of concepts since queries can only contain atomic concepts and roles while general DL concepts can combine any of the constructors in the language. However, this limitation does not necessarily affect the quality of the similarity, since atomic concepts represent the vocabulary chosen by domain experts to describe domain entities, and therefore atomic concepts represent the most important conceptualizations in the domain. Also, many practical optimizations can be performed, such as sorting the query term in such a way that the most restrictive axioms ones are evaluated first.

The other similarity measure for DL (González et al.'s) is much faster, but it is specialized to individual graphs that can be represented as trees, and would not work for individual graphs that contain cycles.

In summary, S_Q is a new practical approach to assess similarity for expressive DL, with a similar classification accuracy or better than existing similarity measures, but more general than González et al.'s, and more efficient than our previous measure $S_{DL\rho}$.

6   Related Work

The work presented in this paper extends our previous work on similarity on Description Logics [20], where we studied how to assess similarity between individuals by transforming them to concepts, and then assessing the similarity of these concepts. The approach presented in this paper is more general (since the language DL queries is common to all DL), more efficient, and more accurate (since we might lose information when converting individuals to concepts).

D'Amato et al. [8] propose to measure concept similarity as a function of the intersection of their interpretations, which is, in fact, an approximation to the semantic similarity of concepts. The approximation is better or worse depending on how good is the sample of individuals used for assessing similarity. Thus, a good sample of individuals is required.

Other approaches have been proposed in order to assess similarity between individuals or concepts without requiring the use of a good sample of individuals. González et al. [10] present a similarity measure for description logic designed for case-based reasoning systems. This similarity measure is based on the idea of hierarchical aggregation, in which the similarity between two instances is computed as an aggregation of the similarity of the values in their roles.

In addition to similarity in Description Logics, there has been a significant amount of work in the more general topic of similarity assessment for other forms of complex data representation formalisms. Hutchinson [12] presented a distance based on the anti-unification of two terms. The Hutchinson distance is the addition of the sizes of the variable substitutions required to move from the anti-unification of two terms to each of these terms. This measure is related to the refinement-based approaches in [20] and [18], but is more coarse grained.

RIBL (Relational Instance-Based Learning) is an approach to apply lazy learning techniques while using Horn clauses as the representation formalism [9]. An earlier similarity measure related to RIBL was that of Bisson [5]. Horváth et al [11] presented an extension of RIBL that is able to deal with lists and terms. The downside of the RIBL approach is that specialized measures have to be defined for different types of data, while other approaches, such as the ones based on refinement operators, do not have this downside.

Bergmann and Stahl [4] present a similarity metric specific for object oriented representations based on the concepts of *intra-class similarity* (measuring similarity among all the common features of two objects) and *inter-class similarity* (providing a maximum similarity given to object classes). This similarity is defined in a recursive way, thus following the same "hierarchical decomposition" idea as RIBL, and limiting the approach to tree representations.

SHAUD, presented by Armengol and Plaza [1], is another similarity measure following the "hierarchical decomposition" approach but designed for feature terms. SHAUD also assumes that the terms do not have cycles, and in the same way as RIBL and Bergmann and Stalh's it can handle numerical values by using specialized similarity measures for different data types.

7   Conclusions and Future Work

This paper has presented a new approach to assess similarity between individuals in Description Logics. Our approach is based on first converting the individuals to conjunctive queries, and then assessing the similarity between the queries. Converting individuals to queries has several advantages with respect to converting individuals to concepts, and then assessing the similarity between the concepts: first, the conjunctive query language is shared among different DLs, and thus, our similarity measure is more generic (although in this paper we focused on the $\mathcal{EL}$ logic). Second, search in the space of queries is more efficient than search in the space of concepts, thus gaining a computational advantage. Our empirical results show that the resulting measure obtains similar results as previous comparable measures, but at a much lower computational cost.

As part of our future work, in addition to evaluation with larger and more complex datasets, we would like to fully explore the applicability of our similarity measure for more expressive description logics. Specifically, we would like to investigate the tradeoffs between relaxing the requirement of having a complete refinement operator (reducing the search space, and thus the computational complexity), and the performance of the resulting similarity measure.

References

1. Armengol, E., Plaza, E.: Relational case-based reasoning for carcinogenic activity prediction. Artif. Intell. Rev. 20(1-2), 121–141 (2003)
2. Ashburner, M.: Gene ontology: Tool for the unification of biology. Nature Genetics 25, 25–29 (2000)
3. Baader, F., Calvanese, D., McGuinness, D.L., Nardi, D., Patel-Schneider, P.F.: The Description Logic Handbook: Theory, Implementation and Applications. Cambridge University Press, New York (2003)
4. Bergmann, R., Stahl, A.: Similarity measures for object-oriented case representations. In: Smyth, B., Cunningham, P. (eds.) EWCBR 1998. LNCS (LNAI), vol. 1488, pp. 25–36. Springer, Heidelberg (1998)
5. Bisson, G.: Learing in FOL with a similarity measure. In: Proceedings of AAAI 1992, pp. 82–87 (1992)
6. Bodenreider, O., Smith, B., Kumar, A., Burgun, A.: Investigating subsumption in SNOMED CT: An exploration into large description logic-based biomedical terminologies. Artif. Intell. Med. 39, 183–195 (2007),
http://portal.acm.org/citation.cfm?id=1240342.1240604

7. Cojan, J., Lieber, J.: An algorithm for adapting cases represented in an expressive description logic. In: Bichindaritz, I., Montani, S. (eds.) ICCBR 2010. LNCS, vol. 6176, pp. 51–65. Springer, Heidelberg (2010)

8. d'Amato, C., Staab, S., Fanizzi, N.: On the influence of description logics ontologies on conceptual similarity. In: Gangemi, A., Euzenat, J. (eds.) EKAW 2008. LNCS (LNAI), vol. 5268, pp. 48–63. Springer, Heidelberg (2008)

9. Emde, W., Wettschereck, D.: Relational instance based learning. In: Saitta, L. (ed.) Machine Learning - Proceedings 13th International Conference on Machine Learning, pp. 122–130. Morgan Kaufmann Publishers (1996)

10. González-Calero, P.A., Díaz-Agudo, B., Gómez-Albarrán, M.: Applying DLs for retrieval in case-based reasoning. In: Proceedings of the 1999 Description Logics Workshop (DL 1999) (1999)

11. Horváth, T., Wrobel, S., Bohnebeck, U.: Relational instance-based learning with lists and terms. Machine Learning 43(1-2), 53–80 (2001)

12. Hutchinson, A.: Metrics on terms and clauses. In: van Someren, M., Widmer, G. (eds.) ECML 1997. LNCS, vol. 1224, pp. 138–145. Springer, Heidelberg (1997)

13. van der Laag, P.R.J., Nienhuys-Cheng, S.H.: Completeness and properness of refinement operators in inductive logic programming. Journal of Logic Programming 34(3), 201–225 (1998)

14. Larson, J., Michalski, R.S.: Inductive inference of VL decision rules. SIGART Bull. 63(63), 38–44 (1977)

15. Lehmann, J., Hitzler, P.: A refinement operator based learning algorithm for the LC description logic. In: Blockeel, H., Ramon, J., Shavlik, J., Tadepalli, P. (eds.) ILP 2007. LNCS (LNAI), vol. 4894, pp. 147–160. Springer, Heidelberg (2008)

16. Motik, B.: Reasoning in Description Logics using Resolution and Deductive Databases. Ph.D. thesis, Univesitat Karlsruhe (TH), Karlsruhe, Germany (January 2006)

17. Ontanón, S., Plaza, E.: Similarity Measures over Refinement Graphs. Machine Learning 87, 57–92 (2012)

18. Ontañón, S., Plaza, E.: On similarity measures based on a refinement lattice. In: McGinty, L., Wilson, D.C. (eds.) ICCBR 2009. LNCS, vol. 5650, pp. 240–255. Springer, Heidelberg (2009)

19. Sánchez-Ruiz, A.A., González-Calero, P.A., Díaz-Agudo, B.: Abstraction in knowledge-rich models for case-based planning. In: McGinty, L., Wilson, D.C. (eds.) ICCBR 2009. LNCS, vol. 5650, pp. 313–327. Springer, Heidelberg (2009)

20. Sánchez-Ruiz, A.A., Ontañón, S., González-Calero, P.A., Plaza, E.: Measuring similarity in description logics using refinement operators. In: Ram, A., Wiratunga, N. (eds.) ICCBR 2011. LNCS, vol. 6880, pp. 289–303. Springer, Heidelberg (2011)

21. Ullman, J.D.: Information integration using logical views. Theor. Comput. Sci. 239(2), 189–210 (2000)

22. Winkler, W.E., Thibaudeau, Y.: An application of the Fellegi-Sunter model of record linkage to the 1990 U.S. decennial census. In: U.S. Decennial Census. Technical report, US Bureau of the Census (1987)

Should Term-Relatedness Be Used in Text Representation?

Sadiq Sani, Nirmalie Wiratunga, Stewart Massie, and Robert Lothian

School of Computing,
Robert Gordon University,
Aberdeen AB25 1HG, Scotland, UK
{s.a.sani,n.wiratunga,s.massie,r.m.lothian}@rgu.ac.uk

Abstract. The variation in natural language vocabulary remains a challenge for text representation as the same idea can be expressed in many different ways. Thus document representations often rely on generalisation to map low-level lexical expressions to higher level concepts in order to capture the inherent semantics of the documents. Term-relatedness measures are often used to generalise document representations by capturing semantic relationships between terms. In this work we conduct a comparative study of common term-relatedness metrics on 43 datasets and discover that generalisation is not always beneficial. Hence, the ability to predict whether or not to generalise the indexing vocabulary of a dataset is important given the computation overhead of generalisation. Accordingly, we present a case-based approach that predicts, given a text dataset, whether or not using generalisation will improve text retrieval performance. The evaluation shows that our approach is able to correctly predict datasets that are likely to benefit from generalisation with over 90% accuracy.

1 Introduction

Large amounts of useful experience and knowledge are captured by organisations in the form of natural language text documents e.g. incident reports, frequently asked questions and error logs. When faced with a new problem (i.e. incident, question or error), the solutions of previous similar problems can be recalled to aid in solving the new problem. Key to the success of this problem solving methodology is the ability to effectively search through and retrieve relevant documents from the organisation's collections. This process involves making use of a query document which captures the description of the new problem, and retrieving relevant documents from the collection using document similarity. The bag-of-words (BOW) model is typically used for representation where documents are represented as un-ordered collections of their constituent terms. However, the BOW model is not able to cope with variation in natural language vocabulary (e.g. synonymy and polysemy) which often requires semantic indexing approaches [13].

The general idea of semantic indexing is to discover terms that are semantically related and use this knowledge to identify conceptual similarity even in the

S.J. Delany and S. Ontañón (Eds.): ICCBR 2013, LNAI 7969, pp. 285–298, 2013.

presence of vocabulary variation. The result is the generalisation of document representations away from low-level expressions to high-level semantic concepts. Different approaches have been proposed for obtaining term relatedness knowledge. These range from using knowledge rich (extrospective) sources (e.g. lexical databases, Wikipedia and the World Wide Web) [8] to knowledge light (introspective) techniques that use statistics of term co-occurrences in a corpus [7]. Despite their simplicity, statistical techniques have so far provided the best performance in text retrieval evaluations [2]. One reason for this is that corpus co-occurrence is particularly helpful for estimating domain specific relationships between terms [3]. Also, statistical approaches are able to capture relationship types other than similarity e.g. the association between 'bank' and 'money'.

Although generalisation has proven quite useful, it remains to be determined whether it always improves text retrieval performance. Thus, the aim of this work is to address two important questions:

1. Does generalisation always improve text retrieval?
2. If no, can we predict when it is likely to work?

We address the first question by investigating the performance of co-occurrence based generalisation on a number of text classification datasets. Text classification using the kNN algorithm is often used to measure text retrieval performance. Given a query document with unknown class, the kNN algorithm uses the class labels of the top k documents in the ranked retrieval set to determine the class of the query document. Thus, higher text classification accuracy signifies a better ranking of the retrieval set with more relevant documents placed closer to the top which indicates better retrieval performance.

To address the second question, we investigate several attributes of text datasets that are predictive of the performance of generalisation. Our aim is to be able to predict when to apply generalisation in both supervised and unsupervised retrieval tasks. Thus, the attributes we consider are completely unsupervised and not dependent on the class labels of datasets. We use these attributes in a case-based system to predict, given any text dataset, whether or not to apply generalisation. Being able to accurately predict when generalisation is not likely to improve performance means that we can conveniently avoid the overhead of having to extract semantic relatedness knowledge in the first place.

The rest of this paper is structured as follows, in Section 2 we describe popular statistical approaches for extracting term relatedness. Section 3 describes the datasets we used in this research. In Section 4 we evaluate the performance of common term relatedness metrics. In Section 5 we present our CBR approach for predicting when to apply generalisation and we evaluate our approach in Section 6. We present related work in Section 7 followed by conclusions in Section 8.

2   Term Relatedness from Corpus Co-occurrence

The general idea of introspective approaches is that co-occurrence patterns of terms in a corpus can be used to infer semantic relatedness such that the more two terms occur together in a specified context, the more related they are. In the following sections, we describe three different approaches for estimating term relatedness from corpus co-occurrence.

2.1 Document Co-occurrence

Documents are considered similar in the vector space model (VSM) if they contain a similar set of terms. In the same way, terms can also be considered similar if they appear in a similar set of documents. Given a standard term-document matrix D where column vectors represent documents and the row vectors represent terms, the similarity between two terms can be determined by finding the distance between their vector representations. The relatedness between two terms, t_1 and t_2 using the cosine similarity metric is given in equation 1.

$$Sim_{DocCooc}(t_1, t_2) = \frac{\sum_{i=0}^{n} t_{1,i} t_{2,i}}{|t_1||t_2|} \tag{1}$$

2.2 Latent Semantic Indexing

Latent Semantic Indexing (LSI) is a technique that uses singular-value decomposition (SVD) to exploit co-occurrence patterns of terms and documents to create a semantic concept space which reflects the major associative patterns in the corpus [7]. In this way, LSI brings out the underlying latent semantic structure in texts.

Given a term-document matrix D, SVD is used to decompose D into three matrices: U, a term by dimension matrix; S a diagonal matrix of singular values; and V, a document by dimension matrix. This decomposition is shown in equation 2. The number of dimensions n is the rank of the original term-document matrix D.

$$D = U \times S \times V \tag{2}$$

The U, S, V matrices are truncated to k dimensions which represent the k most important concepts in the term-document space. A new term-document representation generalised to this concept space can then be obtained by multiplying the rank-reduced U, S, V matrices using equation 2. Term similarities can be obtained from the new term-document space by multiplying term vectors using equation 1. Unlike document co-occurrence however, LSI is able to learn transitive (higher order) relation between terms that do not co-occur within the same document.

2.3 Normalised Positive Pointwise Mutual Information

The use of mutual information to model term associations is demonstrated in [4]. Given two terms t_1 and t_2, mutual information compares the probability of observing t_1 and t_2 together with the probability of observing them independently

288   S. Sani et al.

as shown in equation 3. Thus, unlike document co-occurrence and LSI, PMI is able to disregard co-occurrence that could be attributed to chance.

$$PMI(t_1, t_2) = log_2 \frac{P(t_1, t_2)}{P(t_1)P(t_2)} \tag{3}$$

If a significant association exists between t_1 and t_2, then the joint probability $P(t_1, t_2)$ will be much larger than the independent probabilities $P(t_1)$ and $P(t_2)$ and thus, $PMI(t_1, t_2)$ be greater than 0. Positive PMI is obtained by setting all negative PMI values to 0. The probability of a term t in any context can be estimated by the frequency of occurrence of t in that context normalised by the frequency of all words in all contexts.

$$P(t) = \frac{f(t)}{\sum_{j=1}^{N} \sum_{i=1}^{N} f(t_i, t_j)} \tag{4}$$

PMI values do not lie within the range 0 to 1. Thus we need to introduce a normalisation operation. We normalise PMI as shown in equation 5.

$$Sim_{Npmi}(t_1, t_2) = \frac{PPMI(t_1, t_2)}{-log_2 P(t_1, t_2)} \tag{5}$$

3   Datasets

For evaluation, we obtain a collection of incident reports by crawling the Web. A problem usually encountered when obtaining datasets for evaluating retrieval performance is the absence of user relevance judgements. To address this, we treat our evaluation as a classification task where documents that belong to the same class as the query document are judged to be relevant. A second problem is having access to large amounts of experiential datasets. Because we have decided to conduct our evaluation using text classification, we supplemented our limited incident reports data with standard text classification corpora i.e. Reuters V1, 20 Newsgroups, Ohsumed and Movie Reviews. These corpora are described in detail in the following paragraphs.

Incident Reports corpus was created using incident reports crawled from the Government of Western Australia's Department of Mines and Petroleum website [1]. The corpus contains documents classified according to the types of incidents they report e.g. the class Fire for fire related incidents and Collision for vehicle collisions. Documents in each incident category are further classified into Injury and NoInjury categories depending on whether or not injuries were sustained in the incidents they describe.

Reuters Volume 1 corpus is an archive of 806,791 news stories provided by the global news provider, Reuters. The collection comprises all news stories produced by Reuters journalists within a one year period starting from August,

---

[1] http://dmp.wa.gov.au

1996. Documents within the collection are tagged with descriptive metadata specifying codes for topic, region and industry sector. Topic codes represent the subject area of each news story. Industry codes are used to indicate the type of business or industry referred to by the news story. Region codes indicate the geographical region referred to in the news story. Only topic codes and industry codes where used when creating datasets for our evaluation.

20 Newsgroups corpus is a collection of 20,000 documents collected from Newsnet newsgroups messages. The collection is partitioned almost equally into 20 classes of 1,000 documents each, according to newsgroup topics. For example, the class sci.space contains messages relating to space.

Ohsumed is a subset of MEDLINE, an online database of medical literature, and comprises a collection of 348,566 medical references from medical journals covering a period from 1987 to 1991. The Ohsumed collection is unequally divided into 23 classes according to different disease types e.g. Virus Diseases.

Movie Reviews is a sentiment classification corpus containing 1400 reviews of movies from the Internet Movie Database (IMDB). About half of these reviews are classified as expressing positive sentiment while the other half is classified as negative. Accordingly, the classification task for this dataset is to determine the sentiment orientation of any given review.

4   Performance of Generalisation

Generalisation uses semantic relation between terms to map low-level indexing vocabulary to higher-level semantic concepts. Given a term-document matrix D, a term-relatedness matrix T can be populated using the term-relatedness techniques introduced in Section 2. A new generalised term document matrix D' can then be obtained using equation 6. Further details of this approach are presented in [12].

$$D' = D \times T \tag{6}$$

In this section, we determine if generalisation consistently improves text retrieval performance.

4.1   Experiment Setup

Standard preprocessing operations i.e. lemmatisation and stopwords removal are applied to our datasets. Feature selection is used to limit our term-document space to the top 300 most informative terms for each dataset.

We compare the following algorithms: **BASE**, baseline bag-of-words approach without term relatedness; **DocCooc**, term relatedness estimated from document co-occurrence (see Section 2.1); **NPMI**, term relatedness calculated using Normalised Positive Pointwise Mutual Information (see Section 2.3); **LSI**, term relatedness estimated from latent semantic analysis (see Section 2.2).

We report classification accuracy using a similarity weighted kNN approach (with $k=3$) and using the cosine similarity metric to identify the neighbourhood.

4.2   Results

Classification results are shown in Table 1. Values with the $^+$ sign represent a significant improvement in text classification accuracy compared to the baseline and $^-$ represent a significant decline in classification accuracy. The average difference between DocCooc and Base is 0.96%, between Npmi and Base is 0.84%, and between Lsi and Base is 0.80%.

Although generalisation has resulted in statistically significant improvement in many datasets (42% of the datasets using DocCooc, 37% using Lsi and 47% using Npmi), it has remained neutral on many other datasets and even led to a decline in accuracy in others. Considering the additional cost of acquiring term-relatedness, it is important to empirically determine when it is beneficial to use term relatedness in text retrieval. In the next section, we present a case-based approach for predicting, given any dataset, whether or not apply generalisation.

5   Predicting When to Generalise

We have already established the need to be able to predict when and when not to use generalisation. In this section, we introduce our case-based approach

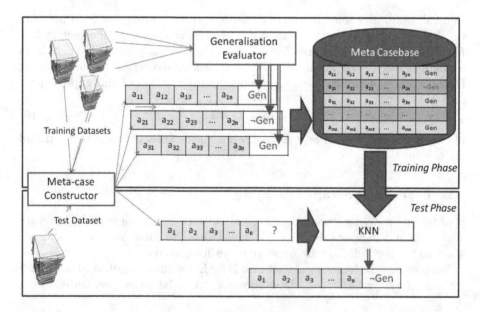

Fig. 1. Case-based approach using dataset meta-data to predict when to generalise

Table 1. Classification accuracy of generalisation techniques

Dataset	Base	DocCooc	Npmi	Lsi
Hardw	89.8	90.9$^+$	91.2$^+$	90.3$^+$
MedSp	95.9	93.8$^-$	95.8	93.6$^-$
CryptE	95.9	90.3$^-$	91.8$^-$	90.6$^-$
ChrisM	88.9	90.5$^+$	89.9$^+$	90.5$^+$
MeastM	95.1	95.3	94.9	95.3
GunsM	93.4	94.0	94.0	94.0
AutoC	94.4	95.1	96.2$^+$	95.0
BaseH	95.9	95.6	96.6$^+$	95.7
StratM	88.8	89.4	83.7$^-$	89.6
EntTour	94.7	95.7$^+$	95.3	95.6$^+$
EqtyB	95.7	95.5	94.8$^-$	95.6
FundA	90.3	92.0$^+$	89.9	92.1$^+$
InRelD	92.6	94.1$^+$	91.7	94.3$^+$
NProdRes	85.9	86.9	80.4$^-$	86.7
ProdNP	87.4	89.3$^+$	88.4	88.9$^+$
MarketA	89.8	89.1	89.4	89.2
MoneyC	94.8	94.5	93.2$^-$	94.5
OilGas	87.8	86.3$^-$	85.7$^-$	86.2$^-$
ElectG	88.7	84.6$^-$	84.0$^-$	84.5$^-$
FinI	86.5	87.0	84.6$^-$	87.0
MovieRev	71.3	78.6$^+$	81.8$^+$	79.3$^+$

Dataset	Base	DocCooc	Npmi	Lsi
NervI	91.4	91.0	92.9$^+$	90.5
BactV	85.1	88.6$^+$	90.0$^+$	87.5$^+$
CardR	90.0	92.2$^+$	93.8$^+$	90.7
MouthJ	89.9	92.2$^+$	92.9$^+$	92.0$^+$
NeopE	91.6	93.8$^+$	94.2$^+$	94.0$^+$
DigNut	87.8	91.3$^+$	93.2$^+$	91.5$^+$
MuscS	83.1	87.0$^+$	91.1$^+$	86.5$^+$
EndoH	91.4	95.8$^+$	96.5$^+$	95.4$^+$
MaleF	92.3	94.9$^+$	95.6$^+$	95.1$^+$
PregN	89.7	90.4	90.9$^+$	90.4
ImmunoV	78.7	82.5$^+$	84.8$^+$	82.7$^+$
NervM	84.5	88.1$^+$	91.0$^+$	87.8$^+$
RespENT	87.2	88.1	91.0$^+$	88.3
SkinN	87.1	88.1	91.2$^+$	87.6
EndoNut	75.2	81.7$^+$	82.7$^+$	81.6$^+$
Fire	84.4	87.0	85.8	86.9
Collision	82.2	80.9	76.8$^-$	81.3
Rollover	79.8	79.1	77.7	78.2
CollRoll	86.5	83.6	80.5$^-$	84.3
MiscInc	84.0	84.1	82.0	84.1
CraneFP	87.5	88.3	82.4$^-$	87.9
ShovFP	88.3	86.6	88.3$^-$	83.8

for doing so. Figure 1 shows both the training and test phases of our case-based system. Given a collection of training datasets, the meta-case constructor creates a meta-case representation for each dataset. The case description of each meta-case comprises a set of nine attributes a_1 to a_n (discussed in Section 5.1) that capture the properties of the dataset. The training datasets are also fed to the Generalisation Evaluator which produces the solutions for the meta-cases. The case solution is a binary judgement of whether or not to apply generalisation to the dataset. A meta-case is labelled with the solution to use generalisation (Gen) if the improvement from generalisation is statistically significant. Otherwise, we label the case with the decision not to use generalisation ($\neg$Gen). For example for the DocCooc technique, generalisation produced a significant improvement on the Hardware dataset of 90.9% compared to BASE (89.9%) (see table 1) and the decision to use generalisation is selected as the case solution for Hardware. On the other hand on the MedSpace dataset, DocCooc produced a decline of 2.1 % in accuracy and thus the solution for this case is not to use generalisation. Similarity between cases is determined using Manhattan distance given in equation 7.

$$Dist(a,b) = \sum_{i=1}^{N}(|a_i - b_i|) \tag{7}$$

In the next section, we discuss the set of features used for case representation.

5.1 Dataset Attributes

Generalisation sometimes fails because of its potential to establish relationships that are too general and hence not very discriminatory. For example the BactViral dataset from Table 1 comprises the classes **Bacterial**, which contains documents on bacteria-related diseases, and **Viral** which contains documents on virus-related diseases. In this particular dataset, the words "biopsy" and "treat" co-occur 10 times which indicates a strong relationship. However, the two words co-occur almost equally across class boundaries which means that the relationship between them is a weak indicator of class membership. In contrast, the words "endoscopy" and "helicobacter" co-occur 5 times, all within the **Bacterial** class which makes this relationship a stronger indicator of class membership. Because of the higher co-occurrence frequency between "biopsy" and "treat", generalisation is likely to treat them as being more related than "endoscopy" and "helicobacter". This indicates that the distribution of terms in the term space is an indicator of the potential performance of generalisation. Accordingly, we utilise attributes that characterise the term-document space of our datasets in order to create our meta-cases.

5.2 Average Terms Per Document

The average term count for the entire dataset is calculated by taking the average term count for all documents in the dataset.

5.3 Document Frequency

The document frequency of a term t_i is a count of the number of documents in which t_i occurs. Document frequency is often used as a feature selection technique under the premise that very rare terms are not informative and thus do not contribute much to document retrieval. At the same time, terms that appear in almost all documents are also not very discriminatory and can be considered noisy in the term document space. Such high frequency terms are also likely to co-occur with almost every other term thus polluting the generalisation process. Hence we utilise three metrics to measure the effect of document frequency: **Maximum DF** which is the maximum document frequency over all terms and **Ave. DF** which is the average document frequency of over all terms.

5.4   Inverse Document Frequency

Inverse Document Frequency (IDF) is a function designed to give a weighting inversely proportional to the document frequency of terms. IDF captures the premise that terms with very high document frequency are less informative than terms that occur less often. The formula for IDF is given in equation 8 where N is the total number of documents and $df(t)$ is the document frequency of t.

$$IDF(t) = log_2 \frac{N}{df(t)} \tag{8}$$

We use the **Maximum IDF** and the **Average IDF** to obtain a measure of rare terms in our datasets.

5.5   Complexity Profile

This measure was originally proposed for measuring the complexity of a case base by looking at the classes of the nearest neighbours of each case [10]. Because we are interested in unsupervised attributes, we instead measure complexity of a dataset using the distance between each document and the other documents in its neighbourhood as shown in Figure 2. Complexity profile of a document d_j is calculated by iteratively retrieving successively larger neighbourhoods k of d_j up to the neighbourhood size K (we use $K = 10$) and computing the proportion $P_k(d_j)$ of documents in that neighbourhood that belong to the same class as d_j as shown in equation 9.

$$Complexity(d_j) = \frac{\sum_{k=1}^{K} P_k(d_j)}{K} \tag{9}$$

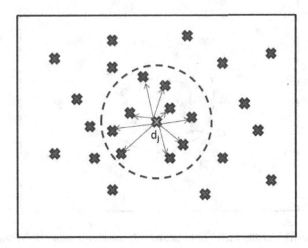

Fig. 2. Complexity calculated using the distance of document d_j to its k nearest neighbours

To convert equation 9 into an unsupervised metric, we ignore the class labels of documents and replace the proportion $P_k(d_j)$ with the similarity between d_j and all documents in its neighbourhood. This is shown in equation 10.

$$P_k(d_j) = \frac{\sum_{i=1}^{k} Sim(d_j, d_i)}{k} \tag{10}$$

Where $Sim(d_j, d_i)$ is the cosine similarity between document d_j and d_i. The final complexity measure for the entire dataset is computed as the average complexity of all documents d_j.

5.6 Neighbourhood Similarity

While Complexity Profile measures the distance between a target document and its nearest neighbours, this metric calculates the average pair-wise similarity between all k nearest neighbours of the target document d_j as shown in Figure 3. We use a neighbourhood size of $k = 10$. We then calculate the average, minimum and maximum nearest neighbour similarity over all documents to obtain the **Average NN Similarity, Minimum NN Similarity** and **Maximum NN Similarity** respectively for that dataset.

The average similarity between the nearest neighbours of a document gives tells us how tightly clustered the neighbourhood of that document is. In turn, the aggregation over all documents provides us with information about how tightly clustered documents are in the entire term document space.

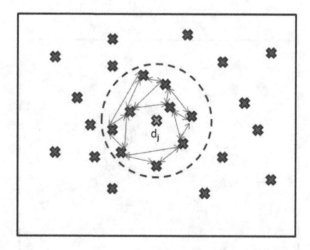

Fig. 3. Neighbourhood similarity of document d_j measures using the distance between k nearest neighbours of d_j

6   Evaluation

The aim of this evaluation is to determine how well our case-based approach (CBR) predicts when and when not to use generalisation for text representation. We compare this with a baseline approach (BASELINE) that always applies generalisation. Our hypothesis is that our case-based approach should be able to identify datasets that are not likely to benefit from generalisation. This allows for applying generalisation to datasets in a systematic fashion. Accordingly, we treat this as a classification task where accuracy is measured as the percentage of test cases that are labelled with the correct decision (to generalise or not). We report the classification accuracy over a leave-one-out validation using a 3-NN approach.

Table 2. Classification accuracy of generalisation prediction

	Gen	DocCooc	Npmi	LSI
BASELINE	55.81	41.86	46.51	37.21
CBR	**79.07**	**81.4**	**88.37**	**72.09**
CBR+	**86.05**	**86.05**	**93.02**	**79.07**

From the results shown in Table 2, it is clear that our case-based predicts when to generalise with high accuracy. The results in the Gen column represent the accuracy of our prediction across all generalisation techniques. That is, deciding to generalise always, we match all datasets that are labelled with the decision to generalise (55.81%) but we also generalise many other datasets (44.19%) that should not be generalised. However, using our case-based apraoch, we correctly match datasets that should/should not be generalised 79.07% of the time. The other columns (DocCooc, Npmi and Lsi) provide a break-down of our performance for each individual generalisation technique respetively. The CBR+ row shows results of the Case-Based approach with optimal weights learnt used for our meta-case attributes using a Genetic Algorithm where the set of weights used range from 0 to 10. A comprehensive review of applying weighting to kNN retrieval is provided in [14]. From these results we can see that our set of attributes are predictive of the effectiveness of applying generalisation for text classification.

The weights learnt for our attributes by the genetic algorithm can be divided into high, **Complexity Profile**; medium, **Maximum IDF, Ave. Tokens Per Doc., Maximum DF** and **Max. NN Sim**; and low, **Ave. DF, Ave. IDF, Max. NN Sim** and **Min. NN Sim**. The high weight assigned to Complexity Profile indicates the importance of the average similarity of cases to their nearest neighbours in determining the performance of generalisation. Note that higher values of Complexity Profile indicate a loosely clustered term document space. Closer investigation reveals datasets with low complexity profile are more likely to benefit from generalisation.

7   Related Work

Much work has been done in the area of meta-learning. For example the approach presented in [1] uses a meta learner to assign classifiers to datasets. Of particular relevance to our work are meta case-based approaches e.g. [6] where a meta case-based technique is used for selecting case-base maintenance algorithms. In this approach, an individual meta-case models a entire case-base where the case solution is the maintenance algorithm that provides the best performance on that case-base and the case description comprises a set of attributes that are derived using complexity measures. While the set of attributes used in [6] are supervised, the attributes used in our approach are completely unsupervised making our approach applicable to both supervised and unsupervised tasks. Also, our approach is concerned with predicting when to generalise while the approach in [6] is concerned with selecting the best algorithm to use for case-base maintenance.

Another case-based approach for selecting the best sentiment lexicon given a sentiment classification dataset is presented in [11]. Here also, a dataset is represented as a single case where the case solution is the best performing sentiment lexicon for the dataset. The case description is modelled as an n-dimensional feature vector derived from document, sentence and term-level statistics of as well counts of part-of-speech information and punctuations. The attributes chosen for case representation are designed to capture the subjectivity of the corresponding dataset. On the other hand, our approach is concerned with trying to predict, given any generic text classification dataset, when to apply generalisation for text representation. Consequently, we use a totally different set of attributes from [11].

The system presented in [9] uses a CBR approach to select the best classification algorithm for a dataset. The datasets considered in this paper are not limited to textual datasets and the attributes used for case representation are designed to capture characteristics of datasets that contain both numeric and symbolic attributes. As such, these features are different from the ones used in our approach and are perhaps not as suited for textual datasets.

A meta case-based approach for obtaining adaptation knowledge for a CBR system is presented in [5]. A meta-case m is made up of a description which encapsulates an adaptation situation and a solution which provides the adaptation action applied. An adaptation situation arises when we have a query case c_q, the best matching case c_i from the casebase, and the adaptation operation (e.g. increment by 10) applied to the solution of c_i to fit c_q. Thus, a meta-case description representing this adaptation situation comprises the description of the query case c_q, the differences between the descriptions of c_q and c_i, and the solution of c_i. Note that this approach uses features from the original casebase for meta-case representation. This contrasts with our approach where a completely different set of features are extracted for meta-case representation. An adaptation action can be a binary judgement of whether or not adaptation is required, or a an update operation (e.g. decrease by 5%) that is applied to the proposed solution.

8   Conclusion

In this paper we investigated whether generalisation is always beneficial for text retrieval. The performance of 3 different co-occurrence based term relatedness techniques on 43 datasets shows that generalisation does not always improve text classification performance and may sometimes even be harmful. Considering that generalisation is an expensive process, we set out to determine when and when not to apply generalisation. Accordingly we presented a case-based approach for predicting when to generalise. Results show that our case-based approach is able to correctly predict the performance of generalisation on a range of datasets with over 90% accuracy.

An important consideration when building a case-based system is the choice of attributes for case representation. The set of attributes we use was obtained from several statistical metrics that capture various important characteristics of text datasets. These range from statistics of document frequencies of terms to measures of clustering of document neighbourhood. The high accuracy achieved in predicting when to generalise indicates that the attributes used for meta-case representation capture characteristics of text datasets that are predictive of the performance of generalisation. We further use a genetic algorithm to learn the relative importance of our attributes. The high weight assigned to the Complexity Profile attribute indicates the importance of the average similarity of cases to their nearest neighbours in determining the performance of generalisation where datasets with low complexity profile are more likely to benefit from generalisation.

References

1. Bensusan, H., Giraud-Carrier, C., Kennedy, C.: A higher-order approach to meta-learning. In: Proceedings of the ECML 2000 Workshop on Meta-Learning: Building Automatic Advice Strategies for Model Selection and Method Combination, pp. 109–117 (2000)
2. Brants, T., Inc, G.: Natural language processing in information retrieval. In: Proceedings of the 14th Meeting of Computational Linguistics in the Netherlands, pp. 1–13 (2004)
3. Chakraborti, S., Wiratunga, N., Lothian, R., Watt, S.: Acquiring word similarities with higher order association mining. In: Weber, R.O., Richter, M.M. (eds.) ICCBR 2007. LNCS (LNAI), vol. 4626, pp. 61–76. Springer, Heidelberg (2007)
4. Church, K.W., Hanks, P.: Word association norms, mutual information, and lexicography. Computational Linguistics 16(1), 22–29 (1990)
5. Craw, S., Wiratunga, N., Rowe, R.C.: Learning adaptation knowledge to improve case-based reasoning. Artificial Intelligence 170(16-17), 1175–1192 (2006)
6. Cummins, L., Bridge, D.: On dataset complexity for case base maintenance. In: Ram, A., Wiratunga, N. (eds.) ICCBR 2011. LNCS, vol. 6880, pp. 47–61. Springer, Heidelberg (2011)
7. Deerwester, S.C., Dumais, S.T., Landauer, T.K., Furnas, G.W., Harshman, R.A.: Indexing by latent semantic analysis. Journal of the American Society of Information Science 41(6), 391–407 (1990)

8. Gabrilovich, E., Markovitch, S.: Wikipedia-based semantic interpretation for natural language processing. Journal of Artificial Intelligence Research 34, 443–498 (2009)

9. Lindner, G., Studer, R.: Ast: Support for algorithm selection with a cbr approach. In: Żytkow, J.M., Rauch, J. (eds.) PKDD 1999. LNCS (LNAI), vol. 1704, pp. 418–423. Springer, Heidelberg (1999)

10. Massie, S., Craw, S., Wiratunga, N.: Complexity profiling for informed case-base editing. In: Roth-Berghofer, T.R., Göker, M.H., Güvenir, H.A. (eds.) ECCBR 2006. LNCS (LNAI), vol. 4106, pp. 325–339. Springer, Heidelberg (2006)

11. Ohana, B., Delany, S., Tierney, B.: A case-based approach to cross domain sentiment classification. In: Agudo, B.D., Watson, I. (eds.) ICCBR 2012. LNCS, vol. 7466, pp. 284–296. Springer, Heidelberg (2012)

12. Sani, S., Wiratunga, N., Massie, S., Lothian, R.: Term similarity and weighting framework for text representation. In: Ram, A., Wiratunga, N. (eds.) ICCBR 2011. LNCS, vol. 6880, pp. 304–318. Springer, Heidelberg (2011)

13. Tsatsaronis, G., Panagiotopoulou, V.: A generalized vector space model for text retrieval based on semantic relatedness. In: Proceedings of the Student Research Workshop at EACL 2009, pp. 70–78 (2009)

14. Wettschereck, D., Aha, D.W., Mohri, T.: A review and empirical evaluation of feature weighting methods for a class of lazy learning algorithms. Artificial Intelligence Review 11(1-5), 273–314 (1997)

Recommending Audio Mixing Workflows

Christian Sauer, Thomas Roth-Berghofer, Nino Auricchio, and Sam Proctor

School of Computing and Technology, University of West London,
St Mary's Road, London W5 5RF, United Kingdom
{first.lastname}@uwl.ac.uk

Abstract. This paper describes our work on *Audio Advisor*, a work-flow recommender for audio mixing. We examine the process of eliciting, formalising and modelling the domain knowledge and expert's experience. We are also describing the effects and problems associated with the knowledge formalisation processes. We decided to employ structured case-based reasoning using the *myCBR 3* to capture the vagueness encountered in the audio domain. We detail on how we used extensive similarity measure modelling to counter the vagueness associated with the attempt to formalise knowledge about and descriptors of emotions. To improve usability we added GATE to process natural language queries within *Audio Advisor*. We demonstrate the use of the *Audio Advisor* software prototype and provide a first evaluation of the performance and quality of recommendations of *Audio Advisor*.

Keywords: CBR, myCBR, audio, mixing, audio engineering, similarity measures, workflow recommendation, knowledge formalisation.

1 Introduction

With automatic composition and improvisation of music expressing the individual style of a human composer as well as the automatic expressive performance of music, the two main steps of music creation are quite well researched [15,16]. There are a variety of approaches to automated composition of expressive music and the expressive performance of music, see e.g., [15,4]. They all need to deal with the problem of formalising emotions in order to relate to the intended emotional effect of a composition and/or performance. The formalisation of affective, emotional statements or descriptive adjectives of an emotion is still a problem [11,6], often encountered by applications dealing with art and deeply linked to emotions and perception of such.

Next to composition and performance, a third important task in professional music production is the mixing of a sound recording. Mixing is the process of applying a set of spectral modifications to sounds in order to achieve a change in timbre or more specifically the emotional effect of the sound on a listener [14]. This process is goal-oriented, with the goal being a desired change in the emotional effect of a sound. The vocabulary describing this effect-change consists of terms that describe the emotion desired to be triggered or altered, i.e., increasing or decreasing an emotional effect. We find queries like 'make it sound more

S.J. Delany and S. Ontañón (Eds.): ICCBR 2013, LNAI 7969, pp. 299–313, 2013.

warm' or 'make it sound less harsh' and onomatopoeia in the language of audio engineers.

The experience of audio engineers is in the linkage between queries containing timbre descriptors such as 'warm' or 'bright' as well as amount descriptors and constrains, and in the choice and application of spectral modifications used to achieve the desired timbre change of the sound. Additionally the effect of such a query is also linked to the context in which it occurs. The modelling and (re-)use of such context embedded queries to recommend the adequate workflows was the main goal of our *Audio Advisor* prototype workflow recommender.

This paper introduces our work on *Audio Advisor*, a workflow recommender system that allows its users to formulate natural language queries for the automatic case-based retrieval of workflows that, when applied to the audio product changes its timbre and/or applies an effect to it. The workflow itself is provided as a sequence of so called presets, where a preset can be described as a selection of frequency descriptors with definite decibel change values for said frequencies. A preset can further contain information on defined effects such as *reverb* or *delay* and the decibel values to be applied to these effects. An example is to define a preset to reduce high frequencies and emphasise lower frequencies while adding a slight echo effect to the sound.

The rest of the paper is structured as follows: We interlink our approach with the current state-of-the-art in the field of artificial music composition and performance in Section 2. Based upon the goals and aims of *Audio Advisor* (Section 3) we examine the domain of audio mixing and its specific knowledge as well as our approaches to elicit and formalise the knowledge in Section 4. In Section 5 we show how we use $GATE^1$ to develop the natural language processing component that enables *Audio Advisor* to 'understand' natural language queries posted to the system. We then demonstrate how we use *myCBR* 3^2 for *Audio Advisor* and examine the overall structure and workflow of the *Audio Advisor* application. Section 6 details on the performed experiments regarding the quality of the workflow recommendations and evaluate the performance of *Audio Advisor*. A summary and outlook on future work then concludes the paper.

2   Related Work

A variety of approaches to formalise emotional annotations and/or descriptive terms that either describe the mood of the music or the way it is to be played [16,9] already exists. Such approaches deal with either playing music in a certain defined way to convey an emotion [7] or to select songs or sounds that are associated with a mood or emotional state [21]. For automated composing, the question of integrating a formal description of the mood the composed music should match is already well researched [16,4].

Emotions or, in our context, the timbre of a sound and its perception are not easy to be a) defined and b) quantised/formalised [13,8,11]. Another problem

---

[1] http://gate.ac.uk/

[2] http://www.mycbr-project.net

we were facing during the domain knowledge formalisation was that we tried to quantify and cluster descriptive adjectives based on very vague data given by the individual descriptions of the emotional effect a sound has on a person describing this effect. The difficulties of capturing a sounds timbre are [10]: "*It is timbre's 'strangeness' and, even more, its 'multiplicity' that make it impossible to measure timbre along a single continuum, in contrast to pitch (low to high), duration (short to long), or loudness (soft to loud). The vocabulary used to describe the timbres of musical instrument sounds indicates the multidimensional aspect of timbre. For example, 'attack quality,' 'brightness,' and 'clarity' are terms frequently used to describe musical sounds.*" The vagueness of the data is based on said variation in the individuals perceptions when they either should describe an emotional effect or perceive something that is annotated with a particular emotion but have a complete different idea of the actual emotion this percept triggers [17,10,12].

The problems caused by the described vagueness of timbre descriptors, which we initially examined in [20], were one of the most prominent ones during the knowledge formalisation process employed for the *Audio Advisor* application's knowledge model. We were able to counter said vagueness by employing complex similarity measures, following the knowledge modelling procedures described for example in [1]. Additionally we also investigated how to extract the meaning, thus the semantics of a natural language query posted to our *Audio Advisor*. We did so mainly by following an approach we developed for a previous Information Extraction (IE) application, *KEWo*, that extracts taxonomies of terms to be used as similarity measures in CBR systems from natural language texts [19].

3   Aims and Opportunities of Our Work

A way to circumvent the lack of quantifiable measures and vagueness is to allow for vagueness and a certain amount of ambiguity within the techniques used for formalising and retrieving problem descriptions based on descriptive adjectives. The vagueness accompanying the formalisation of these descriptive adjectives, mainly the timbre descriptors, can be handled by the use of similarity knowledge in Case-Based Reasoning (CBR) systems [2,22]. The ability of CBR to handle said vagueness has already been used to guide the emotional component of automatic composition as well as performance of music, see e.g., [7,3,18,16].

The aim of *Audio Advisor* is to make audio mixing experience available to its users and to allow them to use and learn the special vocabulary employed by experienced audio engineers. By making the audio engineers experience available through our *Audio Advisor* we thus are able to fulfil the following goals:

- Using *Audio Advisor* in a teaching approach for audio engineering students
- Allowing lay persons to practise / improve mixing skills
- Re-use the knowledge of experienced audio engineers
- Speed up the mixing process and, thus, reduce expensive studio time
- Improve usability to audio mixing software by integrated workflow recommendations

4 The Audio Mixing Domain

The most common mixing task is to change an input sound and consequently an *input* timbre to a desired *target* timbre by a specified *amount*. This basic problem description can be extended by a number of sub timbres to be changed simultaneously and constraints on the desired changes such as 'Make the flute sound more airy but not so breathy.'. Following this basic assumptions about the audio mixing domain, we present in this section our approach to elicit the domain knowledge from experienced audio engineers as well as the knowledge artefacts we were able to elicit. We then consider the problems we faced during the knowledge formalisation process and their influence on our choice of the formalisation techniques that we employed. We then review our resulting initial knowledge model that we modelled using *myCBR 3* and which is currently used as the reasoning component of the *Audio Advisor*.

4.1 Domain Knowledge

The knowledge representing the experience of audio engineers has a high grade of abstraction and is highly encoded. For example the knowledge how to apply a set of frequency changes in a specific order to change a sound in a specific context with a desired effect is simply encoded in a sentence like: 'Make the trumpet a lot fatter and a bit more toppy, like in Jazz music'. This sentence is implicitly associated by the experienced audio engineer with a workflow like: *Increase the 6 kHz frequency in the high shelf segment by 3 dB, then increase the 150 Hz segment by 9 dB with a wide bandwidth and finally reduce the 2.7 kHz segment by 2 dB with a narrow bandwidth.*

Due to this high level of abstraction and encoding we faced the problem of choosing the best suited techniques for the necessary knowledge elicitation. We opted for employing a variety of techniques to minimise the danger of knowledge loss and to maintain a high level of accuracy. To get insight into the audio mixing domain we arranged for several studio sessions where the audio engineers provided actual hands on experience on how to mix an audio product in a studio. Second to these sessions we arranged for a couple of interview sessions with two audio engineers. During these interviews we questioned the experts so they could provide their experience in increasing grades of formalisation.

The knowledge elicitation process also yielded some unexpected artefacts. For example, the audio engineers came up with Venn diagrams classifying the timbre descriptors. Such artefacts were very helpful while building the taxonomic similarity measures for the timbre descriptors.

4.2 Initial Knowledge Modelling

After the elicitation of the described knowledge artefacts that describe a mixing task the next step was to design an initial knowledge model of the audio mixing domain. As we already stated in section 1 one of the most complicated challenges while trying to formalise descriptors for timbres, is the vagueness of said descriptors. We decided to counter this challenge by employing structured CBR as we expected to counter the vagueness of the timbre descriptors and amount descriptors by modelling complex similarity knowledge that describes their relationships. We quickly identified the main domain relationship, presets being applied to timbres, as a perfect candidate to divide the domain into a problem and solution part

The most foreseeable challenge we encountered was the challenge of finding an optimal grade of abstraction. This was of importance as we were, like in any knowledge formalisation task, facing the trade-off between an over engineered too specific knowledge model and the danger of knowledge loss by employing too much abstraction e.g. choosing the abstraction levels too high. Together with the domain experts we chose two additional abstraction levels of frequency segments for the timbre descriptors. We further chose to use a taxonomic order for the timbre descriptors and the amount descriptors, as well as the instruments to be used as structures to model the respective abstraction layers of these knowledge artefacts. Thus we designed taxonomies describing timbres, amounts and instruments from a most abstract root node down to the most specific leafs, see our initial work on this approach [20] for details.

The next modelling step consisted of determining the best value ranges for the numerical attributes we wanted to integrate into our initial knowledge model. Again after discussing this with the domain experts we agreed to use two way to represents *amounts* in our domain. We provide a percentage approach, ranging from 0 to 100% as well as a symbolic approach. The symbolic approach was chosen because the domain experts mentioned that from their experience the use of descriptors for amounts, such as *'a slight bit'* or *'a touch'* were by far more common in audio mixing sessions then a request like *'make it 17% more airy'*. So we integrated, next to the simple and precise numerical approach, a taxonomy of amount descriptors into our initial knowledge model. The taxonomy was ordered based on the amount the symbol described, starting from the root, describing the highest amount down to the leaf symbols describing synonyms of smallest amounts. Additionally to modelling the amounts we also needed to represent the workflow steps, so the application of presets. For this we elicited that the application of spectral modification's is always specified in decibels (dB) and that these settings always follow a certain rasterization, due to the knobs and dials on a mixing board clicking into place with certain amounts of dB being tuned in on this dials. We thus provided the amounts in the workflow descriptions in decibel.

Regarding *myCBR 3* we had to choose between a taxonomic and a comparative table approach. Considering the versatile use of taxonomies in structural CBR [5] we initially opted for the use of taxonomies. Yet regarding the complex similarity relationships between the elicited timbre descriptors we also wanted to investigate whether a comparative table approach for modelling the similarities of the timbre descriptors might yield a more accurate knowledge model, ultimately resulting in better workflow recommendations. So we formalised the similarities of the timbre descriptors also using the comparative table approach.

5   Prototype Implementation

In this section we will detail on how we implemented the *Audio Advisor* application prototype using the GATE framework and *myCBR 3*.

5.1   Using GATE for Natural Language Query Processing

Audio Advisor allows a user to enter a natural language queries such as 'Make the trumpet a bit brighter but not too airy.' Such a query requires the *Audio Advisor* to be able to parse the natural language into settings for the attribute values of the mixing task's problem description. Figure 1 shows the automatically set attribute values based on a real sample query.

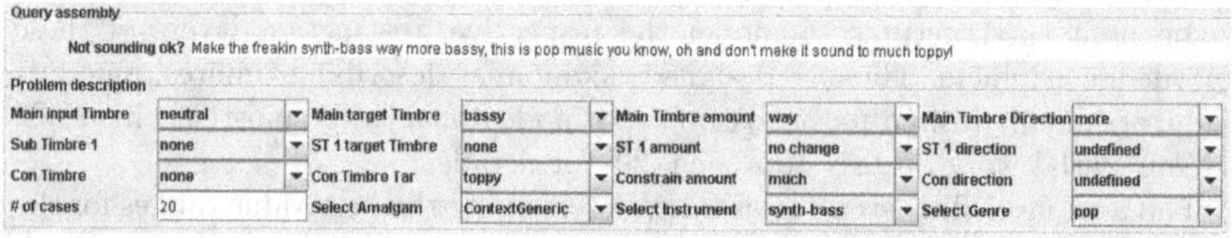

Query assembly								
Not sounding ok? Make the freakin synth-bass way more bassy, this is pop music you know, oh and don't make it sound to much toppy!								
Problem description								
Main input Timbre	neutral	Main target Timbre	bassy	Main Timbre amount	way	Main Timbre Direction	more
Sub Timbre 1	none	ST 1 target Timbre	none	ST 1 amount	no change	ST 1 direction	undefined
Con Timbre	none	Con Timbre Tar	toppy	Constrain amount	much	Con direction	undefined
# of cases	20	Select Amalgam	ContextGeneric	Select Instrument	synth-bass	Select Genre	pop

Fig. 1. Problem description section of the *Audio Advisor* application GUI

   To extract the correct attribute values and their context from the natural language query we employ the GATE Architecture, i.e., a modified version of the ANNIE application[3]. To, for example, distinguish between a query with and without a constraint, we analysed the structure of a number of example queries with the use of the GATE Developer 7.0 GUI application, see Figure 2 for details. After designing the necessary specially built language processing resources, i.e., Gazetteers and Jape grammar rules, we modified the ANNIE Application to allow for the Annotation of the following term categories: amount, constraint, direction, effect, instrument, timbre and timbre-shift. By using these annotations we were able to analyse the query structure as the following figure demonstrates:
   The structural analysis of the queries enabled us to build a classification tree that represents typical semantics formulated in a certain type of query. In this way we can map the queries to reoccurring kinds of problem descriptions and set the values specified within the query to the correct attributes describing specific mixing tasks. Figure 3 shows a section of the classification tree.
   The customised ANNIE is embedded in the *Audio Advisor*. The annotations generated by the customised ANNIE application are stored in an XML file that is then parsed to make the annotations available for the query assembly.

---

[3] http://gate.ac.uk/sale/tao/splitch6.html

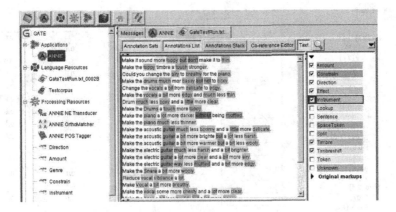

Fig. 2. Query Annotation in GATE

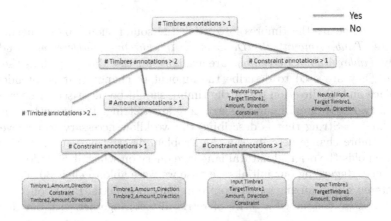

Fig. 3. Classification tree derived from query analysis (excerpt)

5.2   CBR Engine Modelling

myCBR 3[4] provides the knowledge engineer with a variety of graphical user interfaces that allow for rapid prototyping of CBR knowledge models. We used *myCBR 3* Workbench to swiftly transfer our initial knowledge model into a structured CBR knowledge model. Figure 4 provide an insight in the modelling of the local similarity measure for timbre descriptors. The first figure shows the taxonomic modelling on the left and a section from the same similarity measure being modelled in a comparative symbolic table on the right.

The problem description consists of two attributes, *MainInputtimbre* and *Sub-Inputtimbre1* describing the current sound. Additionally the problem description contains the two Attributes *MainTargettimbre* and *SubTargettimbre1* that

---

[4] http://www.mycbr-project.net

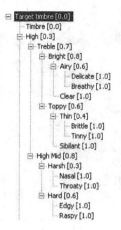

Symmetry ⦿ symmetric ○ asymmetric

	Muffled	High mid	Low	Sibilant
Muffled	1.0	0.0	0.8	0.6
High mid	0.0	1.0	0.1	0.3
Low	0.8	0.1	1.0	0.0
Sibilant	0.6	0.3	0.0	1.0
Timbre	0.0	0.0	0.0	0.0
Dark	0.4	0.6	0.1	0.8
Thin	0.5	0.0	0.2	0.6
Clear	0.0	0.0	0.7	0.1
Treble	0.1	0.1	0.2	0.0
Bright	0.0	0.0	0.5	0.6
Harsh	0.2	0.2	0.0	0.7
Honky	0.4	0.3	0.1	0.0
Hollow	0.0	0.1	0.4	0.6
Airy	0.5	0.1	0.0	0.2

Fig. 4. Timbre descriptor taxonomy

are used to specify the timbres into witch the sound should be changed. The attributes *TimbreAmountEffectDescriptor*, *AmountDirectionDescriptor* and optionally *TimbreAmountPercentage* are available for the Main as well as the Sub timbre. They are used to describe the amount of change that is intended to take place and its direction e.g. if the timbre should be increased or decreased. The last two attributes of our case structure are a String, *WorkflowDescription*, which holds a String that is describing the workflow necessary to achieve the desired timbre change described in the Problem description part. Additionally the Case holds the original natural language query on which it was modelled as a reference. Regarding our initial case base we were able to elicit 30 commonly encountered audio mixing tasks from our domain experts. Based on these 30 initial cases we conducted a series of retrieval tests together with our domain experts.

5.3 *Audio Advisor* at Work

In the following we will give a brief overview over the process of recommending a mixing workflow by *Audio Advisor* (cf. Figure 5). Upon starting the application it initialises a new instance of the Gate framework using the GATE embedded functionalities. Into this instance the customised ANNIE Application is loaded and initialised *Audio Advisor* then uses the *myCBR 3* API to initialise a new mycbr3 project. Loading the case base into the CBR project finishes the initialisation sequence of the *Audio Advisor* and the program is ready to be used.

Input to the *Audio Advisor* application can be provided by either manually selecting values from the drop down menus available in the problem description section of the *Audio Advisor* GUI, or by entering a natural language query into the 'Query Assemble' text field (see Figure 1). If a user decides to enter a natural language query the text is copied into the corpus of the ANNIE application and processed. The ANNIE application returns annotations of timbre

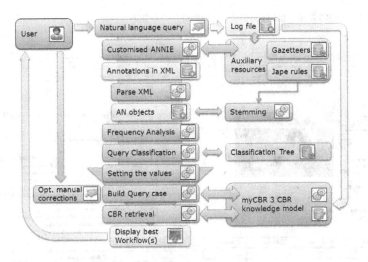

Fig. 5. Recommendation process diagram

descriptors, amount descriptors, timbreshifts, amount directions and constraints in an XML file, which is then read and unmarshalled by an instance of the QueryExtractor class. Within this class the annotations are analysed regarding the frequency of certain types of annotations, e.g. how many timbres are annotated or if there are annotations present annotating a constraint. Based on this analysis the classifier tree is searched for a query structure best matching the query characteristics (annotation frequencies: Number of timbres, number of amount descriptors, presence of constraint annotations) to identify the most likely structure of the natural language input query. Based on the best identified query structure the drop down menus of identified attributes are populated with the extracted values. The user can always adjust the values manually and/or add additional values. By clicking the 'Recommend workflow' button the user triggers the recommendation process.

The GUI of the *Audio Advisor* is quite straight forward (Figure 6). The upper part of the GUI provides all the elements necessary for a user to specify the audio mixing problem at hand. Additionally the user can select which amalgamation function should be used for retrieval. This allows retrieval of mixing tasks in the context of different genres. The query is then analysed and a workflow is recommended by a similarity based retrieval within the CBR Engine. The resulting order of best matching audio mixing workflows is then presented to the user in the lower section of the GUI.

6 Experiments and Evaluation

In this section we explain the aims and setup of the experiments we performed with *Audio Advisor*.

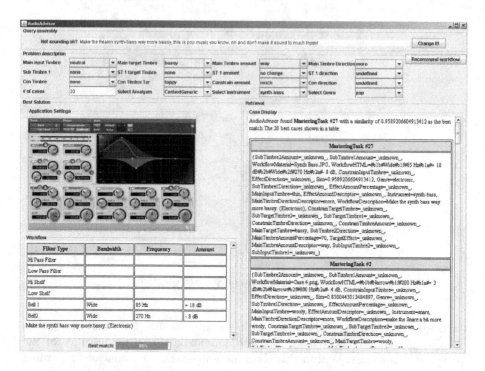

Fig. 6. *Audio Advisor* reading a natural language query containing a constraint

Our knowledge elicitation effort led us to the following knowledge artefacts:

o A set of 39 timbre descriptors with varying grades of abstraction
o A set of 21 amount descriptors
o A set of 15 Direction descriptors
o A set of 20 Effect descriptors with varying grades of abstraction
o Similarity of timbre descriptors in taxonomic and comparative table form
o Similarity of amount descriptors in taxonomic and comparative table form
o Similarity of amount descriptors as Integer function (distance function)
o Similarity of Context of Genre in taxonomic form
o Similarity of Context of Instrument in taxonomic form
o Similarity of the Effect descriptors in taxonomic form
o Global Similarities of the problem description of the mixing task depending
  on the selected Genre and Instrument context
o 30 Screenshots of application settings of the used mixing software
o 30 initial cases describing 30 common mixing tasks

On these artefacts and the knowledge model consequently modelled from them,
we performed experiments to establish the quality of the knowledge model. The
main goal of the experiments was to gain an insight into how good our approach
to formalise experience, from anecdotal into fully formal, worked with regard to
avoiding knowledge formalisation problems. We further aimed to evaluate the

performance/quality of our Software prototype working with our initial knowledge model. Our third goal was to establish the usability and applicability of our overall approach of workflow recommendation in day to day audio mixing work and teaching scenarios. Our fourth goal is it to establish if the use of taxonomies or the use of comparative symbolic tables yields more accurate similarity measures and thus better recommendations.

6.1   Setup of the Experiments

We performed two series of experiments. The first series aims at establishing the usability, quality of recommendation and performance of the *Audio Advisor* application in the day to day use of the software by experienced audio engineers. The second series of experiments aims at establishing the usability of the *Audio Advisor* application for teaching audio engineering students and gather feedback on the quality of the application's recommendations.

At the current time we have conducted experiments from the first series and are currently preparing experiments for the second series. The setup for the first experiment was the following: Two experienced audio engineers were asked to use the *Audio Advisor* software to enter natural language queries into it describing common audio mixing tasks. The engineers were to provide feedback on the usability of the recommended workflows. They were also asked to provide feedback on the correctness of the similarity ordering or sequence of the 5 best matching cases that were retrieved. The data gathering for this experiment was accomplished by logging the natural language queries the audio engineers entered into *Audio Advisor* as well as by providing the audio engineers with questionnaires to provide us with their feedback on *Audio Advisor*'s workflow recommendations. The questionnaire asked for the description of problems encountered with the retrieved workflow, for example, not being applicable for a certain instrument. Further the questionnaire asked for a rating of the quality, applicability of the recommended workflow ranging from 1 (worst) to 5 (best). The third information we gathered was the comparison of the case sequence retrieved to the case sequence deemed optimal by the audio engineers.

The second series of experiments will also use questionnaires to gather feedback from audio engineering students. Students will bring in their own work, consisting of sound samples and songs which they still need to optimise and use the *Audio Advisor* application to get recommendation on how to do so. They will then employ these recommendations on their work (sounds) and rate the actual outcome with regard of the extent the sound has changed as it was intended by the student. Additionally the students are asked for feedback on how fast, in terms of iterations of: Entering query, retrieve workflow, apply workflow in studio, they deem a learning effect to set in. This estimated learning effect will be verified by audio engineering lecturers in the form of a small practical test.

Both series are planned to be repeated with an improved knowledge model that will use symmetric symbol tables as similarity measures rather than the taxonomies used in the first place. This repeated series of experiments aims at providing us with data to compare the performance and accuracy of the two knowledge formalisation approaches we employed.

6.2 Evaluation

Here are first results from our first series of experiments. Each audio engineer was asked to enter 10 queries and provide us with feedback on the applicability of the recommended workflow and the sequence of the first 5 most similar cases retrieved by the *Audio Advisor*. To provide an idea of what kind of natural languages queries were entered by the engineers here is a short excerpt from the actual *Audio Advisor* log file: 'Can you make the drums more toppy?, 'Make the drums more toppy.', 'Change the bass to be more bassy but not toppy.', 'Make the flute more airy but not breathy.', 'The drums need to be way more heavy but not to boomy for a pop song.'

As stated before, the questionnaire we used, asked for the description of problems encountered with the retrieved workflow, for example, not being applicable for a certain instrument. Further the questionnaire asked for a rating of the quality, applicability of the recommended workflow ranging from 1 (worst) to 5 (best). As an informal kind of feedback both engineers reported that if the recommendation was above their rating of 2 it usually was quite useful and perfectly applicable. The third information we gathered was the comparison of the case sequence retrieved to the case sequence deemed optimal by the audio engineers. Table 1 lists the aggregated feedback from both audio engineers with regard to the similarity of the best retrieved case to their query in per cent and the applicability of the recommended workflow:

Table 1. Ratings of results

Rating by Audio Engineer 1	Best	Worst	Average
Match of query case to best retrieved case	95%	69%	79%
Applicability of workflow [1:worst to 5:best]	4	1	2.7

Rating by Audio Engineer 2	Best	Worst	Average
Match of query case to best retrieved case	98%	57%	77%
Applicability of workflow [1:worst to 5:best]	4.5	1	2.13

The second kind of data we gathered from our first set of experiments was the sequence of the five best cases, sorted in a descending order based on their similarity to the query posted. Additionally to his retrieved sequence of cases we asked the audio engineers to provide us with their ordering of the cases and respective mixing workflows with regard of their applicability to the query the engineer entered into the system. We did so to get an insight into the quality of the similarity measures with regard to their effect of "prioritising' the sequence of workflows to recommend in an accurate order. Accurate order meaning the first recommendation (case) being the most applicable and then have the "next best solution" and the 'next best' and so on in a sequence of decreasing applicability. Out of the 20 queries tested 5 were retrieving optimal sequences, the remaining 15 sequences are shown in the following table 2 displays the case sequences with the retrieved sequence in the top row and the engineers suggested optimal sequence in the lower row and starting with the best cases being on the left side of the table:

Table 2. Case retrieval sequence comparisons

	Engineer 1					Engineer 2				
Retrieved sequence	11	3	1	4	7	1	4	2	7	24
Optimal sequence	1	4	11	3	7	24	7	2	4	1
Retrieved sequence	2	7	27	3	1	0	17	12	15	21
Optimal sequence	7	2	27	3	1	17	0	12	21	15
Retrieved sequence	2	1	7	27	4	3	2	1	7	4
Optimal sequence	2	7	27	1	4	4	7	3	2	1
Retrieved sequence	11	26	6	25	13	3	2	1	7	4
Optimal sequence	26	13	11	25	6	2	3	7	1	4
Retrieved sequence	11	3	1	7	2	13	20	26	6	23
Optimal sequence	3	1	11	7	2	20	26	6	13	23
Retrieved sequence	2	1	7	4	11	12	15	17	0	21
Optimal sequence	2	7	1	4	11	17	15	12	0	21
Retrieved sequence	3	11	2	1	7	3	11	4	2	1
Optimal sequence	2	7	3	11	1	4	2	11	3	1
Retrieved sequence	/	/	/	/	/	9	23	13	26	14
Optimal sequence	/	/	/	/	/	9	13	23	26	14

Overall, next to the 25 % of optimal retrieved sequences, the remaining 75% retrieved sequence's orderings were labelled as of sufficient quality by the audio engineers and as a good basis for suggesting alternative workflows. The engineers reported still some flaws in detecting certain amount descriptors. Additionally sometimes the separation of Maintimbre and Subtimbre was not correctly extracted from the natural language query. Both engineers reported that the case structure and interface might still be reduced more to only Input timbre Target timbre an amount descriptor and a constraint on timbre. Overall extraction of the queries from the natural language input was rated as usable, except for the explicitly reported shortcomings which are due to our not yet refined Gazetteers and Jape rules we employ in our ANNIE IE application. The overall feedback from the engineers was quite enthusiastic as they reported to us that once we refine the knowledge model slightly further and made minor changes to the query extraction the *Audio Advisor* actually would be quite powerful in supporting audio mixing and the teaching of audio mixing.

7 Summary and Outlook

In this paper we presented our development of a case-based workflow recommendation system for audio engineering support. We detailed on the entire process of developing the *Audio Advisor* software. We described our approach to formalise the special vocabulary, consisting of vague descriptors for timbres, amounts and directions. We introduced CBR as a methodology to amend the problem of formalising emotions and/or adjectives describing timbres, i.e., the problem of the vagueness of terms and the variance of emotions invoked by the same sound in different humans. We further detailed on our approach to design a Case-based reasoning knowledge model based on the elicited knowledge artefacts. We then described how we designed and implemented the *Audio Advisor* application. While doing so we inspected the *GATE*based natural language processing ability that we integrated into *Audio Advisor* to enable it to process queries posted to it in natural language. We further detailed on the use of *myCBR 3* to rapidly prototype and refine the CBR knowledge model that poses the reasoning component of the *Audio Advisor*. We finished this paper with an overview of our experiments with the *Audio Advisor*and an introduction to a first evaluation of the performance of the *Audio Advisor* and the quality of its recommendations which overall are very promising in both possible roles of the *Audio Advisor* as a support tool for professionals as well as a teaching aid to students.

For the imminent future we plan to refine our knowledge model further based on the evaluated data from the first series of experiments. The next step is the conduction of the teaching related experiments and the further refinement of the knowledge model as well as the GUI employed by *Audio Advisor*.

As a medium future aim we want to investigate the possibility and usefulness of using 'negative similarity values' of the timbre descriptors provided by the domain experts during the knowledge elicitation phase. We therefore want to integrate a negative similarity measure into our knowledge model and perform experiments to establish as this might be useful to be employed as adaptation knowledge as we already suggested in our initial study of the formalisation of knowledge from the audio mixing domain.

References

1. Aamodt, A.: Modeling the knowledge contents of cbr systems. In: Proceedings of the Workshop Program at the Fourth International Conference on Case-Based Reasoning. Citeseer (2001)
2. Aamodt, A., Plaza, E.: Case-based reasoning: Foundational issues, methodological variations, and system approaches. AI Communications 1(7) (March 2007), ftp://ftp.ifi.ntnu.no/pub/Publikasjoner/vitenskaplige-artikler/aicom-94.pdf (letzte Verifikation Juni 11, 2007)
3. Arcos, J., Grachten, M., de Mántaras, R.: Extracting performers behaviors to annotate cases in a cbr system for musical tempo transformations. Case-Based Reasoning Research and Development, 1066–1066 (2003)

4. Arcos, J., De Mantaras, R., Serra, X.: Saxex: A case-based reasoning system for generating expressive musical performances*. Journal of New Music Research 27(3), 194–210 (1998)
5. Bergmann, R., et al.: On the use of taxonomies for representing case features and local similarity measures. In: Proceedings of the 6th German Workshop on Case-Based Reasoning, pp. 23–32 (1998)
6. Broekens, J., DeGroot, D.: Emotional agents need formal models of emotion. In: Proc. of the 16th Belgian-Dutch Conference on Artificial Intelligence, pp. 195–202 (2004)
7. Canamero, D., Arcos, J., de Mántaras, R.: Imitating human performances to automatically generate expressive jazz ballads. In: Proceedings of the AISB 1999 Symposium on Imitation in Animals and Artifacts, pp. 115–120. Citeseer (1999)
8. Darke, G.: Assessment of timbre using verbal attributes. In: Conference on Interdisciplinary Musicology, Montreal, Quebec (2005)
9. De Mantaras, R.: Towards artificial creativity: Examples of some applications of ai to music performance. 50 Anos de la Inteligencia Artificial, p. 43 (2007)
10. Donnadieu, S.: Mental representation of the timbre of complex sounds. In: Analysis, Synthesis, and Perception of Musical Sounds, pp. 272–319 (2007)
11. Fellous, J.: From human emotions to robot emotions. Architectures for Modeling Emotion: Cross-Disciplinary Foundations. American Association for Artificial Intelligence, 39–46 (2004)
12. Halpern, A., Zatorre, R., Bouffard, M., Johnson, J.: Behavioral and neural correlates of perceived and imagined musical timbre. Neuropsychologia 42(9), 1281–1292 (2004)
13. Hudlicka, E.: What are we modeling when we model emotion. In: Proceedings of the AAAI Spring Symposium–Emotion, Personality, and Social Behavior (2008)
14. Katz, B., Katz, R.: Mastering audio: the art and the science. Focal Press (2007)
15. de Mantaras, R.: Making music with ai: Some examples. In: Proceeding of the 2006 Conference on Rob Milne: A Tribute to a Pioneering AI Scientist, Entrepreneur and Mountaineer, pp. 90–100 (2006)
16. de Mantaras, R., Arcos, J.: Ai and music: From composition to expressive performance. AI Magazine 23(3), 43 (2002)
17. Pitt, M.: Evidence for a central representation of instrument timbre. Attention, Perception, & Psychophysics 57(1), 43–55 (1995)
18. Plaza, E., Arcos, J.-L.: Constructive adaptation. In: Craw, S., Preece, A.D. (eds.) ECCBR 2002. LNCS (LNAI), vol. 2416, pp. 306–320. Springer, Heidelberg (2002)
19. Sauer, C., Roth-Berghofer, T.: Web community knowledge extraction for mycbr 3. In: Research and Development in Intelligent Systems XXVIII: Incorporating Applications and Innovations in Intelligent Systems XIX Proceedings of AI-2011, the Thirty-first SGAI International Conference on Innovative Techniques and Applications of Artificial Intelligence, p. 239. Springer (2011)
20. Sauer, C., Roth-Berghofer, T., Auricchio, N., Proctor, S.: Similarity knowledge formalisation for audio engineering. In: Petridis, M. (ed.) Proceedings of the 17th UK Workshop pn Case-Based Reasoning, pp. 3–14. University of Brighton (2012)
21. Typke, R., Wiering, F., Veltkamp, R.: A survey of music information retrieval systems (2005)
22. Watson, I.: Case-based reasoning is a methodology not a technology. Knowledge-Based Systems 12(5), 303–308 (1999)

An Agent Based Framework for Multiple, Heterogeneous Case Based Reasoning

Elena Irena Teodorescu[1] and Miltos Petridis[2]

[1] School of Computing and Mathematical Sciences, University of Greenwich, London, UK
e.i.teodorescu@greenwich.ac.uk
[2] School of Computing, Engineering and Mathematics, University of Brighton, Brighton, UK
m.petridis@brighton.ac.uk

Abstract. This paper investigates the application of Multiple, Heterogeneous Case Based Reasoning (MHCBR) using agents operating on different structures/views of the problem domain in a transparent and autonomous way to retrieve solutions for a new problem from more than one case-base. An MHCBR framework is proposed. This framework includes sub-processes for subscribing of provider case-bases through agents, creating a dynamic structure, and retrieving solutions by using agents and employing a Blackboard communication architecture.

A mechanism based on the competence of a provider case base is introduced to improve MHCBR performance. A negotiation system to support the retrieval process from each source case base of the MHCBR is proposed. ProMHCBR, a MHCBR system employing agents is discussed and an experimental evaluation of this system is presented.

Keywords: Heterogeneous Case Bases, Multiple Case based Reasoning, dynamic CBR structure, Multiple CBR framework, Blackboard Systems, Intelligent agents.

1 Introduction

One of the main problems in modern organisations is that often their data are encapsulated by cases contained in multiple case bases. This reflects the fragmented way in which organisations capture and organise knowledge.

Methods for managing sharing of standardized case bases have been studied in research on distributed CBR (e.g. [4]), as have methods for facilitating large-scale case distribution [1]. Leake and Sooriamuthhi propose a new strategy for MCBR - an agent selectively supplements its own case-base as needed, by dispatching problems to external case-bases with the same representation and using cross-case-base adaptation to adjust their solutions for inter-case-base differences [2], [3], [4]. Ontanon and Plaza [6] looked at a way to "improve the overall performance of the multiple case systems and of the individual CBR agents without compromising the agent's autonomy". They present a framework for collaboration among agents that use CBR and strategies for case bartering [6] - case trading by CBR agents. Nevertheless, they do not consider the possibility of cases having different structures and what impact this will have on applying CBR to heterogeneous case bases.

S.J. Delany and S. Ontañón (Eds.): ICCBR 2013, LNAI 7969, pp. 314–328, 2013.

However, the above approaches bring with them the following challenges:

1. Moving cases into a central case base potentially separates the knowledge from its context and makes maintenance more difficult.
2. Various case bases can use different semantics. There is therefore a need to maintain various ontologies and mappings across the case bases.
3. The knowledge content "value" of individual cases can be related to its origination, or "provenance". This can be lost when merging into a central case base.

Keeping the cases distributed in the form of a Multiple, Heterogeneous Case Based Reasoning system (MHCBR) may have a number of advantages such as increased maintainability and competence and the contextualisation of the cases. Leake states that "an important issue beyond the scope of [their research] is how to establish correspondences between case representations, if the representations used by different case-bases differ." [3]. Sooriamurthi states that if local and external case-bases have different structures and use dissimilar representations, "conversion from one to another [structure] may significantly increase the burden for MCBR" [9].

Given several case bases as the search domain, it is very likely that they have different structures and vocabulary. Ideally, accessing multiple case bases can improve performance by improving competence and coverage. Even if they hold knowledge about the same problem domain, it is very likely that different case bases would hold more varied solutions and also some specialisation would occur. Case bases kept in their initial different structures allow preserving the natural distribution of knowledge present in modern day systems. Autonomy, security and privacy issues are more likely to be solved if systems can provide useful knowledge without changing the provider's structure or accesing more knowledge than needed.

This paper investigates the idea of Multiple, Heterogeneous Case Based Reasoning (MHCBR) and identifies issues that can arise when a CBR system interoperate across multiple case bases with different vocabularies and structures. It presents a MHCBR framework with its sub-processes of subscribing a provider case-bases through agents, creating a dynamic structure, and retrieving solutions by using agents and a blackboard communication architecture.

Furthermore, it introduces a mechanism for improving the MHCBR approach by using a measure for the competence of a provider case base. The competence is to be determined based on the local confidence of a source case base and the trust of the MHCBR system in it. The retrieval process from each source case base of the MHCBR is supported by a negotiation system between a dynamic case-based system retrieval engine and agents associated to heterogeneous provider case bases.

ProMHCBR, a MHCBR system employing agents is proposed and an experimental assessment of this system is presented. The system is developed to automate the MHCBR process which includes: the subscription of a knowledge provider by an agent, maintenance of the Dynamic Structure, the agents' retrieval engine, creation of the Dynamic Case Base and its management by the blackboard component.

The results of the experiments are based on a comparison of the performance of MHCBR against classical CBR systems and a centralised CBR system. These results show that, by changing the number of retrieved cases required from each agent according to their competence, the ProMHCBR system performance improves in time and is comparable to the performance of the centralised CBR system. The experiments have also proven that the ProMHCBR system is capable to learn if case bases are specialised in a particular value of a case attribute. Further experiments show the ProMCHBR system's capability to predict the value of a missing attribute for a target case by looking at sources' knowledge and the results demonstrated the importance of taking the provenance of cases into consideration.

Section 2 discusses the MHCBR framework, architecture and process and the similarity measures used, Section 3 presents the evaluation and experiments conducted as part of this research and section 4 provides conclusions and illustrates areas for further work.

2   Multiple, Heterogeneous Case Based Reasoning Framework

2.1   MHCBR Employing Intelligent Agents and Blackboard Communication Architecture

Blackboards have been used effectively in the past for the construction of hybrid and agent based AI systems [8]. In this research it is proposed that the MHCBR framework employs Intelligent Agents and a Blackboard Communication Architecture.

The MHCBR framework allows the adaptive MHCBR process to take place. This process can be categorised into three main sub-processes: (a) the process of subscribing and managing the heterogeneous structure of a provider knowledge base; (b) the process of retrieval of cases which includes case selection at local provider level and case merging at system level; (c) system level revision of retrieved cases.

The MHCBR framework comprises a Dynamic Case Base System and Intelligent Agents used to register external, heterogeneous case bases to the system (Figure 1).

The MHCBR Dynamic CB System

The Dynamic CB system includes a dynamic case-base structure publisher component and a system level retrieval engine (dynamic smart search engine).

The dynamic case-base structure publisher is in charge of updating and managing the dynamic data structure and makes it available to external agents (for example through Web Services). The dynamic structure makes the self-adaptive multi case base reasoning system possible. By adding a new case base to the existing ones, new attributes are added to a global dynamic structure and new relations linked to these attributes are established by the agents. The dynamic structure is managed by the dynamic CB system which would maintain a data dictionary required to keep all the metadata for the dynamic structure, storing the type and any default value for every single attribute as well as relationships between the Dynamic Case Base Structure attributes themselves. These relationships can be mathematical relationships or look-up tables.

The dynamic "smart" search engine performs solution merging and decides on the final set of solutions in terms of the percentage of selected cases proposed by each local retrieval component. It is based on a blackboard architecture and it contains a blackboard manager component which is in charge of calculating the agents' competence. The blackboard contains the target and retrieved cases from various agents together with similarity calculations and rankings, and also a log of the solution process and the reconciliation strategy followed, thus representing the state of the overall CBR solution process at any point in this process.

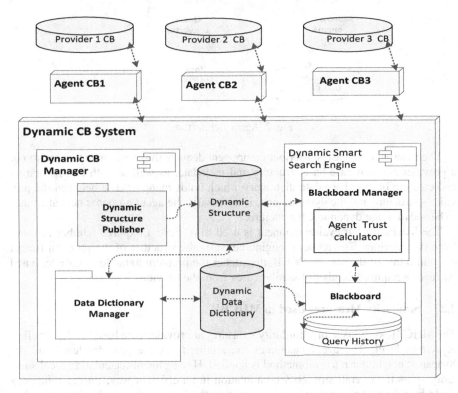

Fig. 1. A general architecture of the MHCBR System employing intelligent agents

Given a new target case, the blackboard manager decides on a strategy for finding similar cases from the CB providers. Based on the agents' confidence and the trust the dynamic CB system has in every agent, the blackboard manager decides the number of cases to retrieve from each provider, as well as other requirements, such as the requirement for diversity, etc. The system then initialises the agents and assigns them a mission. It is their responsibility to translate the problem into the provider's structure and retrieve the best matches. On return, the results (cases) are mapped by the agent to the dynamic data dictionary and sent to the blackboard.

A "global" CBR process is used to decide on the retrieved cases. The system then selects and presents the shortlisted cases after the reconciliation process and provides these to the user, together with links to their original forms for the user to explore and elicit further contextual and explanation knowledge. Finally, the system "reflects" on the process by updating the query history and trust weights for each provider.

HMCBR Intelligent Agents

An Intelligent Agent (Figure 2) has two main components, one in charge with the subscription of a provider CB, and one in charge with the retrieval of cases.

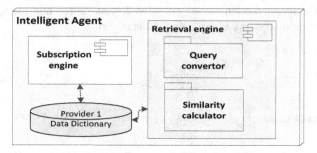

Fig. 2. Agent architecture

The external case base subscriber component deals with the automatic extraction of a provider external case-base structure and metadata, as well as with its registration and creation of case-base data dictionary which holds mapping data between the provider's structure and the dynamic one. If necessary, the agent proposes new attributes to be added to the dynamic CB structure.

The local case retrieval component is used to retrieve a required number of cases from the provider. The retrieval engine converts the structure of the problem (target case) into that of the provider's CB, dispatches the problem to the local case base and retrieves a required number of similar cases from the provider.

2.2   Similarity Measures Used in MHCBR

The MHCBR systems apply similarity measures at provider case base level, as well as at system level. The agent calculates the similarity at the provider level and the K-nearest neighbour retrieval method is applied. Having agents calculating local similarity splits the overall similarity computation in an efficient way; further efficiency could be achieved by applying maintenance methods on each provider case base. For numerical values the agent calculates the distance between the values of a target case attribute and the one of a source case. For some non-numerical values the agents use look-up tables. These look-up tables are part of the metadata of the dynamic structure which is managed by the blackboard manager.

A provider CB structure is always a subset of the dynamic structure and when a query is dispatched to the agents, some of the attributes of the answers will return "null". The absence of a value can also be seen to have meaning. As an illustration, if an attribute of a case is Boolean ("true" means similarity of 1 and "false" means similarity of 0), the absence of a value for this attribute value can itself provide a similarity (an example is provided by Table 1).

Table 1. lookup table to calculate similarity for attributes with Boolean values

	False	True	Null
Garden (of type Boolean)	0	1	0.5

Other ways of mapping values is to use external data, for example using exchange rates from web providers to convert the price of a property into the same currency.

The agent is in charge with the calculation of the similarity at the provider level. To allow for defining locally optimised similarity metrics for different providers, the following similarity measure has been defined:

$$\sigma_{CBy}(C_T, C_s) = \frac{1}{\sum_x \omega_{CBy}(x)} \sum_x \omega_{CBy}(x) * \sigma_{CBy}(C_T, C_s, x) \tag{1}$$

C_T : The target case, C_S : The source case

$\sigma_{CBy}(C_T, C_s)$: The similarity measure for provider CBy.

$\omega_{CBy}(x)$: The weighting from case base provider CBy for attribute x

$\sigma_{CBy}(C_T, C_s, x)$: The local similarity measure for provider CBy for attribute x.

The attributes' similarities are normalised for the calculation of a case similarity.

Using the same similarity metrics for all agents allows for a better aggregation of retrieved knowledge.

2.3   MHCBR Agents Confidence and Trust

As MHCBR is using more than one heterogeneous source, the question of source competence arises: which source is better than another? To answer this question, the idea of agent competence was considered.

Recent research by Manzano et al. explores the idea of agent trust and it proposes a model for the reuse of cases from a base which is divided into two stages: individual reuse and multiagent reuse. If at an individual level the agents produce internally full solution, the multiagent reuse would involve a deliberation process between agents [5] so an agreed final solution, referred to by the term "amalgam" [7], is produced. This paper proposes to preserve the local case similarity at provider level, but the agent is instructed to provide to the blackboard a number of solutions proportional to the trust that the Dynamic CB system has in that particular agent. Two distinct metrics are applied to calculate competence, one for the agent's confidence and one for the system's trust in the agent.

Calculating trust in agents is a long process, as CBR Systems are lazy learner systems and the disadvantage is that trust takes time to build. When a new agent subscribes a provider case base to the MHCBR system, the agent is given an initial default trust value.

In order to obtain the desired cases from the agents, the blackboard manager performs a two-phase retrieval process, which includes a negotiation stage between the dynamic CBR system and the agent.

Given a new problem, the MHCBR system can ask the agents how confident they are in providing good solutions, rather than looking only at the history of agent performance. The agent itself can calculate its confidence for a solving a particular problem and supplies the blackboard with it. The system starts by asking the agent for its confidence in its best required number of cases. Then the system calculates its trust in each agent and decides how many cases it will require for the Dynamic CB from each.

Regarding similarity values, an agent knows only the similarities to the cases from its own provider case base, so it can calculate the confidence based only on its own knowledge: it was decided to use the mean of the highest similarities of the required number of cases (top mean) and the mean of similarities for all cases in the case base (overall mean). The agent confidence is calculated by subtracting the overall mean from the top mean. Equation 2 shows the calculation of an agent's confidence:

$$Conf_{Ai} = 100 - \left(\frac{1}{k} \left(\sum_{i=1}^{k} \sigma(C_T, C_{Si}) \right) - \frac{1}{n} \left(\sum_{j=1}^{n} \sigma(C_T, C_{Sj}) \right) \right) \tag{2}$$

Where:

$Conf\ Ai$: The confidence of agent Ai in k best matches from its provider

n: the number of cases in the case-base

$\sigma(C_T, C_s)$: Overall similarity between C_T and C_S

k: required number of cases most similar to the target   This is a variable number and it is calculated based on the previous trust.

The number of results k from an agent Ai necessary to calculate confidence is calculated by the blackboard and provided to the agent:

$$k_{Ai} = \frac{trust_{Ai} * n}{\sum_{i=1}^{n} trust_{Ai}} \tag{3}$$

Where:

n : The total number of cases required from all agents

$trust_{Ai}$: The system's trust in an agent which was calculated in the previous search

Once an agent provided the confidence, the blackboard manager asks for a particular number of results that an agent will send to the system.

Equation 4 illustrates how the confidence provided by the agent is applied to calculate the new number of solutions an agent is asked to put forward to the blackboard:

$$nrRez_{Ai} = \frac{n * conf_{Ai} * k}{\sum_{i=1}^{n}(conf_{Ai} * k)} \tag{4}$$

Where:

$nrRez_{Ai}$: the number of results that an agent will send to the Dynamic CB System.

$conf\ Ai$: The confidence of agent Ai in k best matches from its provider

n: the number of cases in the case-base ; k: required number of cases most similar to the target

The formula includes the normalisation of the number of results retrieved by an agent, as all agents should send a total of n source cases to the Dynamic CB.

Alternative confidence measures could be introduced to maximize the local agent confidence. Because an agent has only visibly of the local data the confidence can be less accurate. This issue is the cause of introducing the idea of the trust of the dynamic CB system in each agent. The agent trust is calculated by the blackboard manager belonging to the dynamic CB system and it takes into consideration the history of agent performance. Each time when a new query is submitted to the system, the blackboard manager calculates the distance between the overall average of similarity for all agents and the average similarity of the cases returned by each agent. To calculate the new trust for a particular agent the blackboard adds this distance to the old trust.

$$Trust_{Ai}' = Trust_{Ai} + \frac{\left(\sigma_{Ai} - \frac{1}{n}(\sum_{i=1}^{n} \sigma_{Ai})\right)}{\sum_{i=1}^{n} \sigma_{Ai}} * c \tag{5}$$

Where :

    $Trust_{Ai}'$: The trust in agent Ai for the new target case

    n: number of agents

    c: a scaling coefficient used to scale trust values.

    σ_{Ai} : the average similarity of the k nearest neighbour source cases for a particular agent Ai and is calculated as :

$$\sigma_{Ai} = \frac{1}{k}\left(\sum_{i=1}^{k} \sigma(C_T, C_{Si}, Ai)\right) \tag{6}$$

$\sigma(C_T, C_{Si}, Ai)$: average similarity for a particular case of agent Ai

A set of extensive experiments was run for each approach to see how confidence and trust influences in time the overall similarity.

3   Experiments and Evaluation

The case study used for this research is one of searching for a property from many estate agencies without amalgamating their case base structures. Three heterogeneous estate agencies with different case base structures and semantics are used as provider case bases. A detailed representation and of the providers case base structures can be found in the PhD thesis Multiple, Heterogeneous Case base reasoning

3.1   The ProMHCBR System

The purpose of the ProMHCBR system is to allow users to define their target properties and the system is to provide the most similar cases from all provider case bases. A second purpose of the system is to predict a property's price based on the knowledge from the multiple case bases.

The sample data for the case bases was provided by three online estate agents selling holiday properties in Europe. The provider case bases contain 786, 635 and 300 sample cases respectively. Each case base contains cases that describe properties in terms of various attributes. The experiments used repeated random sub-sampling validation.

For comparison reasons, a second CBR system was created, as well, based on a centralised case base that contains all 1721 cases. The centralised CB has the structure of the dynamic CB.

Tests were run on both systems, the centralised CBR one and the MHCBR one.

Each run of the MHCBR system provides a set of data, according to the run setting. This data is kept in the search history by the blackboard. The following is a list of data that can be returned by a run:

- The dynamic case base (cases sent by all three agents);
- The average similarity: the average similarity given by the dynamic case base;
- The agent id: kept to identify the provenance of source cases;
- The agent similarity: the average similarity for all the cases sent by an agent;
- Agent confidence: the confidence of an agent for a required number of cases.
- Agent trust: the trust of an agent for a required number of cases for a particular run. The blackboard keeps two values, the initial trust and trust after a run;
- Number of results: the number of source cases required from an agent;
- The search type (if required): for testing purposes two search types were used, searching for "house" type properties or for "apartment" type ones. It was considered that any property from the source case bases falls into a category or another (according to the similarity values of the attribute). When the search type is set to "house", for example, rules were implemented so the set of targets automatically keeps only properties of this type or very similar to check which agents provides better matches for this category and therefore specialises in this type of properties;
- Predicted price (if required): for testing purposes the initial price of the target case which is removed from each run is also stored.

3.2   First MHCBR Approach: Treating All Agents the Same

This experiment was designed to test the basic form of MHCBR architecture and to assess how it performs when each of the agents requests a fixed number of cases from their providers and returns them to the dynamic case base. The agents calculate the local similarities to the target case at the provider CB level using weights. The basic MHCBR system uses all three agents, but no confidence or trust is applied. Each agent provides its 10 most similar cases.

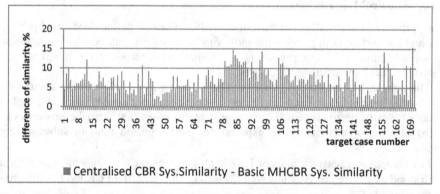

Fig. 3. The difference between the average similarities given by the centralised CBR system and the average similarities given by the MHCBR system for all target cases

The overall average of the differences between the two similarities given by the centralised CBR and the basic MHCBR was only 6.9%. This means that overall the basic MHCBR system performance of 66.01% in finding similar cases to a target is 9.45 % lower than the centralised CBR system of 72.91%.

3.3   Introducing Agent Confidence

Further experiments were conducted to study the effect of introducing agent confidence for solving a problem. Initially the agents are asked to retrieve 10 cases each, but in time, the number of retrieved cases changes according to the confidence. For this test, the system doesn't compare the performance between agents nor does it take into consideration previous agent performance. Instead, the system trusts the agent's confidence for each run.

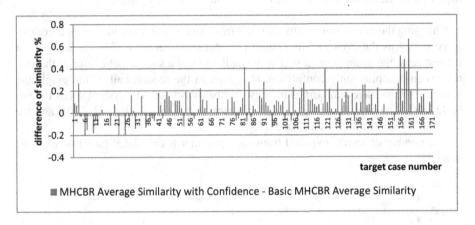

Fig. 4. The difference between the average similarities calculated by the system when applies agent confidence and when it doesn't

Overall, the average similarities improve by 0.16 %, which suggests that applying just confidence brings very little improvement to the values of similarities.

3.4   Applying Agent Trust

In this experiment, the system evaluates the performance of an agent against that of the others.

The system's trust in an agent is calculated based on previous agent performance in comparison to the overall system performance. The following chart shows how the trust in Agent 1 and Agent 2 increases in time in comparison to the starting value, with the performance of Agent 1 being most trustful overall.

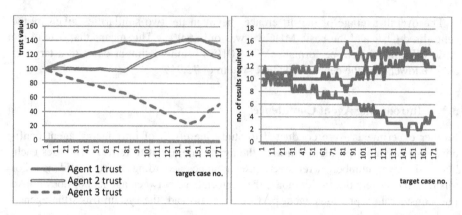

Fig. 5. (1) The values of trust in each agent as they change in time (2) Number of results required from each agent calculated based on the system's trust in agents

Changing the number of results required from each agent according to their competence improves the ProMHCBR system performance over time. For the first part of the test runs, the improvement was very small and had a similar behaviour to the system when it applies only confidence. However, in the second half of the graph, the values of the average similarities increase, probably because the system needs time to build trust. The overall average of the increase of the similarities is 0.42 % and the increase for the last 30 runs has an average of 1.6%.

The number of results required from each provider is calculated based on the trust in agents.

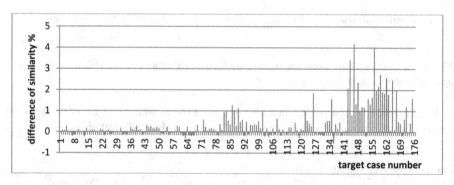

Fig. 6. The difference between the average similarities calculated by ProMHCBR when applies trusts in agents vs when it doesn't

3.5 Finding Out If Agents Are Specialised

For this experiment two different specialised searches were set up, one searching for a property of type house, the other searching for an apartment. The experiment was designed to test if the ProMHCBR system has the capability to train the blackboard manager to decide which agents are specialised in a type of property.

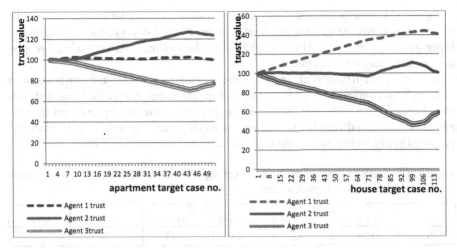

Fig. 7. Evolution of agents trusts when the test set contains only properties of type: (1) apartment and (2) House

Although Agent 3 performs poorly in both types of searches in comparison with the other agents, the results show that it has better matches for apartments than houses. Overall the findings of this experiments show that Agent 2 is specialised in apartments and Agent 1 has a better performance when the system searches only for properties of type "house".

3.6   Predicting a Property Price

For this experiment the attribute "price" was removed, so the case similarity is calculated based on all the other attributes but price. The similarity was therefore used to predict the missing attribute of price. Three methods to calculate the estimated prices were used:(1) use default attribute weights and calculate the final estimated price as the average of the agents' estimated prices; (2) increase the weights of the locality attributes of a property and calculate the final estimated price in a similar way as in Method 1; (3) investigate provenance of cases and decide whether to use the predicted price of only one agent.

In method (1), for each test target case the predicted price was compared against the real price which was removed before running the ProMHCBR process. The difference between the real price and the predicted price was calculated and normalised in a percentage form. The following table shows the spread of the results:

Table 2. Spread of predicted prices according to their distance from the real ones

	0 -20%	20- 30%	30-50%	50- 100%	>100%
Number of cases	32	20	23	35	61

Out of 171 price predictions, only 75 prices were within a 50% distance from the real price and only 30% of the prices were in the desired range (up to 30% higher or lower than the real price). This is explained by the fact that two very similar properties can have very different prices.

In method (2) it was decided to set the locality weights for the attributes such "country", "region" and "city" to 100%. All the other weights for the local attributes were kept as before. The results (see Table 3) demonstrate an improvement in comparison to the initial run. By increasing the locality attribute weight, the number of predicted prices within 30% of the real price distance increased by 8.8%.

Table 3. The spread of the predicted prices according to their distance from the real ones when the locality attributes have highest weight

	0 -20%	20- 30%	30-50%	50- 100%	>100%
Number of cases	32	25	22	37	55

Although the second method of the experiment has given better results than the first one, overall only 33% of the predicted prices are in the desired range, which is up to 30% higher or lower than the real price.

By studying the provider case bases data, it was discovered that the average of the real property prices can vary greatly from country to country. For example, 17% of the properties of the Provider 3 case base are located in United Arab Emirates and the average price is 1,501,287 € . By comparison, the Provider 1 country with the highest number of properties is Italy (16.3 %), for which the average price is 352,886 €.

When a CBR system deals with external source cases and information of how a case base was captured, provenance can provide important knowledge regarding a case's applicability[10]. The ProMHCBR system has the capability to find out which agent supplied a particular solution, and therefore its provider. Method (3) test set of target cases was formed by taking 10 % of the cases out of each provider's case base, so it contained three subsets of cases, one from each provider. For each target case, every agent was supplying a predicted price to the blackboard manager. Those prices were placed on the blackboard for further calculation and stored in the query history. As a new testing method for the estimated prices, it was decided to "believe" the price prediction of the agent in charge of the provider's case base where the target case originally came from.

The new results (Table 4) demonstrate the importance of taking into consideration the provenance of cases and deciding to believe the price estimate of a particular agent in charge of a case base more relevant to a target case.

Table 4. The spread of the predicted prices according to their distance from the real ones when the system takes in consideration provenance

	0 -20%	20- 30%	30-50%	50- 100%	>100%
Number of cases	55	34	21	37	24

The number of predicted prices within 30% of the real price distance increased to 89, which is 52 % of all target cases and 18.7% more than calculating the estimated price based on all agents.

4   Conclusions and Future Work

This paper presents an approach based on agents operating on different structures/views of the problem domain in a transparent and autonomous way. Agents were dynamically added to the system, one agent per case base, thus increasing the search domain and potentially the competence and vocabulary of the system.

The proposed architecture for a self-adaptive MHCBR system involves the use of a dynamic CB system based on a blackboard architecture. This architecture contains the dynamic CB system and intelligent agents that communicate with the blackboard manager. The retrieval process includes a negotiation phase between the dynamic CB system and agents.

The ProMHCBR system was successfully implemented based on the presented architecture.

A centralised CBR system was built for comparison purposes. The structure of its centralised case base was identical to the dynamic structure after the three case bases had been added to the MHCBR system. Because the centralised CB contains all possible cases and it has the structure identical to the dynamic CB, it retrieves the overall best possible matches for a target case. Therefore this system's performance was expected to be better than that of the ProMHCBR. The difference of performance was only 6.9% in comparison to the basic ProMHCBR (competence not taken in consideration). The difference decreased to 5.3 % (over 140 runs) when ProMHCBR continuously adjusted the number of retrieved cases from each case base according to the trust in each agent. When merging case bases there is generally a loss of individual case content value. For example the provenance of a case might be lost, as well as some of the attributes that characterise a particular case base but not necessarily other case bases.

The results showed that agents can specialize in different types of properties (houses or apartments). The conclusion was drawn by observing the trust of the system change according to the type of search.

ProMHCBR was used to predict the value of a property price according to the prices of similar properties form the provider case bases. Initial results were somewhat disappointing with only 30% of the prices being in the desired range. The importance of the attributes that show the locality of a property were increased and it was decided to believe the price prediction of the agent in charge of the provider's case base where the target case originated from. The new results demonstrated the importance of taking into consideration the provenance of cases, with the number of predicted prices in the desired range increasing to 52%.

In the proposed approach, the trust at agent level is determined by the quality of the knowledge from the provider case base it deals with. The research in this area could be taken further by looking into the capability of an agent to search for and discover knowledge elsewhere. It would be interesting to see how an MHCBR system would behave if it was to employ agents with different capabilities. In this situation the calculation of the agent trust should take into account whether an agent is more "hard working" than another.

Another area which could be further investigated is price estimation. In the performed experiments it was decided to believe the price prediction of the agent in charge with the provider's case base where the target case originated from. When the provenance of a target case is not clear, the following question arises: which agent price estimation is best? Further research could identify a way to train the system to believe the estimated price of the agent based on which agent has most properties in the same locality as the target property or a locality in the same price band. Ways of finding price bands and relationships between property location and price could be found by forming clusters of properties according to their prices.

References

1. Hayes, C., Doyle, M., Cunningham, P.: Distributed CBR Using XML. In: Proceedings of the Fourth European Conference on Case-Based (ECCBR), Dublin, Ireland (June 1998)
2. Leake, D.B., Sooriamurthi, R.: When Two Case Bases are Better Than One: Exploiting Multiple Case Bases. In: Aha, D.W., Watson, I. (eds.) ICCBR 2001. LNCS (LNAI), vol. 2080, pp. 321–335. Springer, Heidelberg (2001)
3. Leake, D., Sooriamurthi, R.: Managing Multiple Case Bases: Dimensions and Issues. In: Proceedings of the Fifteenth Florida Artificial Intelligence Research Society (FLAIRS) Conference, Pensacola, Florida (May 2002)
4. Leake, D., Sooriamurthi, R.: Case Dispatching versus Case-Base Merging: When MCBR matters: International Journal on Artificial Intelligence Tools: Architectures, Languages and Algorithms (IJAIT). Special Issue on Recent Advances in Techniques for Intelligent Systems 13(1) (2004)
5. Manzano, S., Ontañón, S., Plaza, E.: Amalgam-Based reuse for multiagent case-based reasoning. In: Ram, A., Wiratunga, N. (eds.) ICCBR 2011. LNCS, vol. 6880, pp. 122–136. Springer, Heidelberg (2011)
6. Ontañón, S., Plaza, E.: A bartering approach to improve multiagent learning. In: AAMAS 2002 (2002)
7. Ontañón, S., Plaza, E.: A formal approach for combining multiple case solutions. In: Bichindaritz, I., Montani, S. (eds.) ICCBR 2010. LNCS, vol. 6176, pp. 257–271. Springer, Heidelberg (2010)
8. Petridis, M., Knight, B.: A blackboard architecture for a hybrid CBR system for scientific software. In: Proceedings of the Workshop Program at the 4th International Conference in Case-based Reasoning, ICCBR 2001, Vancouver (2001)
9. Sooriamurthi, R.: Multi-case-base reasoning. PHD thesis, Department of Computer Science of Indiana University (2007)
10. Leake, D.: Provenance and Case-Based Reasoning. In: Proceedings of the Twenty-First International FLAIRS Conference American Association for Artificial Intelligence 2008 (2008)
11. Teodorescu, E.I.: Multiple, Heterogeneous Case Based Reasoning. PhD Thesis, Department of Computing and Information Systems, University of Greenwich (2012)

Learning-Based Adaptation for Personalized Mobility Assistance

Cristina Urdiales, Jose Manuel Peula, Manuel Fernández-Carmona,
and Francisco Sandoval

ISIS group, ETSI Telecommunications, University of Malaga, 29071 Malaga, Spain
acurdiales@uma.es

Abstract. Mobility assistance is of key importance for people with disabilities to remain autonomous in their preferred environments. In severe cases, assistance can be provided by robotized wheelchairs that can perform complex maneuvers and/or correct the user's commands. User's acceptance is of key importance, as some users do not like their commands to be modified. This work presents a solution to improve acceptance. It consists of making the robot learn how the user drives so corrections will not be so noticeable to the user. Case Based Reasoning (CBR) is used to acquire a user's driving model reactive level. Experiments with volunteers at Fondazione Santa Lucia (FSL) have proven that, indeed, this customized approach at assistance increases acceptance by the user.

1 Introduction

Mobility is of key importance for persons to carry their Activities of Daily Living (ADL). People affected by a disability may require mobility assistance to remain autonomous. Lack of human resources has led to research in mobility assistive devices, like a robotic power wheelchair. These wheelchairs are not a traditional robot in the sense that it is controlled, at least partially, by its user. This approach is known as shared control. Furthermore, doctors and caregivers have reported that excessive assistance may lead to loss of residual skills, whereas an active profile is reported to improve rehabilitation. Hence, in these cases it is desirable to give the user as much control as possible.

There are different approaches to shared control. In *safeguarded navigation*, for example, robots are always under human control, except when a potentially dangerous situation is detected. In these cases, the robot takes over [1][2] and a reactive algorithm is used to avoid such a danger. Other shared control approaches[5][6] rely on a basic set of primitives like *AvoidObstacle*, *FollowWall* and *PassDoorway* to assist the user in difficult maneuvers, either by manual selection or automatic triggering. In extreme, the user just points a destination and the robot does the rest [3]. In order to avoid sharp control switches from human to robot and to prevent loss of residual skills by not letting the user participate at all in complex tasks a third approach to shared control is collaborative control [7][8], where user and robot commands are mixed in a continuous way so that people may contribute their best to any situation.

S.J. Delany and S. Ontañón (Eds.): ICCBR 2013, LNAI 7969, pp. 329–342, 2013.

The authors proposed a collaborative control method in [7] consisting on weighting robot's and user's commands by their respective driving efficiencies at each situation and adding them at reactive level. Thus, the most efficient agent was awarded with more control, yet the least efficient one always contributed to emergent motion. This approach was tested at Fondazione Santa Lucia (FSL), Rome, by a large number of volunteers presenting different degrees of functional disability. Surprisingly, we found out that people with a better functional disability profile actually performed worse than people with worse diagnosis in a significant number of cases. Further analysis proved that this group actually rejected assistance, i.e. tried to counteract robot commands, as soon as they acknowledged that the wheelchair was not doing exactly what they commanded.

To solve this problem, we propose to use a CBR based approach to let the robot learn how the user drives. Rather than choosing the most efficient command, the robot will try to provide the most similar one to what the person would do at each situation within established safety constraints. Since the robot's commands become much more similar to the user's, acceptance is improved and global efficiency grows, as proven by further experiments with volunteers at FSL. To achieve this, the wheelchair learns how a given user drives using CBR.

2   Collaborative Navigation System

Our basic approach to collaborative navigation – fully explained in [7] – is based on reactive navigation. Reactive schemes implicitly deal with several sensors and goals at a time, so we can simply handle user and robot commands as two different goals. Let $\vec{v_U}$ and $\vec{v_R}$ be the user and robot command vectors respectively. $\vec{v_U}$ is extracted from a joystick and $\vec{v_R}$ is calculated via the simplest pure Potential Fields Approach (PFA) [9], where goals and obstacles are modelled as attractors and repulsors, respectively. We can combine $\vec{v_U}$ and $\vec{v_R}$ linearly into a collaborative command $\vec{v_C}$. However, we do not want user and robot to have the same weight in the emerging decision. Instead, we want assistance to adapt to the user's needs. Hence, we weight $\vec{v_U}$ and $\vec{v_R}$ by the local efficiency of user and robot, respectively: the more efficient a user is at solving a given situation, the less assistance he/she receives.

$$\vec{v}_C = \eta_R \cdot \vec{v}_R + \eta_U \cdot \vec{v}_U \tag{1}$$

Motion efficiency η needs to be calculated locally, because in a purely reactive (i.e. memoryless) approach, global factors like trajectory length or completion time cannot be used. We have identified three local factors, ranging from 0 to 1, with an impact on η: *smoothness* (η_{sm} in Eq. 2), *directness* (η_{dir} in Eq. 3) and *safety* (η_{sf} in Eq. 4), corresponding respectively to how smooth the wheelchair is driven, how efficient it is to reach a target and how close it moves to obstacles. Global efficiency η is the average of these three efficiencies, that roughly correspond to the properties of a navigation function:

$$\eta_{sm} = e^{-C_{sm} \cdot |\alpha_{sm}|} \tag{2}$$

$$\eta_{dir} = e^{-C_{dir} \cdot |\alpha_{dir}|} \tag{3}$$

$$\eta_{sf} = 1 - e^{-C_{sf} \cdot |\alpha_{sf}|} \tag{4}$$

α_{sm} being the angle between the heading direction and command direction; α_{dir} being the angle between goal direction and command direction; α_{sf} being the angle between obstacle direction and command direction; and C_{sm}, C_{dir} and C_{sf} constants to decide how much impact have each angle on its respective local efficiency. As most of situations require uniform efficiency factor changes, these constants are set to 1 by default.

The main advantages of our approach [7] are that: i) it tends to preserve curvature and safety, as most PFA-based algorithms do; ii) users contribute to control chair all the time because η_U is never equal to 0; and iii) humans provide (when possible) a deliberation level to the system to avoid local traps.

The main problem with the proposed approach was that $\bar{\eta}_C$ (collaborative command efficiency) turned out to be lower for people with better functional profiles than for people with severe disabilities in a significant number of cases [7]. We observed that a number of these users seemed to be fighting robot control, so we developed a new metric that we called Disagreement. Disagreement is equal to the angle between the user's command $\vec{v_U}$ and the emergent one $\vec{v_C}$ and it represents how similar emergent motion is to what the user expects. Disagreement in our tests for people with good functional profile and low driving efficiency was very high: around 40-45%. Even though there is always a baseline Disagreement when a person drives a vehicle depending on its dynamics and kinematics -we measured it to be around 15% in our wheelchair-, it becomes obvious that it is not comfortable for a person to drive like this.

To solve this problem, we propose to replace commands provided by PFA by commands learnt from previous experience on how the user drives, as described in next section. Thus, the differences between commands proposed by the user and by the robot should be quite lower and acceptance improves.

3 Robot Adaptation to User

3.1 CBR-Based Collaborative Control

CBR has been used in navigation before, typically for global path planning in static environments [10][11] rather than for reactive navigation. There are also approaches for global planning in dynamic environments [12]. However, in [12] new opportunities cannot be discovered when the environment changes unless the topological map, which is based on, is regularly reorganized. Kruusmaa [13] proposed a grid-based CBR global path planning method to overcome the aforementioned problem. However, she concluded that CBR-based global navigation is beneficial only when obstacles are large and dense and only a few solutions exist. Otherwise, the solution space may become too large. Some CBR-based methods focus on reactive navigation [14][15], but they all rely on accumulating experience over a time window while navigating in a given environment to obtain an emergent global goal seeking-behavior. Hence, they are environment-dependent. The authors already proposed a purely reactive navigation layer based on CBR in [7] for autonomous robots. Its original purpose was to create ad-hoc reactive navigation strategies via supervised learning and adapt them to different robot structures via learning by experience. Hence, we could avoid kinematics and dynamics calculations. In the present work, a similar strategy is used to make the robot learn how the user drives to improve acceptance.

Learning a reactive navigation behavior basically consists of associating whatever the driver is doing to the situation at hand. The user is already taking into account the vehicle kinematics and dynamics, as well as the relative position of obstacles and goal and any local consideration, like floor condition, mechanics, etc, that he/she intuitively adapts to through practice. This knowledge is implicitly added to the case base. After a while, all this information is encoded into a set of cases, that can be evaluated with our local metrics: η_{sm}, η_{dir} and η_{sf}. Eventually, the case base stabilizes to the best average solutions the user gives to any input situation. This happens after acquired cases are clustered to obtain valid prototypes, so that duplicates and least efficient cases are removed from the case base as explained below. This case base is not environment-dependent because there are not so many situations one can face from a local point of view [16]: the only relevant information at reactive level is how far we are from close obstacles, where we are heading and where we would like to go.

Obviously, if our system cloned exactly what the user does, it would provide no assistance except to correct punctual errors. However, if we combine a CBR reactive navigation module with the proposed collaborative control approach, advantages are more obvious: we use learnt cases when possible and receive assistance when needed. Each time the robot retrieves a case from the case base, its efficiency (η_{CBR}) is checked. If η_{CBR} is over a given threshold U_η (in our case, 0.7), the retrieved case solution becomes $\vec{v}_R$. Otherwise, $\vec{v}_R$ is obtained from PFA, even if η_{CBR} is bigger than η_{PFA}. Then $\vec{v}_R$ and $\vec{v}_U$ are weighted by their respective efficiencies as usual to calculate $\vec{v}_C$, which is stored in the case base for future reference.

3.2   CBR Implementation

The number of different situations that a robot can locally find is not too large [16], so the number of cases to acquire a given motion strategy is not large either (150-200 cases in our previous tests [7]). Hence, we can use a flat structures and the usual feature-value vector representation. Our cases are compared using a nearest-neighbour (NN) algorithm. After several tests, we chose to work with the Tanimoto distance T_s. For cases C_1 and C_2:

$$T_s(C_1, C_2) = \frac{C_1 \cdot C_2}{|C_1|^2 + |C_2|^2 - C_1 \cdot C_2} \tag{5}$$

We found that similar environment arrangements result in a lower Tanimoto distance, whereas other metrics, e.g. Euclidean distance, reward higher partial similarities in nearby obstacles than the environment as a whole [7]. Our input instance includes all commented reactive factors: goal position, laser readings (nearby environment) and wheelchair heading. Our output includes the user motion command ($\vec{v}_{CBR}$) and our evaluation measure is its global efficiency (η_{CBR}). Fig. 1 shows our case structure. Since minor differences between sensors readings usually correspond to slight robot shiftings rather than to different situations, we have discretized the space sampled by the robot into 8 equal sized arc regions and also discretize laser readings into 5 non equal intervals that have proven to be valid for a typical indoor environment [7]: i) critical (0-20 cm); ii) near (20-50 cm); iii) medium (50-100 cm); iv) far (100-150 cm); and v) no influence (more than 150 cm).

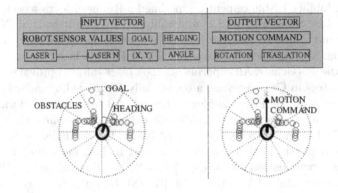

Fig. 1. CBR case structure and example

Our system uses both *learning by observation* and *learning by own experience*. If there is any recorded trace of a given user driving the wheelchair without assistance, it is used to initialize the case base. As soon as the user starts driving, his/her commands are combined with the robot's and learnt by the system. If the user is unable to provide an efficient way of solving a problem, the solution will be mostly based on PFA. However, the robot's commands tend to be more and more similar to the user's as a whole. It needs to be noted that we include no case adaptation stage in our CBR cycle to preserve user's commands as much as possible. Instead, adaptation is implicitly provided by collaborative control.

Finally, a MaxMin clustering algorithm is applied to the case base on a regular basis to group similar cases into a single cluster prototype (CP). The resulting CP is equal to the average of all cases in the cluster weighted by their own efficiencies. Thus, a given CP is not the most frequent response to a situation, but the most efficient response the user may produce on a regular basis. Also, low efficiency cases weight too little in CP calculation, so in practice they are removed unless there is no learnt alternative to cope with the related situations.

Our final case base is composed only of CPs and we only compare new instances with those Cps, but we preserve all learnt cases for future rearranging. This whole process is performed offline each time the number of newly acquired cases exceeds a threshold. Typically, it needs to be performed at least twice, once for the learning by observation stage and another for the first run while learning by own experience. Later on, the number of acquired cases decreases significantly and it is no longer necessary to cluster cases after each experiment. This process has, mainly, four targets: i) to remove duplicated cases, ii) to bind the number of cases in the database, iii)to avoid oscillations between similar cases, and iv) to clean spurious solutions from the database.

4   Experiments and Results

The proposed CBR system was built on CARMEN (Collaborative Autonomous Robot for Mobility ENhancement), a modified a Runner Meyra wheelchair (Fig. 2.a),donated by Sauer Medica S.L. and equipped with an industrial PC running Linux OS and a frontal Hokuyo laser URG04-RX for localization and obstacle detection. Moving backwards was not allowed due to lack of rear sensors and mirrors in the wheelchair. All experiments took place -after approval by FSL Ethical Committee- in Casa Agevole, a 60 m2 fully furnitured, standard-compliant test house built in the FSL complex in Rome[1]. All tests were performed by 18 volunteering inpatients presenting different degrees of disability (Left or Right Hemiparesis, Ischemic Stroke, Spinal cord injury, Cerebral Hemorrhage). Their cognitive and physical skills ranged from good to low, according to the minimental state examination (MMSE)[17] (1-30), the Barthel Index [18] (0-100) and the Instrumental ADL (IADL) [19] (0-5/8). In our volunteers, these scales were MMSE: 3-29, Barthel: 8-100, IADL: 0-8. Volunteers were divided into 3 groups depending on their diagnosed profile: 1) minor; 2) mild; and 3) severe physical/cognitive disabilities .Fig. 2.b shows the approximate path that volunteers were asked to perform. It can be observed that it involves door crossing, narrow areas and significant turns. This path was suggested by our medical staff because it includes most situations faced in ADL.

Each volunteer performed at least three runs (autonomous mode, shared control using PFA, and shared control using CBR). First of all, they drove the wheelchair without assistance (for benchmarking). In this mode, only a safeguard layer is active to prevent collisions. Many volunteers did not manage to complete the path in this mode. Our second mandatory run was PFA-based collaborative control navigation. Most users managed to do the run in assisted mode at first attempt, but one group of inpatients – group 3 – had notable exceptions. During this run, CBR was active and cases were acquired for the next run. Finally, all volunteers tried at least a CBR-based collaborative control runs, using their own case bases to assist navigation. During these runs, the case base kept acquiring new cases on a need basis.

---

[1] http://www.progettarepertutti.org/progettazione/
  casa-agevole-fondazione/index.html

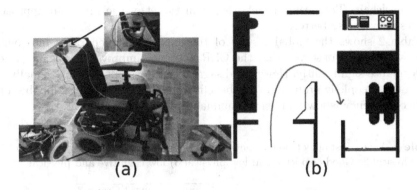

Fig. 2. a) Robot wheelchair (CARMEN) b) Proposed path at Casa Agevole

Table 1. Average results for all inpatients in collaborative control tests

	PFA Tests		CBR Tests	
	Mean	Dev.	Mean	Dev.
Global efficiency (%)	65.23	20.10	71.42	18.74
Smoothness (%)	64.18	27.80	69.05	25.54
Directness (%)	39.31	28.93	53.60	26.73
Safety (%)	92.25	17.60	91.73	16.06
Intervention Level (%)	79.17	-	75.59	-
Disagreement (%)	41.83	27.75	26.11	21.87
Joystick variation (%)	1.62	4.69	1.64	4.47
Inconsistency (%)	9.53	10.93	6.59	9.28
Completion time (sec)	48.45	-	43.46	-

Table 1 shows the average results of our experiments. Standalone results are omitted, as they are possibly biased by the users' learning curve. Our task metrics include all efficiency factors η, η_{sm}, η_{sf} and η_{dir} and total time. Our psych metrics include: *Intervention Level*, defined as the portion of time that the user moves a joystick [20] and showing if the user presents an active profile; ii) *Disagreement*, as previously commented, related to effort and frustration; it needs to be noted that due to mechanical issues like inertia, response time, joystick sensitivity, etc., our lowest wheelchair disagreement seemed to settle around 20% in standalone mode for our wheelchair; iii) *Inconsistency*, defined as the variation of the user's commands when facing similar situations; and iv) *Joystick variation*, which measures changes over 10% in the position of the stick and has been used as an indirect measure of workload [21][22].

It can be observed that efficiency in CBR-based collaborative mode is higher – specially in terms of directness – and its deviation is lower. The most important issue, however, is that *disagreement* decreases from 41.83% (PFA-based mode) to 26.11% (CBR-based mode), meaning that users are more comfortable with the wheelchair. This was, in fact, the main target of the proposed approach. Consistency is also better.

Table 2 shows the global results of the experiments separated into our 3 groups. It can be observed that the CBR approach improves and homogenizes performance significantly. However, disagreement does not decrease equally for all groups. People in group 3 do not benefit from this method, probably because their consistency is low and learnt commands are not too efficient.

Table 2. Anova Test of global efficiency and disagreement for groups: 1) Good cognitive and physical 2) Good cognitive and low physical 3) low cognitive and physical

	Group 1	Group 2	Group 3
PFA Global efficiency (%)	66.77	70.75	68.86
CBR Global efficiency (%)	74.50	76.20	76.33
ANOVA (pvalue)	0.000	0.018	0.000
PFA Disagreement (%)	32.87	31.85	25.66
CBR Disagreement (%)	25.10	19.40	31.8
ANOVA (pvalue)	0.000	0.004	0.000
Patients in group	3	3	12

4.1   A Case in Detail

The problem our CBR approach meant to solve is clearly represented by volunteer 1, a 56 years old female affected by multiple sclerosis with good physical and cognitive skills (MMSE=26, Barthel=100, IADL=8). This person could move with the help of a walker, but had no previous experience with power wheelchairs. Table 3 briefs her performance in her five tries: standalone, PFA-based (x3) and CBR-based collaborative control. This person repeated the PFA test three times and her third try was the worst of all. In fact, she failed to finish her last two PFA-based paths and reported that "the wheelchair was not working" while trying to move it into a wall.

In her first standalone run, only the safeguard layer was active – to prevent imminent collisions –, hence the minor differences between human and collaborative performance in this mode (table 3). Her standalone global efficiency was equal to 67.56%, and her worst feature was directness, probably due to lack of practice with power wheelchairs. Intervention Level (table 3) was very high because the wheelchair did not move unless there was some human input (99.73%). Besides, her Joystick variation was very low (0.09 %). Her standalone trajectory was quite smooth and efficient, even at door crossing, except at the second turn, when she got too close to the walls and had to steer right sharply. In brief, her standalone run was quite good and she reached her goal in just 30.67 seconds. In order to observe what this volunteer actually lacked, we clustered her commands according to the relative position of wheelchair, goal and obstacles and realized that she had trouble adjusting turns.

Table 3. Results for inpatient 1 trials in all modes. In the CBR column, data without parenthesis represents the PFA value and data with them the CBR value.

Control type		User	PFA	PFA	PFA	CBR
η_{global} (%)	Robot	—	63.12	65.73	58.97	68.61(69.23)
	User	67.51	65.17	60.7	57.05	70.6
	Shared	67.56	68.4	67.88	64.0	74.5
η_{sm} (%)	Robot	—	56.45	57.37	36.81	52.2(78.73)
	User	66.4	65.91	56.4	61.26	66.62
	Shared	66.33	65.1	63.17	61.96	74.6
η_{dir} (%)	Robot	—	43.85	51.6	48.75	61.98(34.13)
	User	44.28	37.46	38.42	22.0	48.61
	Shared	44.39	45.75	44.85	37.0	52.73
η_{sf} (%)	Robot	—	88.9	88.24	91.39	91.81(94.92)
	User	91.95	92.34	87.41	87.94	96.55
	Shared	92.03	94.45	95.59	92.96	96.22
Intervention Level	%	99.73	70.73	77.85	97.95	82.83(82.56)
Disagreement	%	25.18	21.37	29.87	47.31	20.89(25.11)
	dev	25.18	14.96	18.02	27.7	18.72(26.63)
Joystick variation	%	0.09	0.2	0.07	0.06	2.38
	dev	1.12	2.37	1.08	0.99	3.65
Inconsistency	%	6.73	7.7	8.58	4.91	6.69
	dev	6.73	7.7	8.58	4.91	6.69
Total Length	m	6.7	6.68	4.98	4.8	6.4
Completion time	sec	30.67	38.2	31.61	32.22	30.63

Her next 3 runs were performed with PFA-based collaborative control: not only did η_C not increase but even decreased in the last run whereas the volunteer tried to collide into a wall to check if the wheelchair obeyed her. Obviously, the wheelchair tried to correct her and they struggled for a couple of seconds. Her *directness* dropped to 22% and η_C decreased sharply. In this particular run, *disagreement* became as high as 47.31% – meaning that half the time, user and emergent commands were in conflict –.

Fig. 3 shows η_H, η_R and η_C for the third PFA-based run. We have chosen to represent η_{sm}, and η_{dir} and η_{sf} as the red, green and blue channels of the RGB colorspace, respectively, for visibility in our efficiency plot. Hence, pink efficiency means loss of directness (steering areas) and purple-blue means loss of directness and smoothness (sharp direction changes). Fig. 4.b shows how *disagreement* was highly correlated with η_R, i.e. it grew when the robot had more impact on emerging commands. Fig. 4.a shows how disagreement grew after crossing the door – when η_R began to increase and the user became aware of the robot intervention– . After that, the rest of the trajectory was a fight for control. Although the robot tried to compensate the user's motion commands (cyan η_R area), emergent ones were affected and, as a result, she failed to reach the goal. The inpatient's *inconsistency* in this run shows that she was driving far from her usual skills: from the 43 different commands clusters that she typically used

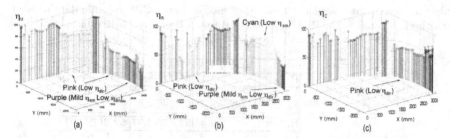

Fig. 3. Inpatient 1 results for PFA-based collaborative mode: a) human efficiency; b) robot efficiency; c) collaborative efficiency

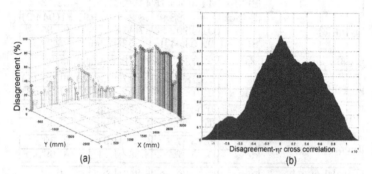

Fig. 4. Inpatient 1 results for PFA-based collaborative mode: a) *disagreement* for PFA-based collaborative control; b) cross correlation disagreement/η_R

to solve the proposed trajectory in standalone mode, 25 – corresponding mostly to steering decisions – had a very large standard deviation in this run. This basically means that her decisions were statistically erratic during the struggle.

Fig. 5 shows two clusters corresponding to locations involving steer correction and strong *disagreement*. The robot -in the center of the plot- is heading in the 0 degrees direction, obstacles are represented with circles and the goal is marker with an x. On the right of each plot, we can see the joystick shadow for human and robot. Cluster 16 corresponds to the beginning of the trajectory, when the first steering decision needed to be taken. As usual, inpatient 1 delayed her steering command, so the robot initiated it, and provoked a noticeable turn to the right to avoid getting too close to the wall. The emerging command, no longer equal to the user's, was practically a 90^0 right turn, only much slower than the user's command (shorter vector). At this point, the user became fully aware of the robot's influence and tried to fight it, but, eventually, the struggle became so intense that we got clusters like cluster 34, with a goal on the right side of the wheelchair, obstacles on the left and yet, the user pushing the joystick hard into the obstacles. Since her efficiency was very low at the point, the robot was dominant and the combination was a slow forward motion. This eventually got the wheelchair so close to an obstacle that further safe maneouvre was not possible and inpatient 1 failed to finish the trajectory.

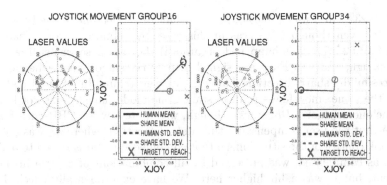

Fig. 5. Clusters representation of two situations with a strong *disagreement* between inpatient 1 and robot in PFA-based collaborative control

Fig. 6.a shows the path in CBR-based collaborative control. As commented, the case base was filled with user's data coming from all the previous runs, but in this case mostly included data from the standalone and first PFA runs, because efficiency grew worse later. During execution, if a situation is not similar enough to the output case or η_{CBR} is too low, we use a PFA command instead in collaborative control at that specific location (areas marked in Fig. 6.b with a dot). We can observe that *disagreement* was specially high when PFA was used instead of CBR. Nevertheless, *disagreement* was no longer correlated with η_R because most learnt cases were efficient enough to be extensively used through the trajectory. In this case, *inconsistency* was similar in average to standalone mode and there were less, more homogeneous command groups than in PFA mode.

Results of this test are briefed in the last column of table 3. As commented, we calculate what PFA would do all the time, but PFA commands are not used unless η_{CBR} goes under a safety threshold or the retrieved case instance is too far from the current situation. Table 3 shows how η_C in the CBR-based mode case increased to 74.5%, higher than in any of the previous runs, including standalone mode, and higher than each of the components separately: PFA (68.61%),

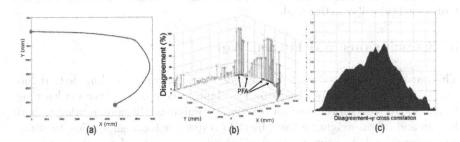

Fig. 6. Inpatient 1 results for CBR-based collaborative mode: a) path; b) disagreement (PFA and CBR); c) correlation between disagreement and system efficiency

CBR (69.23%) and user (70.6%). More specifically, *smoothness* was boosted up to 74.6%, even though CBR was in this case, by definition, worse than PFA (78.73% to 52.2%). This happened because: i) the case base was still not complete enough; and ii) PFA are designed to preserve smoothness, whereas CBR produces mildly sharp direction changes at case switching. Inpatient 1 had practically the same smoothness than in standalone mode, i.e. she was not fighting the machine anymore. *Directness*, though, was quite low for CBR with respect to PFA, but the user compensated this and, at locations where η_C was too low, PFA took control of the situation, so that combined *directness* raised to 52.73%. *Safety*, as commented, was preserved by all combinations of control in the experiment, but it was a bit higher here. We have experimentally checked that this happens when users are comfortable with control and drive smoothly, as, in these cases, they tend not to get too close to obstacles.

To illustrate how CBR decisions are closer to human commands, Fig. 7 shows two clusters corresponding to approximately the same situations regarding the relative position of obstacles and goal. Fig. 7.a corresponds to PFA-based collaborative mode, whereas Fig. 7.b is obtained in CBR-based collaborative mode. It can be observed that PFA-based collaborative mode corrected *directness* and reduced variability, but CBR-based collaborative mode reduced *disagreement* as well.

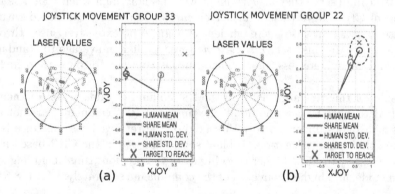

Fig. 7. Inpatient 1 cluster comparative for the same location for PFA-based (a) and CBR-based (b) collaborative mode

5   Conclusions and Future Work

This paper has presented a CBR-based collaborative control technique to reduce user's stress and assistance rejection by adapting help to the user via learning. Collaborative control is based on reactively combining the contribution of both human and robot, weighting them by their respective local efficiencies, to obtain an emergent collaborative navigation behavior. The robot learns how the user drives via CBR and contributes to control with more familiar commands. Thus, differences between user and emergent commands are less perceivable by the user. However, if CBR commands are not efficient enough, and due to safety reasons, PFA is used instead of CBR.

The system was tested by 18 inpatients at FSL in a home-like environment. Average efficiency was higher in CBR-based collaborative mode than in PFA-based one or standalone mode. Besides, all users managed to finish a mildly complex trajectory in CBR-based collaborative mode. As expected, CBR-based navigation mimicked the user's way of driving and, in most cases – good or mild cognitive skills –, reduced disagreement between user and machine. Persons with very low cognitive skills did not provide enough efficient patterns to build a valid user model, so in those cases the system typically behaved like a PFA-based collaborative one.

The main drawback of the proposed system is that most users agreed that the wheelchair moved a bit brusque. This is provoked by case swapping, since the case base has a limited number of cases, and also by swaps from CBR to PFA-based collaborative control. This problem can be solved by adding some temporal inertia and future work will focus on this.

Future work will focus too on checking if the contents of a given user's case base can be correlated with the person's condition and, if so, on predicting the amount of help required by a specific user at a given situation instead of just providing it in a completely reactive fashion. This would allow us to include temporal inertia and reduce the commented problems related to case switching.

Acknowledgement. This work has been partially supported by the Spanish Ministerio de Educacion y Ciencia (MEC), Project TEC2011-29106-C02-01. The authors would like to thank Santa Lucia Hospedale and all volunteers for their kind cooperation and Sauer Medica for providing the power wheelchair.

References

1. Parikh, S.P., Grassi, V., Kumar, V., Okamoto, J.: Usability study of a control framework for an intelligent wheelchair. In: Proc. of the 2005 IEEE International Conference on Robotics and Automation, Barcelona, Spain, pp. 4745–4750 (April 2005)
2. McLachlan, S., Arblaster, J., Liu, D.K., Valls, J., Chenoweth, L.: A multi-stage shared control method for an intelligent mobility assistant. In: Proc. of the 2005 IEEE 9th International Conference on Rehabilitation Robotics, Chicago, USA, pp. 426–429 (July 2005)
3. Frese, U., Larsson, P., Duckett, T.: A multigrid algorithm for simultaneous localization and mapping. IEEE Transactions on Robotics 21(2), 1–12 (2005)
4. Mandel, C., Huebner, K., Vierhuff, T.: Towards an autonomous wheelchair: Cognitive aspects in service robotics. In: Proceedings of Towards Autonomous Robotic Systems (TAROS 2005), pp. 165–172 (2005)
5. Bruemmer, D.J., Few, D.A., Boring, R.L., Marble, J.L., Walton, M.C., Nielsen, C.W.: Shared understanding for collaborative control. IEEE Transactions on Systems, Man and Cybernetics - Part A: Systems and Humans 25(4), 494–504 (2005)
6. Horiguchi, Y., Sawaragi, T.: Effects of probing to adapt machine autonomy in shared control systems. In: Proc. International Conference on Systems, Man and Cybernetics, Hawaii, USA, vol. 1, pp. 317–323 (October 2005)

7. Urdiales, C., Fernandez-Carmona, M., Peula, J., Annicchiaricco, R., Sandoval, F., Caltagirone, C.: Efficiency based modulation for wheelchair driving collaborative control. In: Proc. of 2009 IEEE Conf. on Robotics for Rehabilitation (ICRA 2010), Anchorage, USA (2010)
8. Carlson, T., Demiris, Y.: Human-wheelchair collaboration through prediction of intention and adaptive assistance. In: IEEE International Conference on Robotics and Automation, ICRA 2008, pp. 3926–3931 (May 2008)
9. Khatib, O.: Real-time obstacle avoidance for manipulators and mobile robots. International Journal of Robotics Research 5(1), 90–98 (1986)
10. Branting, L.K., Aha, D.W.: Stratified case-based reasoning: reusing hierarchical problem solving episodes. In: IJCAI 1995: Proceedings of the 14th International Joint Conference on Artificial Intelligence, Montreal, pp. 384–390. Morgan Kaufmann, San Mateo (1995)
11. Fabrizi, E., Oriolo, G., Panzieri, S., Ulivi, G.: Mobile robot localization via fusion of ultrasonic and inertial sensor data. In: Proc. of the Sixth Midwest Artificial Intelligence and Cognitive Science Conference, Carbondale, USA, pp. 32–36 (1995)
12. Haigh, K.Z., Veloso, M.: Route planning by analogy. In: Aamodt, A., Veloso, M.M. (eds.) ICCBR 1995. LNCS, vol. 1010, pp. 169–180. Springer, Heidelberg (1995)
13. Kruusmaa, M.: Global navigation in dynamic environments using Case-Based Reasoning. Autonomous Robots 14, 71–91 (2003)
14. Likhachev, M., Arkini, R.C.: Spatio-temporal case-based reasoning for behavioral selection. In: Proc. of the IEEE Int. Conf. on Robotics and Automation (ICRA), pp. 1627–1634 (2001)
15. Santamaria, J., Ram, A.: A multistrategy case-based and reinforcement learning approach to self-improving reactive control systems for autonomous robotic navigation. Tech. Rep. (1993)
16. Minguez, J., Osuna, J., Montanor, L.: A divide and conquer strategy based on situations to achieve reactive collision avoidance in troublesome scenarios. IEEE Trans. on Robotics (2009)
17. Crum, R., Anthony, J., Bassett, S., Folstein, M.: Population-based norms for the mini-mental state examination by age and educational level. Journal of the American Medical Association 269(18), 2386–2391 (1993)
18. Mahoney, F., Barthel, D.: Functional evaluation: the barthel index. Maryland State Medical Journal (14), 56–61 (1965)
19. Lawton, M., Brody, E.: Assessment of older people: self-maintaining and instrumental activities of daily living. Gerontologist (9), 179–185 (1969)
20. Cooperstock, J., Pineau, J., Precup, D., Atrash, A., Jaulmes, R., Kaplow, R., Lin, N., Prahacs, C., Villemure, J., Yamani, H.: Smartwheeler: A robotic wheelchair test-bed for investigating new models of human-robot interaction. In: Proc. of the IEEE Conference on Intell. Robots and Systems (IROS), San Diego, USA (2007)
21. Clarke, D., Yen, S., Kondraske, G.V., Khoury, G.J., Maxwelle, K.J.: Telerobotic network workstation for system performance and operator workload monitoring. NASA JSC, Houston, TX. Tech. Rep. 91-013R (1991)
22. Khoury, G.J., Kondraske, G.V.: Measurement and continuous monitoring of humanworkload associatedwith manual control devices. NASA JSC, Houston, TX. Tech. Rep. 91-011R (1991)

Biological Solutions for Engineering Problems: A Study in Cross-Domain Textual Case-Based Reasoning

Swaroop S. Vattam and Ashok K. Goel

Design & Intelligence Lab, School of Interactive Computing,
Georgia Institute of Technology Atlanta, GA 30332, USA
{svattam,ashok.goel}@cc.gatech.edu

Abstract. Textual Case-based Reasoning (TCBR) is a powerful paradigm within CBR. Biologically inspired design – the invention of technological systems by analogy to biological systems - presents an opportunity for exploring cross-domain TCBR. Our *in situ* studies of the retrieval task in biologically inspired design identified findability and recognizability of biology articles on the Web relevant to a design problem as major challenges. To address these challenges, we have developed a technique for semantic tagging of biology articles based on Structure-Behavior-Function models of the biological systems described in the article. We have also implemented the technique in an interactive system called Biologue. Controlled experiments with Biologue indicate improvements in both findability and recognizability of useful biology articles. Our work suggests that task-specific but domain-general model-based tagging might be useful for TCBR in support of complex reasoning tasks engaging cross-domain analogies.

1   Introduction

Textual case-based reasoning (TCBR) entails the use of unstructured cases in the form of textual documents (Weber, Ashley & Bruninghaus 2006). TCBR has become especially important with the advent of the Web that provides access to a large number of textual documents containing potential cases. A major question then becomes how do we access the right cases from the Web for a given query or problem? Thus, research into TCBR is closely intertwined with research on information retrieval (IR) and text mining (TM) (Rissland & Daniels 1996). According to Weber, Ashley & Bruninghaus (2006), the major differences between TCBR and IR/TM are that the former (1) is more explicitly interested in supporting complex reasoning, and (2) thus uses task-specific and domain-specific knowledge to access the right case for supporting the reasoning. For example, Burke *et al.* (1997) describe a technique that uses task-specific and domain-specific knowledge to answer FAQ questions in a specific domain; Bruninghaus & Ashley (2001) describe a different technique that too uses task-specific and domain-specific knowledge to access textual cases in the legal domain. Lenz (1998) describes knowledge layers for supporting TCBR; Raghunandan *et al.* (2008) propose evaluation measures for TCBR systems.

S.J. Delany and S. Ontañón (Eds.): ICCBR 2013, LNAI 7969, pp. 343–357, 2013.

In this work, we are interested in a related but slightly different question: how might TCBR work if the target problem and the textual cases are from different domains? That is, we are interested in *cross-domain* TCBR. We have encountered this problem in the context of *biologically inspired design (BID)* – the invention of new technological products, processes and systems by analogy to biological systems. The needed biological knowledge typically is found in the form of unstructured textual documents, typically on the Web. Due to its growing importance, we posit that BID presents a great opportunity for exploiting and exploring cross-domain TCBR.

In general, BID entails all the major tasks of CBR such as retrieval, adaptation, evaluation and storage (Kolodner 1993). In this paper, we focus on the *retrieval* task. Given a target design problem, one of the first tasks in the BID process involves finding the right biological system to emulate in order to generate a design solution. Designers, including expert designers, typically are novices in biology and thus are aware of only a small fraction of the vast space of biological systems. Thus, designers typically rely on external information environments such as the Web for finding biological cases relevant to their design problems. Most biological cases on the Web are available only in unstructured forms such as textual documents. Thus, the retrieval task in BID takes a query in a design domain, such as engineering, as input, and has the goal of returning as output textual documents in the domain of biology that are relevant to the query. The retrieval task is challenging not only because of the unstructured nature of the cases, but also because the retrieval process cannot rely on domain-specific knowledge and conventional techniques for retrieving textual documents lead to poor precision and recall.

Our *in situ* studies of the retrieval task in BID identified findability and recognizability of biology articles relevant to a design problem as major challenges (Vattam & Goel 2011); we define findability and recognizability below. To address these challenges, we have developed a technique for semantic tagging of biology articles based on Structure-Behavior-Function (SBF) models of the biological systems described in the article. We have also implemented the technique in an interactive system called Biologue. Controlled experiments with Biologue indicate improvements in both findability and recognizability of useful biology articles. In this paper, we describe the design and evaluation of Biologue.

2   Background

The growing movement of biologically inspired design (BID) or biomimicry views nature as a large library of sustainable designs that could be a powerful source of technological innovation (e.g., Benyus 1997). Recent examples of BID include the design of wind turbines inspired by the tubercles on the pectoral fins of humpback whales, and fog harvesting devices inspired by the arrangement of hydrophilic and hydrophobic surfaces found on the back of Namibian beetles, etc.

2.1   Related Research

Research on computational methods and tools for supporting BID can be categorized into three broad approaches. The first approach uses digital libraries of functional models of biological systems (Chakrabarti *et al.* 2005). For example, the DANE system provides access to a functionally indexed digital library of SBF models of biological systems (http://dilab.cc.gatech.edu/dane/; Goel *et al.* 2012). The difficulty with this approach is scalability: it takes expertise, time and effort to build such a library.

The second approach uses text mining techniques (Shu 2010), including syntax-level heuristics customized to BID (Chiu & Shu 2007), and enhanced by engineering-to-biology thesaurus (Nagle, Stone & McAdams 2010). Although more scalable than the first approach, this technique could be subject to the usual limitations of keyword-based search; the efficacy of this approach is still being explored.

The third approach uses semantic indexing for accessing biological information. For example, Biomimicry 3.8 Institute's AskNature (Biomimicry 2008) is a popular Web portal that provides access to a functionally indexed digital library of biology articles. Our work on TCBR in BID presented here takes a similar approach. We posit that our approach is more human-centered, emphasizing (1) first gaining a deep understanding of TCBR in BID as it naturally occurs in the real world, (2) grounding our system design in that understanding, and (3) rigorously evaluating our claims using controlled experiments.

This work both builds on and differs from our previous work on case-based design. In earlier work, we grounded the process of case-based design in SBF models of physical systems (Goel, Bhatta, Stroulia 1997), exploited TCBR for understanding design problems stated in natural language (Peterson, Mahesh & Goel 1994), and explored TCBR for acquiring SBF models of everyday devices from textual documents. In more recent work, we have formalized SBF models (Goel, Rugaber & Vattam 2009), conducted *in situ* studies of BID (Vattam, Helms & Goel 2008; Vattam & Goel 2011), and developed digital libraries of SBF models of biological systems in support of BID (Goel *et al.* 2012). In this paper we describe the design and evaluation of Biologue, our interactive system for addressing the findability and recognizability challenges of cross-domain TCBR in the retrieval task of BID.

3   Human-Centered TCBR in the Context of BID

We conducted our *in situ* studies of BID in the context of ME/ISyE/MSE/PTFe/BIOL 4740, a senior-level, interdisciplinary, project-based course at Georgia Tech. Yen *et al.* (2011) describe this course in detail. The two studies described below were conducted in Fall 2006 and Fall 2008 sections of the class, respectively. In these studies we observed a total of ten interdisciplinary teams of designers engaged in open-ended, semester-long BID projects that led to novel *conceptual* designs of technological systems such as the design of a new levee for New Orleans inspired in part by Iron Snail. While details of the studies can be found in other sources (Vattam 2012), here we summarize our findings related to TCBR in BID.

3.1   Characteristics

CBR in BID can be characterized as follows:
• *Cross-domain analogies*: The design problem originates in a design domain such as engineering but the cases for addressing the problem are in the domain of biology.

• *Compound analogies*: A single design solution may often require multiple biological source cases (Vattam, Helms & Goel 2008).

• *Textual cases*: Cases that are retrieved and used by the human designers mostly are found in the form of textual biological articles.

• *Cases distributed across multiple online environments*: Designers use a range of online information environments to seek biological cases, including (1) digital libraries like Web of Science, Google Scholar, ScienceDirect, etc., (2) online encyclopedia like Wikipedia, (3) popular life sciences blog sites like Biology Blog, (4) biomimicry portals like AskNature, and (5) general web search engines like Google.

• *Human-in-the-loop retrieval process*: In our observations, the designers used search results from a design query to formulate new queries for online search in an iterative process of formulating queries, searching online, finding biology articles, reading the articles, formulating new queries and so on. We call this process *interactive analogical retrieval* (Vattam 2012).

3.2   Challenges

We found that designers faced three major challenges in accessing biology articles on the Web relevant to their design problems (Vattam & Goel 2011): *findability, recognizability* and *understandability*. These difficulties were encountered irrespective of the specific type of information environment used and made the retrieval process quite inefficient.

Findability: designers often went for long periods without finding a single relevant biological case in a retrieval process that typically extended over several weeks and often was tedious and frustrating for the designers. Thus, the relative frequency of encountering relevant articles containing biological cases was very low, suggesting that the match between the retrieval task and the information environment was not very good. A rough calculation suggests that designers spent approximately three person-hours of search time on the Web in order to find a single relevant article.

Recognizability: designers were prone to making errors of judgment about the true utility of articles that they encountered in the search process. In almost all online environments, search queries brought back a ranked list of search results (a set of articles). One important aspect of the search process was assessing and selecting promising articles from this list for further consumption. But, this decision had to be made based on *proximal cues* – hyperlink titles and snippets of text that are intended to represent the distal documents. In many instances, designers picked up on low-utility articles, only to realize later that it was not actually very useful (false positives). False positives lead to wasted time and effort (resource cost). Conversely, consider situations where designers might dismiss an article they encounter during the search as having low utility even though it might have contained a potential biological source (false negatives). False negatives represent lost opportunities.

Understandability: Since designers typically are novices in biology, they often have difficulty understanding the biological systems described in the textual documents they retrieve from the Web. While this challenge is covered in detail in other sources (Vattam 2012), here we focus on addressing the findability and recognizability challenges.

4   Addressing the Challenges of TCBR

Let us consider the issue of *findability*. According to the ACME theory of analogy (Holyoak & Thagard 1989), in order to retrieve source cases analogous to a target problem, the retrieval mechanism should simultaneously satisfy three constraints: semantic similarity (the overlap in terms of the number of similar concepts between the target and potential sources), structural similarity (the overlap in terms of the higher-order relationships between the target and potential sources), and pragmatic similarity (the overlap in terms of the pragmatic constraints or goals surrounding the target and potential sources). It is these three constraints acting simultaneously that distinguish analogical retrieval from other kinds of information retrieval mechanisms.

However, keyword-based search mechanisms found in common current online information environments support access to cases based on literal similarity (word-for-word matching), or at most semantic similarity to a limited degree, while ignoring structural and pragmatic similarity. As a result, each attempt at access can contain a large number of spurious articles that contain systems that are superficially similar to the target design as opposed to analogically similar. This results in low precision and recall.

Alternate methods of indexing and accessing biological articles in online environments may help address this challenge. Literature on case-based reasoning suggests guidelines for the alternate method (Kolodner 1993). (1) Indexing at storage time should anticipate the vocabulary the reasoner might use at retrieval time. (2) Indexing should use concepts and relations described at a level of abstraction that is justified from the perspective of the reasoning task.

In our *in situ* studies we found that designers' vocabulary used concepts and relations like functions, structures, physical principles, and operating environments. This vocabulary significantly overlaps with the vocabulary of Structure-Behavior-Function (SBF) models (Goel, Rugaber & Vattam 2009). Briefly, SBF models are a family of conceptual functional models. In SBF models, *Structure* pertains to components of a system; *Behaviors* describe causal processes or mechanisms in the system; and *Functions* specify outcomes of the system. In past work, SBF models have proved to be useful for design (Goel *et al.* 2012), understanding (Helms, Vattam & Goel 2010; Vattam, *et al.* 2011) as well as TCBR (Goel *et al.* 1996; Peterson, Mahesh & Goel 1994). Therefore, we posit that semantically indexing and accessing biology articles using concepts and relations derived form SBF ontology may better address the issue of findability in BID.

Now consider the second issue of *recognizability*. Information foraging theory (Pirolli 2007), which explains human information-seeking behavior in online information environments, claims that navigation towards useful information is guided by perception of *information scent* based on the proximal cues available in these environments. The issue of recognition errors in BID is attributable to the affordances - or lack thereof - of the proximal cues for accurately perceiving the information scent of the biology articles in the information environments.

One way to address this problem is by enhancing the proximal cues with additional information to help designers perceive the analogical similarity between the target design problem and the contents of the biological cases in the textual documents. We posit that enhancing the proximal cues of the distal articles by including visual overviews derived from the SBF models of the biological cases described in the articles may lower the rate of recognition errors.

5   Biologue

Biologue is an interactive information system that embodies the two main claims discussed in the previous section for addressing the challenges of the retrieval task in BID. Biologue represents a social approach to establishing an online corpus of biology articles annotated by their corresponding SBF models. It is based on the principle of social bookmarking (Sen *et al.* 2006) and is aimed to promote the sharing of biology articles in the BID community.

As one posts a reference to an article in Biologue, one can also manually add tags for annotating and organizing that reference. However, instead of keyword-based tags, the semantic tags in Biologue are linked to the ontology and schema of SBF models. As more and more people tag a particular reference, partially-structured SBF models of biological systems emerge in a socially-distributed fashion and get associated with that article. As Figure 1 illustrates, Biologue leverages these models to: (1) index an article and provided access to it based on the SBF schema and ontology, (2) offer visual overviews of the SBF models of biological cases described in the article.

As a use-case scenario, consider a situation where a designer, in the course of her day-to-day work, comes across a relevant online article on rat intestine and how that organ passively transports water across an osmotic gradient. The designer uses Biologue to: (1) bookmark this article in her personal library, (2) enter the article's bibliographic information, (3) tag the article with **Function:Transport(Water)**, where Transport **Is-A** Function in SBF ontology, and (4) share this tagged article with a teammate. The teammate reads the article and understands that the intestine achieves this function using the *three-chamber* mechanism, which uses a combination of *forward-* and *reverse-osmosis* principles. The teammate then adds a new tag to this article, **Behavior:Three-Chamber-Method**, and links it to the **Function:Transport(Water)** tag, where Behavior **Is-A-Part-Of** Function in SBF ontology. The teammate further adds two mores tags **Principle:Osmosis** and **Principle:Reverse-Osmosis** and links them to the **Behavior:Three-Chamber-Method** tag, where Principle **Is-A-Part-Of** Behavior in SBF ontology. Assuming that this article is read, tagged and retagged by many people, a conceptual model of how the rat intestine works emerges through negotiation and gets associated with this article.

Biologue implements an auto-complete feature to encourage tag reuse and minimize proliferation of user-generated tags, and a simple drag-and-drop interface for linking one tag to another and for linking tags to parts of the document.

Now, consider a use-case scenario where some other designer, completely unrelated to the first set who tagged the article, is trying to design a bio-inspired, energy-efficient, seawater desalination technique. This user logs into Biologue and proceeds to search the collection of articles in Biologue for a relevant biological source. Biologue currently allows users to search for articles based on features derived from SBF ontology, including Function, Principle, Structure, and Operating environment.

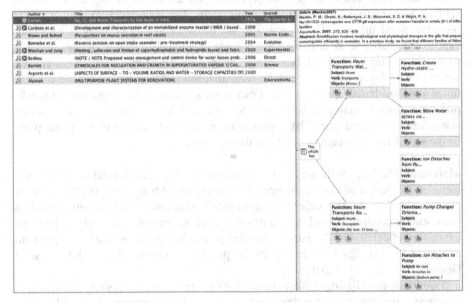

Fig. 1. Search results in Biologue; a selected article, and the visual overview of the SBF model tagging the article

Let us assume that this designer chooses to search the collection based on Function "remove salt from water" and the Principle "reverse osmosis". Because of the partial match across Principle, Biologue returns a non-ranked set of articles that includes the rat intestine article tagged by the first set of users. When this designer clicks on this article in the result set, she can not only view the traditional information that one would expect, but also the SBF model associated with the article as illustrated in Figure 1. Studying this model gives her a gist of how the rat ileum works from a designer's perspective, which also helps her make the analogical connection between the target desalination technology and the rat ileum system and allows her to decide whether it is worth pursuing the article. The model also gives her a high-level conceptual schema that she can refer back and forth to guide her development of understanding of the article. Upon reading the article and understanding how the rat ileum works in detail, she may choose to use this biological system as a source of inspiration and develops a novel desalination technique that removes salt from water by a combination of forward- and reverse-osmosis.

6   Findability Study

Hypothesis: In the context of BID, indexing and accessing biological articles using concepts and relations derived from their corresponding SBF models will lead to higher rate of findability when compared to keyword-based indexing and retrieval.

Procedure: This was a between-subject study conducted in Spring 2011. Sixteen subjects were recruited to participate in this study. A preselected BID challenge was presented to all subjects. The goal was to use Biologue to find as many articles relevant to the challenge as possible (and provide a rationale for their choice) within a stipulated amount of time. Eight subjects comprising the control group were given a version of Biologue where the articles were indexed and accessed using the conventional keyword-based approach. The other eight subjects in the experimental condition were given a version of Biologue where the articles were indexed and accessed using features derived from SBF representation. The performance of the participants on the search task was compared across the two groups.

Materials: The BID challenge involved a technology for solar thermal collectors, and included the design of (1) a bio-inspired reflective panel that could be fitted onto an existing absorber and was capable of dynamically changing its reflectivity, and (2) a bio-inspired feedback control system that regulates the temperature of glycol by regulating the reflexivity of the panel. This design challenge was an authentic problem attempted by a design team in one of the previous offerings of the BID course at Georgia Tech (Yen *et al.* 2011).

Two versions of Biologue were created for this study. In one version, the articles in Biologue's repository were indexed by keywords. Consequently, the articles were accessible only through keyword search. The search panel consisted of a single text box similar to Google and a search and a clear button. In the second version, the articles in Biologue's repository were indexed by concepts and relationships that were part of the SBF models associated with those articles. Biologue's repository had more than 200 articles in it. Fourteen designated articles that were known to be relevant to the target design challenge were included in the repository. If a participant's information seeking was efficient and accurate, then nearly all these 14 articles would be found and reported by the participant.

Data: Three kinds of data were collected for this study. First, the video of each participant's entire search process was captured using screen-capture software. This provided the bulk of the data to understand and analyze the retrieval process of the participants. Second, every article found by the participant in the course of their search was collected, along with his or her stated rationale for selecting that article. Third, participant demographic data was also collected.

Analysis: First, the participants in the two groups were compared to establish the equivalency of the two treatment groups. There was no statistically significant difference between the two groups with respect to participants' gender, their biology background, extent of design experience, extent of interdisciplinary research experience, or extent of use of scholarly articles in their work practices.

Second, the found-article data was analyzed to determine: (1) the total number of articles found by a participant, (2) the number of designated articles within that total number. The rationale provided by the participant for selecting an article was also analyzed to ensure that participants were meaningfully undertaking the task.

Third, the video data obtained for each participant was coded using a coding scheme. The origin of the coding scheme lies in information foraging studies conducted by researchers in the human-information interaction community (Pirolli, 2007). Their coding scheme coded and visualized the behavior of a person engaged in online information activity in the form of *web behavior graphs*. From these web-behavior graphs, collecting interesting statistics about the information seeking behavior in our experiments becomes possible. Two coders independently coded the videos (inter-coder reliability was 87.93%; in those cases where there was no consensus, the coding of the experienced researcher was included).

From the coded video data we were able to derive four retrieval performance characteristics for each participant: (1) *find period* (minutes per article found), (2) *mean between-patch foraging time* (minutes spent searching and navigating), (3) *mean number of regions foraged* (number of hops from one search page to another), and (4) *mean information yield per region* (ratio of actually relevant articles in an information region to all the articles encountered in that region). These four dependent variables, related to the findability issue, indicate the efficiency with which participants were able to search and retrieve the articles that they found.

Results: In the experimental condition, we expected that find period would be lower, that between-patch foraging time would be lower, that number of regions foraged would be smaller, and that information yield would be larger. Table 1 shows the actual results that confirm our predictions.

Table 1. Participant performance on the findability task

Treatment	Find period (avg)	Between-patch foraging (avg-mean)	Num. of regions foraged (avg-mean)	Yield per region (avg-mean)
Control	11.48 mins	11.68 mins	4.3	0.07
Experimental	5.85 mins	2.96 mins	2.45	0.212

The data collected in this study was submitted to a statistical significance test. An *a priori* power analysis suggested a sample size of 27 (power=0.7, d=0.5, α=0.05) for utilizing the t-test. Since our sample size of 16 was less than the recommended number, the data could not be subjected to the t-test. Therefore a non-parametric version of the t-test, known as *Mann-Whitney U* test was used.

There was *no* statistically significant difference between the two groups with respect to the number of articles found. The median *total articles found* in the control and experimental groups were 9 and 7.5 respectively; the distributions in the two groups did not differ significantly (Mann-Whitney U = 23.5, n1 = n2 = 8, P = 0.369 two tailed). Similarly, the median *total designated articles found* in the control and experimental groups were 3 and 3.5 respectively; the distributions in the two groups did not differ significantly (Mann-Whitney U = 27, n1 = n2 = 8, P = 0.595 two tailed).

Although the number of articles found by the two groups was comparable, there was a statistically significant difference between the two groups with respect to the cost incurred to find those articles.

In experimental condition, the average *find period* was 52% less as compared to the control condition and the difference was statistically significant (median1 = 92.5, median2 = 39, Mann-Whitney U = 1.5, n1 = n2 = 8, P = 0.001 two tailed). In the experimental condition, the average *mean between-patch foraging time* was 74.63% less compared to the control condition and the difference was statistically significant (median= 587, median2 = 157, Mann-Whitney U = 2, n1 = n2 = 8, P = 0.002 two tailed). In the experimental condition, the average *mean number of information regions foraged* was 43% less compared to the control group and the difference was statistically significant (median1 = 3.1, median2 = 1.8, Mann-Whitney U = 16, n1 = n2 = 8, P = 0.093 two tailed). Finally, in the experimental condition, the average *mean information yield per region* was 67% more compared to the control group and the difference was statistically significant (median1 = 0.063, median2 = 0.221, Mann-Whitney U = 5, n1 = n2 = 8, P = 0.005 two tailed).

Discussion: The above results suggest that both the treatment groups were similar with respect to the quantity and quality of articles that they found during this task. But, the experimental group took significantly less time and effort compared to the control group. In other words, for a similar output, the cost of retrieval in the experimental group was significantly lower. The differences between the two groups with respect to the four measurements taken together indicate that participants in the experimental condition more frequently encountered relevant biology articles when compared to the control condition. This implies that SBF-based indexing and access to biology articles has greater affordance for dealing with the findability issues, thus validating our proposed hypothesis.

7   Recognizability Study

Hypothesis: In the context of BID, enhancing proximal cues to include visual overviews derived from corresponding SBF models will lead to lower rate of recognition errors when compared to traditional proximal cues that do not include such overviews.

Procedure: This too was a between-subject study conducted in Spring 2011. The same sixteen subjects from Study 1 also participated in this study, but a sufficient time gap was provided between the two studies. A second preselected BID challenge was presented to all subjects. Biologue was used to then present a set of eight biology articles' proximal cues to the participants. The goal was to judge the relevancy of each of the eight biology articles for the given design challenge. The relevancy was reported on a five-point scale; the subjects were also asked to provide a rationale for their ratings. The articles were chosen such that four of them were relevant to the given design challenge and the other four were not relevant.

It is important to note that rather than entire articles, Biologue presented them with just proximal cues associated with articles. Eight subjects in the control group were given a version of Biologue that presented conventional proximal cues (containing information like the title, publication information, and an abstract of the article). The other eights subjects in the experimental condition were given a version of Biologue that presented them proximal cues which were additionally enriched with visual overviews derived from the articles' associated SBF models. All subjects were given a stipulated amount of time to complete the task. Because the researchers knew beforehand which four articles were relevant and which four were not relevant, we were able to calculate the extent of correct and incorrect classifications for each participant. We compared the classification accuracy of the participants across the two groups.

Materials: The BID challenge given to the participants in this study involved the design of a bio-inspired desalination technique such that: (1) the salinity of output fresh water should be fit for human consumption (specifically drinking), and (2) the energy footprint of the new technique must be less than the existing industry-standard techniques. This design challenge was subject to the following simplifying assumptions: (1) the feed water is already filtered and pre-treated to remove all other unwanted contents, leaving designers to deal with only pure saline water, and (2) the design will not actively control for other parameters like pH and alkalinity, free residual chlorine, boron, etc. The subjects were also given information about two existing industry-standard techniques for doing desalination, namely flash distillation method and reverse osmosis method. Some of the energy-related problems associated with the industry-standard techniques were also presented. To sum up, the subjects were given enough information so that a novice could be brought up to speed on the problem and had a rich enough mental model of the problem to be able to read an article and make a determination about its relevancy. They were also tested on their knowledge about this problem before they proceeded to perform the rating task. Again, this was an authentic problem addressed by one of the design teams (Team FORO) in the BID course at Georgia Tech in Fall 2008.

A total of eight biology articles were chosen for this task. These articles were selected from a pool of articles that Team FORO had researched in Fall 2008. Four of those articles were noted by the team as being relevant to solving the problem, and four as being irrelevant and leading to dead ends.

Biologue's repository for the purposes of this study consisted of only those eight articles. Biolgue for this study was instrumented such that as soon as a participant launched it, she would be instantly presented with a list of these eight articles (proximal cues only). This was meant to simulate a snapshot in the information seeking process where the seeker has just entered an information region and then needs to prioritize the order in which these articles would be visited based on the perceived relevance of each article to the target problem.

Two versions of Biologue were created for this study. In the control condition version, participants saw the traditional version of proximal cues, consisting of articles' title, abstract, and publication information. In the second experimental condition version, participants saw the SBF-augmented version of the proximal cues consisting of visual model overviews in addition to the other elements.

To minimize research bias, the primary researchers recruited another researcher to build the SBF models of biological systems discussed in the eight articles. This model builder had not encountered the desalination problem and was not aware of the purpose to which the SBF models would be put to use. Therefore, he could not introduce bias by tailoring the SBF models to match the desalination problem. These SBF models were then entered into Biolgoue and made available as part of the cues in one of the treatment groups.

All participants were required to rate the eight articles on a scale of 1 to 5. This was achieved by asking the participants to take an online survey when they were ready to rate the articles. The survey contained eight questions, one for each article they were required to rate. The rating was couched as a recommendation question: what would be their recommendation for the article to a team doing the desalination project on a scale of 1 to 5, where 1 represented "completely irrelevant (skip reading the article altogether)" and 5 represented "absolutely relevant (mimic the biological system in the paper and you will have solved the problem)." The middle value 3 represented "may be relevant, may not be relevant, can't say which."

Data: Participant demographic data was one of the data points used for this study. But the primary data for this study came from the online survey, which contained participants' article classification data, including the rationale for their classification.

Analysis: Although the participants were the same in the two studies, their distribution across the treatment groups was different. Therefore, a group equivalency test had to be performed in this study as well. We found no statistically significant difference between the two groups with respect to participants' gender, biology background, design experience, interdisciplinary research experience, and the use of scholarly articles in their everyday work practices.

Participant classification data, which was on a 5-point scale, was converted into a 3-point scale. A value of 1 or 2 was classified as "irrelevant," a value of 4 or 5 was classified as "relevant," and a value of 3 was classified as "unclassified." For each participant and for each article, the participant classification was compared against the actual classification of the article. Based on this comparison, a determination was made as to whether it the classification was correct, false positive, false negative, or null (no classification).

Table 2. Participant performance data on the recognizability task

Treatment		Correct classification	False +ve	False -ve	Undecided
Control	avg	3.75	1.88	0.88	1.50
	stdev	1.04	0.83	0.64	0.93
Experimental	avg	5.50	0.38	0.88	1.25
	stdev	1.31	0.52	0.64	1.16

Results: Table 2 summarizes the data from this study. Again, the data collected in this study was submitted to Mann-Whitney U test. This data shows that in the experimental condition, the *average recognition error* was 41.18% less compared to the control condition and the difference was statistically significant (median1 = 4, median2 = 2.5, Mann-Whitney U = 9, n1 = n2 = 8, P = 0.015 two tailed).

In the experimental condition, the average *false positives* was 79% fewer when compared to the control condition and the difference was statistically significant (median1 = 2, median2 = 0, Mann-Whitney U = 4.5, n1 = n2 = 8, P = 0.003 two tailed). But, in the experimental condition, there was 0% difference in *false negatives*. Finally, in the experimental condition, there was 16.67% fewer *undecided classifications*, but the differences was *not* statistically significant median1 = 1.5, median2 = 1.5, Mann-Whitney U = 28, n1 = n2 = 8, P = 0.66 two tailed).

Discussion: The above results shows that in the context of this study, the group that worked with redesigned proximal cues containing SBF information did significantly better in terms of number of false positives. This difference in false positives heavily contributed towards the difference observed in the total error rate between the two groups. It is not clear why there was no change in the number of false negatives or undecided classifications. More fine-grained studies are required to determine the affordance of proximal cues vis-à-vis the different kinds of recognition errors.

8   Conclusions

We know from past work that TCBR is a powerful paradigm within CBR especially with the advent of the Web (e.g., Bruninghaus & Ashley 2001; Burke *et al.*, 1997; Lenz 1998; Rissland & Daniels 1996; Raghunandan *et al.* 2008; Weber, Ashley & Bruninghaus 2006). However, on one hand complex reasoning tasks such as BID require access to cross-domain analogies, and, on the other, conventional search engines on the Web do not support easy access of cross-domain analogies. This gap between the demand and lack of supply for cases creates both a challenge and an opportunity for TCBR: how to accomplish cross-domain TCBR?

Our *in situ* studies of the retrieval task in BID identified findability and recognizability as two of the main challenges for cross-domain TCBR. There are potentially several approaches that one can use to mitigate the identified challenges, including engineering a structured case-base of biological and engineering systems using a domain-general knowledge representation language, semantically tagging documents in a socially-distributed fashion, using natural language processing to automatically extract the semantic tags from textual documents, using machine learning to learn semantic tags for the biology articles. While in this work we have chosen to use the social semantic approach, in other threads we exploring alternative approaches.

We have developed a technique for model-based semantic tagging of biology articles based on SBF ontology of systems because SBF models have proved to be useful for design, understanding as well as TCBR. We have also implemented, fielded and evaluated the technique in an interactive system called Biologue. From the point of view of TCBR, this is a new approach not only because of the cross-domain nature of the cases, but also because of the social dimension of semantic tagging of the cases that is part of the solution. From the point of view of BID, Biologue represents a new class of technological aids; earlier technologies relied on fully structured knowledge-bases that typically entailed a high cost of knowledge engineering, or employed most-ly-syntactic bottom-up natural language processing techniques that often are prone to poor precision and recall.

Our controlled experiments with Biologue indicate that when users do indeed adopt the social semantic approach, the improvements in retrieving cross-domain textual cases are significant. In particular, improvements were noticed with respect to both findability and recognizability issues. In our other work (Helms, Vattam & Goel 2010), we have found that SBF annotations also help design teams better *understand* biological systems.

Technological aids such as Biologue however raise the issue of reducing the chances of serendipitous encounter with fringe information that can sometimes lead to creative design solutions. The additional focus brought in by semantic search can inadvertently have the result that designers do not spend as much time browsing through articles and accidentally finding information that might be useful. Therefore, in practice, a more directed search feature in such a tool should be accompanied by other features that allow users to browse articles using a different set of criteria so that serendipity too is supported.

We know from past work that TCBR uses task-specific and domain dependent knowledge to retrieve and reuse textual cases for complex reasoning tasks (Weber, Ashley & Bruninghaus 2006). Our work indicates that task-specific but domain-general knowledge might be useful for TCBR in support of complex reasoning tasks engaging cross-domain analogies.

Acknowledgements. We are grateful to our research partners Michael Helms and Bryan Wiltgen for many discussions about biologically inspired design. We thank Professor Jeannette Yen, the coordinator of the ME/ISyE/MSE/PTFe/BIOL 4740 class that acts as a teaching and research laboratory for us. We are grateful to the US National Science Foundation that has generously supported this research through an NSF CreativeIT Grant (#0855916) entitled "Computational Tools for Enhancing Creativity in Biologically Inspired Engineering Design."

References

1. Benyus, J.: Biomimicry: Innovation Inspired by Nature. William Morrow (1997)
2. Biomimicry Institute (2008), Ask Nature – The Biomimicry Design Portal, http://www.asknature.org/
3. Burke, R., Hammond, K., Kulyukin, V., Lytinen, S., Tomuro, N., Schoenberg, S.: Question answering from frequently-asked questions files: experiences with the FAQ Finder system. AI Magazine 18(1), 57–66 (1997)
4. Brüninghaus, S., Ashley, K.: The role of information extraction for textual CBR. In: Aha, D.W., Watson, I. (eds.) ICCBR 2001. LNCS (LNAI), vol. 2080, pp. 74–89. Springer, Heidelberg (2001)
5. Chakrabarti, A., Sarkar, P., Leelavathamma, B., Nataraju, B.: A functional representation for aiding biomimetic and artificial inspiration of new ideas. AIEDAM 19, 113–132 (2005)

6. Chiu, I., Shu, L.: Biomimetic design through natural language analysis to facilitate cross-domain analysis. AIEDAM 21, 45–59 (2007)
7. Goel, A., Bhatta, S., Stroulia, E.: Kritik: An Early Case-Based Design System. In: Maher, Pu (eds.) Issues and Applications of Case-Based Reasoning in Design, pp. 87–132 (1997)
8. Goel, A., Mahesh, K., Peterson, J., Eiselt, K.: Unification of Language Understanding, Device Comprehension and Knowledge Acquisition. In: Proc. Cognitive Science Meeting 1996 (1996)
9. Goel, A., Rugaber, S., Vattam, S.: Structure, Behavior & Function of Complex Systems: The SBF Modeling Language. AIEDAM 23, 23–35 (2009)
10. Goel, A., Vattam, S., Wiltgen, B., Helms, M.: Cognitive, collaborative, conceptual and creative - Four characteristics of the next generation of knowledge-based CAD systems: A study in biologically inspired design. Computer-Aided Design 44(10), 879–900 (2012)
11. Helms, M., Vattam, S., Goel, A.: The Effects of Functional Modeling on Understanding Complex Biological Systems. In: Proc. 2010 ASME IDETC/CIE, Montreal, Canada (2010)
12. Holyoak, K., Thagard, P.: Analogical Retrieval by Constraint Satisfaction. Cognitive Science 13(3), 295–355 (1989)
13. Kolodner, J.: Case-Based Reasoning. Morgan Kaufmann Publishers, San Mateo (1993)
14. Lenz, M.: Defining knowledge layers for textual case-based reasoning. In: Smyth, B., Cunningham, P. (eds.) EWCBR 1998. LNCS (LNAI), vol. 1488, pp. 298–309. Springer, Heidelberg (1998)
15. Nagel, J., Stone, R., McAdams, D.: An engineering-to-biology thesaurus for engineering design. In: Proc. ASME 2010 IDETC/CIE, Montreal, Canada (2010)
16. Peterson, J., Mahesh, K., Goel, A.: Situating Natural Language Understanding in Experience-Based Design. IJHCS 41, 881–913 (1994)
17. Pirolli, P.: Information foraging theory: Adaptive interaction with information. Oxford University Press, Oxford (2007)
18. Raghunandan, M.A., Wiratunga, N., Chakraborti, S., Massie, S., Khemani, D.: Evaluation Measures for TCBR Systems. In: Althoff, K.-D., Bergmann, R., Minor, M., Hanft, A. (eds.) ECCBR 2008. LNCS (LNAI), vol. 5239, pp. 444–458. Springer, Heidelberg (2008)
19. Rissland, E., Daniels, J.: Using CBR to Drive IR. In: Procs. International Joint Conference on Artificial Intelligence, vol. 14, pp. 400–407 (1995)
20. Sen, S., Lam, S., Rashid, A., Cosley, D., Frankowski, D., Osterhouse, J., Harper, F., Riedl, J.: Tagging, communities, vocabulary, evolution. In: Procs. CSCW 2006, Banff, Canada, pp. 181–190 (2006)
21. Shu, L.H.: A natural-language approach to biomimetic design. AIEDAM 24(4), 483–505 (2010)
22. Vattam, S.: Interactive Analogical Retrieval: Practice, Theory and Technology, Doctoral Dissertation, Georgia Institute of Technology (2012)
23. Vattam, S., Goel, A.: Foraging for inspiration: Understanding and supporting the information seeking practices of biologically inspired designers. In: Proc. ASME DETC Conference on Design Theory and Methods, Washington, DC (August 2011)
24. Vattam, S., Helms, M., Goel, A.: Compound Analogical Design: Interaction Between Problem Decomposition and Analogical Transfer in Biologically Inspired Design. In: Proc. DCC 2008, Atlanta, pp. 377–396. Springer (June 2008)
25. Weber, R., Ashley, K., Bruninghaus, S.: Textual Case-Based Reasoning. Knowledge Engineering Review 20(3), 255–260 (2006)
26. Yen, J., Weissburg, M., Helms, M., Goel, A.: Biologically Inspired Design: ATool for Interdisciplinary Education. In: Bar-Cohen, Y. (ed.) Biomimetics: Nature-Based Innovation. Taylor & Francis (2011)

Similarity Measures to Compare Episodes in Modeled Traces

Raafat Zarka[1,2], Amélie Cordier[1,3], Elöd Egyed-Zsigmond[1,2],
Luc Lamontagne[4], and Alain Mille[1,3]

[1] Université de Lyon, CNRS
[2] INSA-Lyon, LIRIS, UMR5205, F-69621, France
[3] Université Lyon 1, LIRIS, UMR5205, F-69622, France
[4] Department of Computer Science and Software Engineering,
Université Laval, Québec, Canada, G1K 7P4
{raafat.zarka,amelie.cordier,elod.egyed-zsigmond,
alain.mille}@liris.cnrs.fr, luc.lamontagne@ift.ulaval.ca

Abstract. This paper reports on a similarity measure to compare episodes in modeled traces. A modeled trace is a structured record of observations captured from users' interactions with a computer system. An episode is a sub-part of the modeled trace, describing a particular task performed by the user. Our method relies on the definition of a similarity measure for comparing elements of episodes, combined with the implementation of the Smith-Waterman Algorithm for comparison of episodes. This algorithm is both accurate in terms of temporal sequencing and tolerant to noise generally found in the traces that we deal with. Our evaluations show that our approach offers quite satisfactory comparison quality and response time. We illustrate its use in the context of an application for video sequences recommendation.

Keywords: Similarity Measures, Modeled Traces, Recommendations, Edit Distance, Human Computer Interaction.

1 Introduction

Recently, there has been a growing interest in the analysis of user activity on the web. Indeed, from the observation of human activity, one can learn a lot about behaviors of users and use these findings for improving the quality of services.

As part of a collaboration with the company Webcastor [1], we are working on a web application called Wanaclip[2]. This application allows users to compose video clips from different audio-visual sources. Wanaclip has a built-in recommendation system that guides users in both the selection of videos, and the actions to perform in order to make a nice clip. The recommendation engine is fed by interaction traces left by previous users of the application.

---

[1] www.webcastor.fr
[2] www.wanaclip.eu

S.J. Delany and S. Ontañón (Eds.): ICCBR 2013, LNAI 7969, pp. 358–372, 2013.
© Springer-Verlag Berlin Heidelberg 2013

To build recommendations, we use a classic Case-Based Reasoning cycle. First, the system identifies the current situation of the user (the target problem). Then, it searches for similar situations in memory (source cases). Once the source cases are retrieved, the system adapts them to fit the current situation, e.g. to make recommendations relevant and appropriate to the current context of the user. The whole process is itself traced. As a consequence, the experience base of the system is incrementally enriched as the system is used. We are aware that the size of the trace base will rise scalability problems. We plan to implement trace "forgetting" methods, but these are out of scope for this paper.

In the following, we call the source cases **episodes**. An episode is a subsequence of **modeled traces** which structure is often complicated. A modeled trace is a structured record of **observed elements**, denoted **obsels**, captured from users' interactions with a computer system. Therefore, when computing similarity between episodes, we have to face new challenges that are not addressed by well-known CBR similarity measures. We have to take into account the fact that episodes are sequences of elements, but more importantly, we have to compare obsels which are complex objects. A formalization of the modeled traces will be fully explained in Section 3.

In this paper, we focus on the problem of assessing the similarity between two episodes identified in a modeled trace. For this, we defined a new similarity measure that is based on two main components: a similarity measure used to compare obsels having rich structures, and an algorithm combines the obsel similarity to compare episodes. The algorithm we propose is an adaptation of the *Smith-Waterman Algorithm* [1]. We implemented our proposal as a web service of $\mathcal{T}$Store, a Trace-Based Management System that handles the storage, processing and exploitation of traces [2]. We applied the proposal to Wanaclip in order to provide users with contextual recommendations.

The rest of the paper is organized as follows. Section 2 describes related work in the field of similarity measures for sequential data. Section 3 recalls definitions of $\mathcal{M}$-Traces and the $\mathcal{M}$-Trace Model. Similarity Measures between obsels are presented in Section 4. In Section 5, we describe the similarity algorithm between the episodes of $\mathcal{M}$-Traces. We report our experiments and performance study in Section 6, and conclude our work in Section 7.

2 Similarity Measures for Sequential Data

There exist several approaches for comparing strings and defining similarity measures over sequential data. A detailed comparison between three major classes of such similarity measures (*i.e. edit distance, bag-of-word models* and *string kernals*) has been studied in [3]. In this section, we present some similarity measures methods and their usages in different domains.

2.1 Methods for Defining Similarity Measures

String distances of Hamming [4] and Levenshtein [5] are among the first approaches. They are originated from the domain of telecommunication for detection of erroneous data transmissions. These approaches enable calculating the minimum edit distance between two strings which is the minimum number of editing operations (insertion, deletion and substitution) needed to transform one sequence into another. Needleman-Wunsch [6] is a global alignment method which attempts to align every residue in every sequence. Smith-Waterman Algorithm [1] is a local alignment method which is more useful for dissimilar sequences that are suspected to contain regions of similarity within their larger sequence context.

A different approach to sequence comparison is the vector space (or bag-of-words) model which originates from information retrieval and implements comparison of strings by embedding sequential data in a vector space [7]. This concept was extended to *n-grams* for approximate matching [8]. An *n*-gram is a contiguous sequence of *n* items from a given sequence of text. The vector space approach has been widely used for analysis of textual documents.

Kernel-based learning is a recent class of similarity measures derived from generative probability models. Various kernels have been developed for sequential data, starting from the original ideas of [9] and extending to domain-specific variants, such as string kernels designed for natural language processing [10] and bio-informatics [11].

2.2   Similarity Measures in the Case-Based Reasoning Field

Similarity measures applying on complex sequences are also developed in the CBR field. The Episode-Based Reasoning framework [12] provides mechanisms to represent, retrieve and learn temporal episodes. In the CR2N system [13], the similarity assumption is used to determine reusable textual constructs. A confidence measure for a workflow adaptation based on introspection of the case base has been introduced in [14]. The CeBeTA system [15] combines a sentence similarity metric based on edit distance with a reuse approach based on text-transformation routines, in order to generate solutions for the text modification problem. A proper case structure and a new distance measure have been proposed in [16], that are exploited to retrieve traces similar to the current one. They use a graph edit distance definition by focusing on traces of executions. For them, it was guaranteed that the actions in the traces always matched reality. However, our approach is based on similarity measures between obsels by comparing their contents, *i.e.* their time-stamps, users, types, and values.

Most of these approaches enable comparison of homogeneous elements (such as characters or symbols). In this paper, we focus on modeled traces, which are sequential records of complex elements. As the elements are complex, they cannot be directly compared. Therefore, we propose an approach inspired by the Smith-Waterman Algorithm [1], but it takes into account the richness of the compared elements. We chose this algorithm because, by combining it with our similarity measures between obsels, it has all the properties expected for the comparison of episodes, namely: processing of sequential data, tolerance to variations in representation, noise resistance, high degree of customization and satisfactory response time.

3 $\mathcal{M}$-Traces Definitions and Formal Representation

An interaction trace is a rich record of the actions performed by a user on a system. Therefore, traces enable capturing users' experiences. $\mathcal{M}$-**Traces** (short for Modeled Trace) differ from logs in the sense that they come with a model and their observed elements (called **obsels**) are highly connected. An $\mathcal{M}$-Trace includes both the sequence of temporally situated obsels and the model of the trace which gives the semantics of obsels and the relations between them. The notations and definitions of $\mathcal{M}$-Trace and its model are defined in detail in [17]. In the following, we provide a simplified formalization fitting our needs.

3.1   Trace Model Definition ($\mathcal{M}_T$)

Definition 1. *A Trace model is defined as a tuple* $\mathcal{M}_T = (T, C, R, A, p)$

- T: *temporal domain representing the period during which the $\mathcal{M}$-Traces of this model occurred. Usually, it contains two time-stamps: start and end.*
- C: *finite set of obsel types.*
- R: *finite set of relation types between the obsel types in C, including (but not limited to) inheritance relation.*
- A: *hierarchical set of attributes for each obsel type in C. $A(c)$ is the set of attributes for a c obsel type.*
- p: *parent trace model that represents the hierarchy between the trace models.*

3.2 $\mathcal{M}$-Trace Definition

Definition 2. *An $\mathcal{M}$-Trace is a tuple* $\mathcal{M} - Trace = (\mathcal{M}_T, O, \leq_O, u, s_t, e_t, v, \lambda_R, \lambda_C)$

- $\mathcal{M}_T$: *trace model $\mathcal{M}_T = (T, C, R, A, p)$.*
- O: *sequence of obsels of this trace, where $|O|$ is the number of obsels in the trace.*
- $\leq_O$: *partial order defined on the obsels, it represents their chronological order.*
- u: *identification of the user executing the actions associated with this trace.*
- s_t and e_t: *starting and ending timestamps of the trace.*
- v: *visibility of the trace. It can be: public, private or custom. It is used for defining security properties of a trace.*
- λ_R: *function describing relation between two obsels $\lambda_R : O \times O \rightarrow R.$, where R is the set of relations type defined in $\mathcal{M}_T$.*
- λ_C: *total function that associates each obsel with its type. $\lambda_C : O \rightarrow C$.*

3.3   Obsel Definition

Definition 3. *We define an obsel as* $o = \{\mathcal{M}\text{-}Trace, c, A_o, u, s_t, e_t\}$ *where:*

- $\mathcal{M}$-*Trace: trace containing this obsel.*
- c: *type of the obsel. Each obsel type has a predefined set of attributes.*

- A_o: set of attributes of the obsel o and their values. $A_o = \{(a_i, v_i)\}_{i=1,|A_c|}$. Note that a_i is an obsel attribute type, v_i is an obsel attribute value, and A_c is the set of attributes of an obsel type c.
- u: user executing the action associated with the obsel. It is obtained from the $\mathcal{M}$-Trace of this obsel.
- s_t and e_t: starting and ending timestamps of the obsel.

Note that obsel attributes are hierarchical so only the leafs can have values. These values are not mandatory.

3.4   What Is an Episode?

An episode is a temporal pattern composed of an ordered set of events corresponding to a specific task defined by a task signature. The concept of task signature has been introduced in [18] as a set of event declarations, entity declarations, relations, and temporal constraints.

Definition 4. *Given an $\mathcal{M}$-Trace, O is the sequence of its obsels. We define an episode $\mathcal{E}_k = O_{n_k}$ as a sub-sequence of O where $(n_1 < n_2 < \ldots)$ is a temporally coherent sequence of obsels of O.*

An episode $\mathcal{E}$ can be derived from O by deleting some obsels without changing the order of the remaining obsels.

4   Similarity Measures between Obsels

In order to define the similarity between the obsels, we need to define several local similarity measures for the obsel types, users, attributes and timestamps, which are the significant components of an obsel.

Definition 5. *We define $sim_{obs}(o_1, o_2)$ as a similarity measure between the obsels $o_1 = \{c_1, A_{o_1}, u_1, s_{t1}, e_{t1}\}$ and $o_2 = \{c_2, A_{o_2}, u_2, s_{t2}, e_{t2}\}$ as:*

$$sim_{obs}(o_1, o_2) = \alpha \times sim_{obstype}(c_1, c_2) + \beta \times sim_{obsattr}(A_{o_1}, A_{o_2})$$

$$+ \gamma \times sim_{obsuser}(u_1, u_2) + \delta \times sim_{obstime}(s_{t1}, e_{t1}, s_{t2}, e_{t2}) \qquad (1)$$

where:

- $sim_{obstype}(c_1, c_2)$: is the obsel type similarity,
- $sim_{obsattr}(A_{o_1}, A_{o_2})$: is the obsel attribute similarity,
- $sim_{obsuser}(u_1, u_2)$: is the obsel user similarity,
- $sim_{obstime}(s_{t1}, e_{t1}, s_{t2}, e_{t2})$: is the obsel time-stamp similarity,
- $\alpha, \beta, \gamma, \delta$: are weights, with $(\alpha + \beta + \gamma + \delta) = 1$ to keep this measure normalized.

The similarity measure between obsels $sim_{obs}(o_1, o_2)$ is a normalized value $\in [0, 1]$ since all its sub measures $(sim_{obstype}, sim_{obsattr}, sim_{obsuser}, sim_{obstime})$ produce normalized values and the sum of the weights $(\alpha + \beta + \gamma + \delta) = 1$. It is application experts who define these values.

4.1   Obsel Type Similarity $sim_{obstype}(c_1, c_2)$

Basically, obsel types are similar if they are identical. Thus, we propose to define a substitution matrix over obsel types $S_{obstype}(|C| \times |C|)$. In bio-informatics and evolutionary biology, a substitution matrix describes the rate at which one character in a sequence changes to other character states over time. We consider that the substitution matrix has normalized values between 0 and 1. The simplest possible substitution matrix would be one in which each obsel type in $(c \in C)$ is considered maximally similar to itself, but not similar to any other obsel type. This matrix would look like:

$$S_{obstype} = \begin{bmatrix} 1 & 0 & \cdots & 0 & 0 \\ 0 & 1 & \cdots & 0 & 0 \\ \vdots & & \ddots & & \vdots \\ 0 & 0 & \cdots & 1 & 0 \\ 0 & 0 & \cdots & 0 & 1 \end{bmatrix}$$

Definition 6. *For two obsel types $c_1, c_2 \in C$, we define their similarity as:*

$$sim_{obstype}(c_1, c_2) = S_{obstype}(c_1, c_2) \in [0, 1] \tag{2}$$

Where:

- *C is the set of all obsel types*
- *S is a substitution matrix $|C| \times |C|$*
- *Both of the rows and columns of S are obsel types. Each cell in this matrix has a normalized value representing the distance between a pair of obsel types.*

The substitution matrix depends on the obsels defined on the application. By default, it is filled manually by an expert.

4.2   Attribute Similarity $sim_{obsattr}(A_{o_1}, A_{o_2})$

When two obsels are of the same type, it is easy to compare them. However, when two obsels are of different types, a more complex comparison method has to be defined. For example in Wanaclip, we have the obsel types "playVideo" and "infoVideo". Both of them contain the meta-data of the video. To compare their attributes we need to know the common attributes between these obsel types.

Definition 7. *We define $cA(A_{o_1}, A_{o_2}) = \{(ca_{1,i}, ca_{2,i})\}_{i=1,|cA(A_{o_1}, A_{o_2})|}$ the set of common type attribute couples of the attribute sets A_{o_1} and A_{o_2} where o_1, o_2 are two obsels that can have different obsel types. An obsel can not have two attributes of the same type unless they are not in the same hierarchy level.*

Definition 8. *For two obsels $o_1 \in O$ and $o_2 \in O$, we define the similarity between the two sets of attributes of o_1 and o_2 as:*

$$sim_{obsattr}(A_{o_1}, A_{o_2}) = \sum_{i=1}^{|cA(A_{o_1}, A_{o_2})|} w_{importance}(i) \times sim_{attr}(ca_{1,i}, ca_{2,i}) \tag{3}$$

where:

- $sim_{attr}(ca_{1,i}, ca_{2,i}) \in [0, 1]$. *For each attribute type, there has to be at least one similarity function provided to compare their values.*
- $sum_{i=1}^{|cA(A_{o_1}, A_{o_2})|} w_{importance}(i) = 1$. *The different weights are defined separately and possibly modified by the user or the system. It ensures that the similarity measure between two sets of attributes is normalized.*

4.3   Obsel User Similarity $sim_{obsuser}(u_1, u_2)$

When comparing $\mathcal{M}$-Traces, we prefer to give more importance to the traces which belong to the same user or to users members of the same group. Social media give a lot of importance to this measure. Indeed users having similar profiles and interests are more likely to do similar activities, and therefore, to produce similar $\mathcal{M}$-Traces [19].

Definition 9. *The similarity measure between two users u_1, u_2 is defined as:*

$$sim_{obsuser}(u_1, u_2) = \begin{cases} 1 & u_1 = u_2 \\ \lambda \times sim_{profile}(u_1, u_2) + \mu \times sim_{groups}(u_1, u_2) & u_1 \neq u_2 \end{cases} \tag{4}$$

where:

- $sim_{profile}(u_1, u_2)$ *is the similarity between the profiles of the users according to their interests and activities. We will consider that this value is between 0 and 1.*
- $sim_{groups}(u_1, u_2)$ *is the Jaccard similarity coefficient [20] between user groups. By considering G_1 as the groups that the user u_1 belongs to, G_2 for the user u_2, then we define the similarity measure between these groups as:*

$$sim_{groups}(u_1, u_2) = \frac{|G_1 \cap G_2|}{|G_1 \cup G_2|} \tag{5}$$

- λ, μ: *are weights, where $\lambda + \mu = 1$.*

This measure is to be considered in a future work. For the moment in Wanaclip, we only consider similar, identic users.

4.4   Obsel Timestamp Similarity $sim_{obstime}(o_1, o_2)$

According to the $\mathcal{M}$-Trace model, each obsel has two timestamps (begin and end). At the moment, similarity is only defined with regard to the duration of the obsel. However, this measure could easily be extended (for example, by using Allen's interval relations [21]).

Definition 10. *For two obsels o_1, o_2 having time-stamps (s_{t1}, e_{t1}) for o_1 and (s_{t2}, e_{t2}) for o_2, we define the obsel timestamps similarity measure as:*

$$sim_{obstime}(o_1, o_2) = \begin{cases} 1 & |(e_{t1} - s_{t1}) - (e_{t2} - s_{t2})| = 0 \\ \frac{min(e_{t1} - s_{t1}, e_{t2} - s_{t2})}{max(e_{t1} - s_{t1}, e_{t2} - s_{t2})} & |(e_{t1} - s_{t1}) - (e_{t2} - s_{t2})| \neq 0 \end{cases} \tag{6}$$

where $(e_{t1} - s_{t1})$ is the duration of o_1. We calculate the fraction between the minimum and maximum duration of o_1, o_2 to get a normalized value. For example, if $t_1(2, 9), t_2(4, 14)$ then $sim_{obstime}(o_1, o_2) = 0.7$

5   Similarity between Episodes

In the previous section, we defined a similarity measure for comparing two obsels. In this section we present an approach for comparing episodes based on the minimum edit distance. This approach makes use of the similarity measure introduced earlier. This approach relies on a dynamic programming algorithm that solves the smaller problems optimally and uses the sub-problem solutions to construct an optimal solution for the original problem. We use the Smith-Waterman Algorithm [1] to compare the episode. We extend the algorithm by introducing similarity measures between obsels.

5.1   The $\mathcal{M}$-Trace Smith-Waterman Algorithm

The Smith–Waterman algorithm is a well-known algorithm for performing local sequence alignment. Instead of looking at each sequence in its entirety, it compares segments of all possible lengths and optimizes the similarity measure. We adapted the algorithm using the similarity measure described in the previous section to make it more accurate in terms of temporal sequencing and tolerant to noise generally found in the traces that we deal with. For that, we build a substitution matrix using the similarity measures between obsels and a gap function to determine the reduced score according to the number of indels (insertions/deletions) between obsels. It represents the cost of replacing an obsel by another one. This method not only evaluates the similarity between two episodes but also the transformations needed to go from one episode to the other (alignment), which is particularly useful for issuing the recommendations.

Definition 11. *Gap penalty function determines the reduced score according to the number of indels (insertions/deletions) in the sequence alignment. It determines the lost value when replacing an obsel by a gap. For the moment, for any obsel o, we consider that $gap(o) = -1$.*

Definition 12. *For two episodes $A, B, \forall a \in A \cup \{'-'\}, b \in B \cup \{'-'\}$, we define the substitution matrix of obsels as:*

$$S(a, b) = \begin{cases} gap(a) & b =' -' \\ gap(b) & a =' -' \\ (sim_{obs}(a, b) - 0.5) \times 2 & a \in A \wedge b \in B \end{cases} \quad (7)$$

where:

- $gap(a) \in [-1, 0]$ *is the gap penalty function for an obsel $a \in A$*
- $'-'$ *is the gap-scoring scheme*
- $sim_{obs}(a, b) \in [0, 1]$ *is the obsel similarity measure between a, b*
- $(sim_{obs}(a, b) - 0.5) \times 2$ *means converting the values from $[0, 1]$ to $[-1, +1]$*

Algorithm 1. The $\mathcal{M}$-Trace Smith-Waterman Algorithm

---

Data: A, B are episodes to compare

Result: The similarity measure $sim_{episode}(A, B)$ and the local alignments L_1, L_2

1: $\left.\begin{array}{l} H_{Score}(i,0) \leftarrow 0 \quad \forall\, 0 \le i \le |A| \\ H_{Score}(0,j) \leftarrow 0 \quad \forall\, 0 \le j \le |B| \end{array}\right\}$ ▷ Initialization of the similarity-score matrix

2: $\left.\begin{array}{l} S(A_i, -) \leftarrow gap(A_i) \quad \forall\, 1 \le i \le |A| \\ S(-, B_j) \leftarrow gap(B_j) \quad \forall\, 1 \le j \le |B| \end{array}\right\}$ ▷ Compute gap penalties of obsels

3:

4: **for** $i = 1 \to |A|$ **do**

5:   **for** $j = 1 \to |B|$ **do**

6:     $S(A_i, B_j) \leftarrow (sim_{obs}(A_i, B_j) - 0.5) \times 2$ ▷ Fill the substitution matrix

7:     $H(i,j) \leftarrow max \begin{cases} \textbf{Score} & \textbf{Pointer} \\ 0 & None \\ H(i-1, j-1) \;\; + S(A_i, B_j) & Substitution \\ H(i-1, j) \;\; + S(A_i, -) & Deletion \\ H(i, j-1) \;\; + S(-, B_j) & Insertion \end{cases}$

8:   **end for**

9: **end for**

10:

11: $L_1, L_2 \leftarrow [\,]$

12: $i, j = max(H_{i,j}(i,j))$

13: **while** $(i > 0) \wedge (j > 0) \wedge (H_{pointer}(i,j) \ne "None")$ **do**   ▷ Trace-back

14:   **if** $(H_{pointer}(i,j) = "Substitution")$ **then**

15:     $push(L_1, A_i)$

16:     $push(L_2, B_j)$

17:     $i \leftarrow i - 1$

18:     $j \leftarrow j - 1$

19:   **else if** $(H_{pointer}(i,j) = "Insertion")$ **then**

20:     $push(L_1, -)$

21:     $push(L_2, B_j)$

22:     $j \leftarrow j - 1$

23:   **else if** $(H_{pointer}(i,j) = "Deletion")$ **then**

24:     $push(L_1, A_i)$

25:     $push(L_2, -)$

26:     $i \leftarrow i - 1$

27:   **end if**

28: **end while**

29:

30: $sim_{episode}(A, B) \leftarrow max(H_{score}(i,j)$

31: **return** $L_1, L_2, sim_{episode}(A, B)$

The $\mathcal{M}$-Trace Smith-Waterman Algorithm 1 compares two episodes A, B. It computes the episode similarity measure $sim_{episode}(A, B)$ and the local alignments L_1, L_2 of A, B. The similarity-score matrix (H) between a suffix of $A[1 \cdots i]$ and a suffix of $B[1 \cdots j]$ with a size of $(|A| + 1 \times |B| + 1)$, is built during the algorithm. Firstly, the first row and the first column are initialized to zero. During the algorithm, a gap penalty is computed for each obsel in $(A \cup B)$. As shown in Figure 1, the algorithm iterates for each cell in H from the top-left cell to the bottom-right cell (see lines 4 to 8). At each time, it computes the similarity measure between the two obsels A_i, B_j according to Definition 5. The score of the current cell $H(i, j)$ is the highest score of three other scores in the similarity-score matrix (see line 7): the up-left neighbor cell $H(i - 1, j - 1)$ added to the similarity of obsels A_i, B_j, the left neighbor cell added to the gap penalty and the up neighbor cell added to the gap penalty. We keep pointers to the highest cells to keep directions of movement used to construct the matrix. This will be used in the trace-back step to obtain the best alignment. If we want the best local alignment, we find $H_{opt} = max_{i,j} H(i, j)$ and we trace-back. The cell of highest score can be anywhere in the array. Otherwise, if we want all local alignments, we find $H(i, j) > threshold$ for all i, j and we trace-back. To obtain the optimum local alignment, we start with the highest value in the matrix $H(i, j)$ (see line 12). Then, we go backwards to one of the positions $H(i - 1, j)$, $H(i, j - 1)$, and $H(i - 1, j - 1)$ depending on the direction of the movement used to construct the matrix. (see lines 13 to 28) We keep the process until we reach a matrix cell with zero value, or the cell $H(0, 0)$. Once we have finished, we reconstruct the alignment as follows: Starting with the last value, we reach $H(i, j)$ using the previously calculated path. A diagonal jump implies a replacement. A top-down jump implies a deletion. A left-right jump implies an insertion. The trace-back step stops when the score is 0 or when we reach the first row or column. The algorithm returns the maximum similarity score H_{opt} and the alignments L_1, L_2 found. Sometimes, we have different possible alignments especially when we have more than one maximum cell.

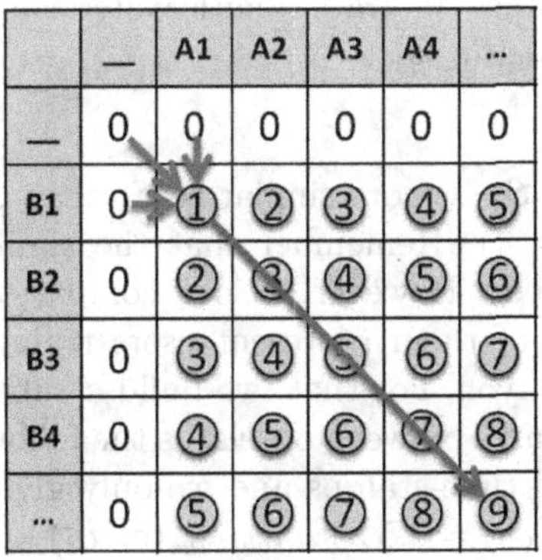

Fig. 1. The iterations of the Smith-Waterman Algorithm

5.2   Example

The following example comes from the web application Wanaclip. Let us consider that we have an episode $\mathcal{E}_1 = (S_8 P_1 A_1 A_2 X_1 A_3 G_7)$. It represents a user searching for videos about "Lyon" (S_8: S is the obsel type, and 8 is the value). Then, he starts playing (P_1) and adding videos (A_1 and A_2) to his selection. After that, he adds the text X_1 to the first selected video and adds another video A_3. Finally, he generates the clip G_7. This episode is collected and stored in $\mathcal{T}$Store. Later, two users are using Wanaclip. The first one selects the tag "Lyon", plays and adds a video, then adds text to it. This episode is noted $\mathcal{E}_2 = (T_8 P_1 A_1 X_1)$. The second user is searching for a different keyword "Paris". He plays and adds other videos. His episode is noted $\mathcal{E}_3 = (S_9 P_3 A_3 A_5)$. Using the stored $\mathcal{M}$-Traces, an assistant recommends the next actions to do. For that, the assistant compares the current episodes with the stored ones and recommends the most similar ones. Figure 2 shows how to compare the stored episode $\mathcal{E}_1$ with the current episodes $\mathcal{E}_2$ and $\mathcal{E}_3$ using Smith-Waterman Algorithm without using the similarity measures between obsels (The default values: gap=-1, matching = 1, mismatch=-1).

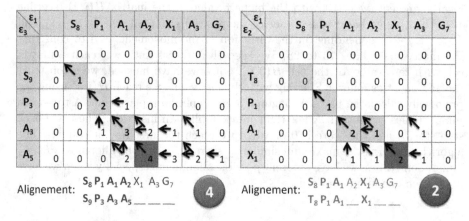

Fig. 2. An example of the $\mathcal{M}$-Trace Smith-Waterman Algorithm to compute $sim_{episode}(\mathcal{E}_1, \mathcal{E}_2)$, $sim_{episode}(\mathcal{E}_1, \mathcal{E}_3)$

The results show that $\mathcal{E}_3$ is more similar to $\mathcal{E}_1$ than $\mathcal{E}_2$ ($sim_{episode}(\mathcal{E}_1, \mathcal{E}_2) = 2 < sim_{episode}(\mathcal{E}_1, \mathcal{E}_3) = 4$) (Remember that the similarity measure between episodes is not normalized). However, in our context, $\mathcal{E}_2$ is more likely to be similar to $\mathcal{E}_1$ since both of them represent users trying to add similar videos to generate a clip about Lyon. For that, we build a substitution matrix defined using the similarity measures between obsels, as the following: we ignore the similarity between users and time-stamps and we only give weights to obsel types and attributes $sim_{obs}(o_1, o_2) = 0.5 \times sim_{obstype}(c_1, c_2) + 0.5 \times sim_{obsattr}(A_1, A_2)$. We also consider that the similarity between the obsel types (search and select-Tag) is $sim_{obstype}(S, T) = 0.5$. According to that, the substitution matrices for

$\mathcal{E}_1, \mathcal{E}_2$ and $\mathcal{E}_1, \mathcal{E}_3$ are computed as shown in Figure 3. For example, to compute the substitution value for the obsels S_8, T_8 we compute the similarity measure between them $sim_{obs}(S_8, T_8) = 0.5 \times 0.5 + 0.5 \times 1 = 0.75$. Then we convert the value to be in the range $[-1, +1]$ as $S(S_8, T_8) = (sim_{obs}(S_8, T_8) - 0.5) \times 2 = 0.5$.

$\mathcal{E}_1$ / $\mathcal{E}_3$	—	S_8	P_1	A_1	A_2	X_1	A_3	G_7
—		-1	-1	-1	-1	-1	-1	-1
S_9	-1	0	-1	-1	-1	-1	-1	-1
P_3	-1	-1	0	-1	-1	-1	0	-1
A_3	-1	-1	-1	0	0	-1	1	-1
A_5	-1	-1	-1	0	0	-1	0	-1

$\mathcal{E}_1$ / $\mathcal{E}_2$	—	S_8	P_1	A_1	A_2	X_1	A_3	G_7
—		-1	-1	-1	-1	-1	-1	-1
T_8	-1	0.5	-1	-1	-1	-1	-1	-1
P_1	-1	-1	1	0	-1	0	-1	-1
A_1	-1	-1	0	1	0	0	0	-1
X_1	-1	-1	0	0	-1	0	-1	-1

Fig. 3. The substitution matrices for $\mathcal{E}_1, \mathcal{E}_2$ and $\mathcal{E}_1, \mathcal{E}_3$

Figure 4 shows the $\mathcal{M}$-Trace Smith-Waterman Algorithm after applying the substitution matrix based on the similarity measures between obsels. We see that $\mathcal{E}_2$ is more similar to $\mathcal{E}_1$ than $\mathcal{E}_3$ ($sim_{episode}(\mathcal{E}_1, \mathcal{E}_2) = 2.5 > sim_{episode}(\mathcal{E}_1, \mathcal{E}_3) = 1$) which is more reasonable and accurate than the past results. Using the alignments of the episodes, we can recommend the next actions. The detailed description of the recommendation mechanism is out of the scope of this paper.

$\mathcal{E}_1$ / $\mathcal{E}_3$	S_8	P_1	A_1	A_2	X_1	A_3	G_7
	0	0	0	0	0	0	0	0
S_9	0	0	0	0	0	0	0	0
P_3	0	0	0	0	0	0	0	0
A_3	0	0	0	0	0	0	1	0
A_5	0	0	0	0	0	0	0	0

$\mathcal{E}_1$ / $\mathcal{E}_2$	S_8	P_1	A_1	A_2	X_1	A_3	G_7
	0	0	0	0	0	0	0	0
T_8	0	0.5	0	0	0	0	0	0
P_1	0	0	1.5	0.5	0	0	0	0
A_1	0	0	0.5	2.5	1.5	0.5	0	0
X_1	0	0	0	1.5	1.5	2.5	1.5	0.5

Alignement:
S8 P1 A1 A2 X1 **A3** G7
_ _ _ S9 P3 **A3** A5   (1)

Alignement:
S8 P1 A1 A2 X1 A3 G7
T8 P1 A1 __ X1 __ __   (2.5)

Fig. 4. An example of the $\mathcal{M}$-Trace Smith-Waterman Algorithm to compute $sim_{episode}(\mathcal{E}_1, \mathcal{E}_2)$, $sim_{episode}(\mathcal{E}_1, \mathcal{E}_3)$ after applying obsels similarity measures

6 Implementation and Evaluation

We have implemented our local similarity measures and the $\mathcal{M}$-Trace Smith-Waterman Algorithm as a PHP web service, within the $\mathcal{T}$Store framework. Any client connected to $\mathcal{T}$Store can execute the service to compare obsels, episodes of $\mathcal{M}$-Traces or other types of sequences. These implementations allow us to customize the similarity measures by specifying all the weights used in the similarity measures between obsels defined in Definition 5. In addition, we can customize the $\mathcal{M}$-Trace Smith-Waterman Algorithm by changing the substitution matrix. We also customize the gap penalty function and change the range of values of the similarity scores between obsels. We can extend it to be in $[-\infty, +\infty]$. For example, decreasing the score of similar obsels will make the episodes more likely to be similar by giving more score to the matched obsels. The same thing applies for mismatching obsels. If we give a negative value for the substitution between two obsels, it means that a mismatch between them will decrease the probability of matching the episodes.

The $\mathcal{M}$-Trace Smith–Waterman algorithm is fairly demanding in time: to compare two episodes $\mathcal{E}_1, \mathcal{E}_2$, $O(|\mathcal{E}_1||\mathcal{E}_2|)$ time is required. In addition, improvements can easily be made. Indeed, through observation of the similarity matrix calculation process in Figure 1, we found that we can optimize the calculation time of the similarity scores. For each iteration, every element on an anti-diagonal line marked with the same number could be calculated simultaneously, taking into consideration the elements that could be calculated at the same time. For example, in the first cycle, only one element marked as (1) could be calculated. In the second cycle, two elements marked as (2) could be calculated. In the third cycle, three elements marked as (3) could be calculated, *etc.*

We have implemented a collection process in Wanaclip. The $\mathcal{M}$-Trace handler captures users' actions, and stores them as obsels in $\mathcal{T}$Store. In order to illustrate the efficiency of the proposed algorithm, we conducted run-time experiments on episodes regenerated randomly. Both of $\mathcal{T}$Store and Wanaclip are running on the same computer. We performed several experiments changing the weights used in the similarity measures between obsels and the similarity values in the substitution matrix. We compared the run-time of the $\mathcal{M}$-Trace Smith–Waterman algorithm over 100 episodes of different lengths from 2 to 20 obsels. We used about 50 obsel types (10 between them were the most frequently). Each obsel has in average 3 attributes. The procedure was repeated 10 times for various weights, and the run-time was averaged over all runs. As shown in Figure 5, the run-time of the algorithm shown logarithmic growth. It is composed of the time required for the iterations of comparison and the time of the alignment part. We are not obliged to compute the alignments but it is important for the recommendations. In the right chart, we see that about the half of the similarity scores of the comparison test we performed were between 0 and 5. Where 25% were totally mismatched which helps to eliminates many episodes while retrieving the similar ones for the recommendation.

The most important thing was the obsel types, the length of the compared episodes (the number of obsels in each episode) and the number of common

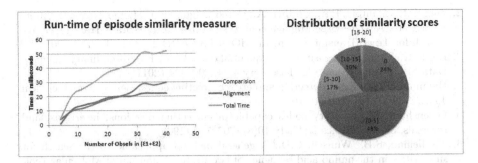

Fig. 5. Evaluation of the $\mathcal{M}$-Trace Smith-Waterman Algorithm

attributes between compared obsels. We have noticed also that the alignments of episodes is sometimes better when expanding the range of the substitution matrix. However, the run-time for the $[-1, +1]$ range is better than the expanded range because there are more (0) values which decrease the number of pointers in the score matrix and that leads to less paths to find the alignments.

7 Conclusion

In this paper, we introduced a similarity measure for comparing episodes belonging to $\mathcal{M}$-Trace. Comparing such episodes is a difficult task, not only because they are sequences, but also because they are composed of obsels (observed elements) which are complex objects therefore difficult to compare. Our method is based on two major components: a similarity measure for comparing obsels, and an adaptation of the Smith-Waterman Algorithm, using the similarity measure described above, to determine the similarity between two episodes. This method not only evaluates the similarity between two episodes but also the transformations needed to go from one episode to the other (alignment), which is particularly useful given the recommendations.

Evaluations of the method showed that the algorithm is not time-consuming and that it has all the properties that we expected: ability to compare complex objects, calculate the similarity score, many customization options, *etc*. We implemented this measure in the platform $\mathcal{T}$Store [2]. We have experimented with the measure within the context of the Wanaclip application (www.wanaclip.eu) in order to provide Wanaclip users with contextualized recommendations bases on traces of previous users. In future work, we will perform a thorough evaluation of the quality of the recommendations in order to better tune the similarity measure. We will work also in the optimization of the $\mathcal{M}$-Trace Smith-Waterman Algorithm and its application to different domains. In addition, we will use users' feedback for the computation of the similarity measures.

References

1. Smith, T.F., Waterman, M.S.: Identification of common molecular subsequences. Journal of Molecular Biology 147(1), 195–197 (1981)

2. Zarka, R., Champin, P.A., Cordier, A., Egyed-Zsigmond, E., Lamontagne, L., Mille, A.: TStore: A Trace-Base Management System using Finite-State Transducer Approach for Trace Transformation. In: MODELSWARD 2013. SciTePress (2013)
3. Rieck, K.: Similarity measures for sequential data. Wiley Interdisciplinary Reviews: Data Mining and Knowledge Discovery 1(4), 296–304 (2011)
4. Hamming, R.W.: Error detecting and error correcting codes. Bell System Technical Journal 29(2), 147–160 (1950)
5. Levenshtein, V.I.: Binary codes capable of correcting deletions, insertions, and reversals. Soviet Physics Doklady 10(8), 707–710 (1966)
6. Needleman, S.B., Wunsch, C.D.: A general method applicable to the search for similarities in the amino acid sequence of two proteins. Journal of Molecular Biology 48(3), 443–453 (1970)
7. Salton, G., Wong, A., Yang, C.S.: A vector space model for automatic indexing. Communications of the ACM 18(11), 613–620 (1975)
8. Damashek, M.: Gauging Similarity with n-Grams: Language-Independent Categorization of Text. Science 267(5199), 843–848 (1995)
9. Watkins, C.: Dynamic Alignment Kernels. Advances in Large Margin Classifiers, 39–50 (January 1999)
10. Lodhi, H., Saunders, C., Shawe-Taylor, J., Cristianini, N., Watkins, C.: Text Classification using String Kernels. Journal of Machine Learning Research 2(3) (2002)
11. Cuturi, M., Vert, J.P., Birkenes, O., Matsui, T.: A Kernel for Time Series Based on Global Alignments. In: 2007 IEEE International Conference on Acoustics Speech and Signal Processing, ICASSP 2007, vol. 2(i), pp. II-413–II-416 (2006)
12. Sánchez-Marré, M., Cortés, U., Martínez, M., Comas, J., Rodríguez-Roda, I.: An Approach for Temporal Case-Based Reasoning: Episode-Based Reasoning. In: Muñoz-Ávila, H., Ricci, F. (eds.) ICCBR 2005. LNCS (LNAI), vol. 3620, pp. 465–476. Springer, Heidelberg (2005)
13. Adeyanju, I., Wiratunga, N., Lothian, R., Sripada, S., Lamontagne, L.: Case Retrieval Reuse Net (CR2N): An Architecture for Reuse of Textual Solutions. In: McGinty, L., Wilson, D.C. (eds.) ICCBR 2009. LNCS, vol. 5650, pp. 14–28. Springer, Heidelberg (2009)
14. Minor, M., Islam, M. S., Schumacher, P.: Confidence in Workflow Adaptation. In: Agudo, B.D., Watson, I. (eds.) ICCBR 2012. LNCS, vol. 7466, pp. 255–268. Springer, Heidelberg (2012)
15. Valls, J., Ontañón, S.: Natural Language Generation through Case-Based Text Modification. In: Agudo, B.D., Watson, I. (eds.) ICCBR 2012. LNCS, vol. 7466, pp. 443–457. Springer, Heidelberg (2012)
16. Montani, S., Leonardi, G.: Retrieval and clustering for supporting business process adjustment and analysis. Information Systems (December 2012)
17. Settouti, L.S.: M-Trace-Based Systems - Models and languages for exploiting interaction traces. PhD thesis, University Lyon1 (2011)
18. Champin, P.A., Prié, Y., Mille, A.: MUSETTE: a framework for Knowledge from Experience. In: EGC 2004, RNTI-E-2, Cepadues Edition, pp. 129–134 (2004)
19. Kietzmann, J.H.: Social media? Get Serious! Understanding the Functional Building Blocks of Social Media 54 (2011)
20. Lipkus, A.H.: A proof of the triangle inequality for the Tanimoto distance 26 (1999)
21. Allen, J.F.: Maintaining knowledge about temporal intervals. Communications of the ACM 26(11), 832–843 (1983)

Author Index